Aqueduct Hunting in the Seventeenth Century

Aqueduct Hunting in the Seventeenth Century

Raffaello Fabretti's *De aquis et aquaeductibus veteris Romae*

by Harry B. Evans

THE UNIVERSITY OF
MICHIGAN PRESS
Ann Arbor

Published in the United States of America by
The University of Michigan Press
Manufactured in the United States of America
♾ Printed on acid-free paper

2005 2004 2003 2002 4 3 2 1

A CIP catalog record for this book is available from the British Library.

Library of Congress Cataloging-in-Publication Data

Evans, Harry B.
Aqueduct hunting in the seventeenth century : Raffaello Fabretti's De aquis et aquaeductibus veteris Romae / by Harry B. Evans.
p. cm.
Includes bibliographical references and index.
ISBN 0-472-11248-1 (cloth : alk. paper)
1. Fabretti, Raffaele, 1618–1700. De aquis et aquaeductibus veteris Romae. 2. Water-supply—Italy—Rome. 3. Aqueducts—Italy—Rome. 4. Rome (Italy)—Antiquities. I. Title.
TD280.R6 E93 2002
628.1'5'0937—dc21 2002000562

UXORI OPTIMAE

Acknowledgments

My debts in completing this study of Fabretti's *De aquis* are many and varied. First, I am greatly indebted to the Library of the American Academy in Rome, where most of the research was done, and to its ever helpful staff, as well as to the indispensable Biblioteca Apostolica Vaticana with its incomparable collection of books and manuscripts, most important among them for my study being the 1788 Barbiellini edition of Fabretti's text. I am most grateful to the administration of the Vatican Library for kindly permitting me to reproduce Fabretti's map of the Roman Campagna, as well as to the Beinecke Rare Book and Manuscript Library at Yale University for permission to photograph and reproduce the other figures from the original edition of 1680.

Fordham University generously supported my work through a Faculty Fellowship during the 1996–97 academic year, a reduction in teaching load in the spring of 1999, and several grants to offset travel and research expenses. I am indeed grateful to my home institution for this commitment to scholarly research.

Many scholars and colleagues on both sides of the Atlantic have also been most helpful. Robert Rodgers very kindly made available to me the text and commentary of his forthcoming edition of Frontinus's *De aquaeductu;* aqueduct hunters all over the world have been eagerly awaiting its publication, and it was a tremendous help to have a sneak preview of it. In addition, I am greatly indebted to Giovanna Tedeschi Grisanti, who encouraged me in this undertaking some six years ago; to Lorenzo Quilici, who generously shared with me his extensive knowledge of both Rome's aqueduct system and Fabretti's topographical stomping grounds; and to the readers for the University of Michigan Press and my expert copy editor Jill Butler Wilson, who caught many infelicities and errors in the manuscript I submitted. Of course, I take full responsibility for the mistakes that remain.

After this book had already gone to page proof, Christer Bruun very kindly made available to me a copy of a recently published article directly relevant to my study, "Frontinus, Pope Paul V, and the Aqua Alsietina/ Traiana Confusion," *BSR* 69 (2001): 299–315. I am happy to say that we are in substantial agreement about the seventeenth-century confusion Fabretti attempted to clarify in his first dissertation, and it is a pleasure to acknowledge this assistance, one more example of the valuable help I have received from colleagues in completing this work.

My greatest debt is to my wonderful wife, Fran, to whom this book is lovingly dedicated. Although hardly an ardent fan of aqueduct hunting, she has kindly put up with one. In addition, as an expert editor, she read through my final draft, catching all sorts of embarrassing mistakes and improving my translation and prose style at many points. Her affectionate help and support in bringing this project to its conclusion mean more to me than I can say.

Contents

Figures

Figures 1–36 are courtesy of the Beinecke Rare Book and Manuscript Library, Yale University.

Abbreviations

Abbreviations in the text and notes follow closely those listed in the *Oxford Classical Dictionary* (3d ed., 1996) and *AJA* 95 (1991): 1–16. Wherever possible, works are cited from reprint editions indicated in Works Cited. Works cited frequently are abbreviated as follows:

Aicher	Peter J. Aicher. *Guide to the Aqueducts of Ancient Rome*. Wauconda, Ill., 1995.
Angeloni	Francesco Angeloni. *L'istoria augusta da Giulio Cesare a Costantino il Magno illustrata con la verità dell'antiche medaglie*. Rome, 1641.
Aringhi	Paolo Aringhi. *Roma subterranea*. Rome, 1651.
Ashby	Thomas Ashby. *The Aqueducts of Ancient Rome*. Oxford, 1935. Published in Italian as *Gli acquedotti dell'antica Roma* (Rome, 1991).
Ashby, *Campagna*	Thomas Ashby. *The Roman Campagna in Classical Times*. London, 1927.
Barbiellini	Natale Barbiellini, pub. *Raph. Fabretti Gasparis F. Urbinatis de aquis et aquaeductibus veteris Romae dissertationes tres, editio secunda romana adnotationibus illustrata*. Rome, 1788.
Barrington Atlas	R. J. A. Talbert, ed. "*Barrington Atlas of the Greek and Roman World*" and "*Map-by-Map Directory*." 3 vols. Princeton and Oxford, 2000.

Blackman and Hodge	Deane R. Blackman and A. Trevor Hodge. *Frontinus' Legacy: Essays on Frontinus' "de aquis urbis Romae."* Ann Arbor, 2001.
Bruun	Christer Bruun. *The Water Supply of Ancient Rome: A Study in Roman Imperial Administration*. Commentationes Humanarum Litterarum 93. Helsinki, 1991.
Clüver	Philip Clüver. *Italia antiqua, auctoris methodo, verbis et tabulis geographicis retentis*. Wolfenbüttel, 1659.
DBI	Alberto M. Ghisalberti, ed. *Dizionario biografico degli Italiani*. Rome, 1960–.
Donati	Alessandro Donati, S.J. *Roma vetus ac recens utriusque aedificiis ad eruditam cognitionem expositis*. Rome, 1638. (= Graevius, 3:541–938.)
DSB	Charles Coulston Gillispie, ed. *Dictionary of Scientific Biography*. New York, 1970–.
EHCA	Nancy T. de Grummond, ed. *An Encyclopedia of the History of Classical Archaeology*. 2 vols. Westport, Conn., 1996.
Fabricius	Georg Fabricius [Georgius Fabricius Chemnicensis]. *Descriptio urbis Romae*. Rome, 1550. (= Graevius, 3:457–540.)
FI 1.10	Lorenzo Quilici. *Collatia*. Forma Italiae 1.10. Rome, 1974.
Frutaz	Amato Pietro Frutaz. *Le piante di Roma*. 3 vols. Rome, 1962.
Fulvio	Andrea Fulvio. *Dell'antichità della città di Roma*. Venice, 1543.
Graevius	Johannes Georgius Graevius. *Thesaurus antiquitatum romanarum congestus*. 2d ed. 12 vols. Venice, 1732.
Gruter	Janus Gruter. *Inscriptiones antiquae totius orbis Romani in corpus absolutum simul redactae*. Heidelberg, 1603.
Helbig^4	W. Helbig. *Führer durch die öffentlichen Sammlungen klassicher Altertümer in Rom*. Ed. H. Speier. 4th ed. 4 vols. Tübingen, 1963–72.
Hodge	A. Trevor Hodge. *Roman Aqueducts and Water Supply*. London, 1992.
Holste	Lucas Holste [Lucas Holstenius]. *Annotationes in Geographiam sacram Caroli a S. Paolo, Italiam antiquam Cluverii, et Thesaurum geographicam Ortellii*. Rome, 1666.
Jordan	H. Jordan. *Topographie der Stadt Rom im Alterthum*. 2 vols. in 4. Berlin, 1871–85.
Kircher	Athanasius Kircher, S.J. *Latium: Id est, nova et parallela Latii tum veteris tum novi descriptio*. Amsterdam, 1671.

Lanciani	Rodolfo Lanciani. *I commentarii di Frontino intorno le acque e gli acquedotti*. MemAccLinc 3d ser., 4 (1881): 213–616.
Lanciani, *FUR*	Rodolfo Lanciani. *Forma Urbis Romae*. Milan, 1893–1901.
Ligorio, *Paradosse*	Pirro Ligorio. *Libro delle antichità di Roma, nelquale si tratta de circi, theatri e anfitheatri, con le paradosse del medesimo autore, quai confutano la commune opinione sopra varii luoghi della città di Roma*. Venice, 1553.
LIMC	*Lexicon Iconographicum Mythologiae Classicae*. 9 vols. in 18. Zurich, 1981–97.
Lipsius	Justus Lipsius. *Admiranda sive, de magnitudine Romana libri quattuor*. Ed. ultima. Antwerp, 1617.
LTUR	E. Margareta Steinby, ed. *Lexicon Topographicum Urbis Romae*. 6 vols. Rome, 1993–2000.
L'urbs	*L'urbs: Espace urbain et histoire (Ier siècle avant J-C.–IIIe siècle après J-C.)*. CÉFR 98. Rome and Paris, 1987.
Nardini	Famiano Nardini. *Roma antica*. Rome, 1666.
Nash	Ernest Nash. *A Pictorial Dictionary of Ancient Rome*. 2d ed. 2 vols. London, 1968.
Nolli	Giambattista Nolli. *Rome, 1748: The Pianta grande di Roma of Giambattista Nolli in facsimile*. Highmount, N.Y., 1984.
Oisel	Jacques Oisel. *Thesaurus selectorum numismatum*. Amsterdam, 1677.
Pace	P. Pace. *Gli acquedotti di Roma e il "De aquaeductu" di Frontino: Con testo critico, versione, e commento*. Rome, 1983.
Panimolle	Giuseppe Panimolle. *Gli acquedotti di Roma antica*. Rome, 1963.
Peto	Luca Peto [Lucas Paetus]. *De mensuris et ponderibus Romanis et Graecis cum his quae hodie sunt collatis libri V*. Venice, 1573. (= Graevius, 11:1608–73.)
Pfeiffer	Rudolf Pfeiffer. *History of Classical Scholarship from 1300 to 1850*. Oxford, 1976.
Pökel	Wilhelm Pökel. *Philologisches Schriftsteller-Lexicon*. Leipzig, 1882.
Poleni	Giovanni Poleni. *Sex. Iulii Frontini de Aquaeductibus Urbis Commentarius*. Padua, 1722.
RIC^2	Harold Mattingly, Edward Allen Sydenham, and Robert Andrew Glindinning Carson. *The Roman Imperial Coinage*. 6 vols. London, 1923–61.

Richardson	L. Richardson, jr. *A New Topographical Dictionary of Ancient Rome*. Baltimore and London, 1992.
Richmond	Ian Archibald Richmond. *The City Wall of Imperial Rome*. Oxford, 1930.
Roma sotterranea	R. Luciani, ed. *Roma sotterranea*. Rome, 1984.
Sandys	John Edwin Sandys. *A History of Classical Scholarship*. 3d ed. 3 vols. Cambridge, 1921.
Tomassetti	Giuseppe Tomassetti. *La campagna romana antica, medioevale e moderna*. 4 vols. Rome, 1910–26.
Trionfo	Giuseppina Pisani Sartorio and Anna Maria Liberati Silverio, eds. *Il trionfo dell'acqua: Acque e acquedotti a Roma, IV. sec. a.C.–XX. sec.* Rome, 1986.
Trionfo II	Giuseppina Pisani Sartorio and Anna Maria Liberati Silverio, eds. *Il trionfo dell'acqua: Atti del convegno "Gli antichi acquedotti di Roma: Problemi di conoscenza, conservazione, e tutela."* Rome, 1992.
Van Deman	Esther Boise Van Deman. *The Building of the Roman Aqueducts*. Washington, D.C., 1934.
VZ	Roberto Valentini and Giuseppe Zucchetti. *Codice topografico della città di Roma*. 4 vols. Rome, 1940–53.
Wikander, *Handbook*	Orjan Wikander, ed. *Handbook of Ancient Water Technology*. Technology and Change in History 2. Leiden, 2000.

Introduction

Since the early days of the Renaissance, aqueduct hunting in and around Rome has been a favorite outdoor sport for both casual visitors to the city and scholars interested in the topography and monuments of ancient Rome itself. The activity is documented as early as the mid–fifteenth century, when Flavio Biondo describes an expedition of Pope Pius II and his court in September 1461 to the valley of the Empiglione east of Tivoli and provides an account of the aqueducts seen there and along the Via di Carciano.[1] Aqueduct hunting continues to this day: the recent publication of a detailed guide to the ruins, both inside and outside the modern city of Rome, is clear indication that interest in these imposing monuments to Roman greatness is not flagging and will no doubt continue to flourish.[2] Indeed, ongoing concerns about the preservation

1. See Ashby, 1 (19–20). Throughout, citations of Ashby's *The Aqueducts of Ancient Rome* give the pagination of both the original edition (Oxford, 1935) and, in parentheses, the Italian translation (Rome, 1991).

2. Peter Aicher has recently provided a detailed and up-to-date guide, with a full bibliography and accurate maps and directions to the remains, effectively superseding the guide of E. M. Winslow, *A Libation to the Gods: The Story of the Roman Aqueducts* (London, 1963). Filippo Coarelli's *Dintorni di Roma* (Rome, 1981) in the Laterza series of archaeological guides includes a section on the aqueducts (227–30). The recent publications of Italia Nostra, particularly Lorenzo Quilici's *La Via Prenestina: I suoi monumenti e suoi paesaggi*, Passeggiate nel Lazio 2 (Rome, 1969), and the publication of *Trionfo* and *Trionfo* II show that interest in the aqueducts is still strong. See also L. Quilici, "Gli acquedotti di Roma," *Archeo. attualità del passato* 53 (July 1989): 51–97.

and protection of the conduits in the city and Campagna point to their critical importance as part of the patrimony of Roman civilization.[3]

Serious aqueduct hunting had its beginning in the seventeenth century, with Raffaello Fabretti's *De aquis et aquaeductibus veteris Romae dissertationes tres*, our earliest topographical work focused on the aqueducts of the ancient city. Since its first publication in 1680, Fabretti's *De aquis* has received ample praise from topographers and ancient historians. Rodolfo Lanciani, who cites Fabretti often, describes him as "oculatissimo ricercatore del vero" [a most observant investigator of the truth].[4] Thomas Ashby, in his own magisterial study of the Roman aqueduct system, an irreplaceable work recently republished in Italian translation, describes Fabretti's treatise as "among the most valuable works on the subject."[5] In addition, Lorenzo Quilici, in his comprehensive and closely focused study of Collatia, has praised Fabretti's work as "opera titanica . . . per noi la prima grande opera di ricerca storica-topografica del territorio" [a titanic work, for us the first grand work of historical-topographical research of the territory], citing his "attente descrizioni e vagliate notizie, spesso oggi irrepetibili" [careful descriptions and closely examined reports, often not to be repeated today].[6]

In marked contrast to much of the topographical study of the *ager Romanus* (Roman Campagna) in the sixteenth through eighteenth centuries, Fabretti's work on the aqueducts had an enormous impact on later scholarship; all commentators on Frontinus and the aqueducts, from the early eighteenth century to the present, cite him frequently.[7] A complete reprint of the original 1680 edition of the *De aquis* is available within the series The Printed Sources of Western Art, chosen no doubt for inclusion within that collection because of the many carefully detailed plans and maps it presents.[8] Yet to date, no scholarly assessment of Fabretti's treatise

---

3. See, for example, the many discussions of the problem presented in different articles in *Trionfo* II, 175–259.

4. Lanciani, 352.

5. Ashby, 1 (20).

6. FI 1.10, 18.

7. The practice began with Giovanni Poleni's famous edition, *Sex. Iulii Frontini de Aquaeductibus Urbis Romae Commentarius* (Padua, 1722), and continues to the present day; a full listing is unnecessary.

8. R. Fabretti, *De aquis et aquaeductibus veteris Romae dissertationes tres* (Rome, 1680; reprint, 1788; original edition reprinted in the series The Printed Sources of Western Art, ed. T. Besterman, no. 3 [Portland, Oregon, 1972]).

as a whole has been attempted, and no translation of his original Latin has appeared in any modern language.

In some ways, it is not surprising that no translation of the original has been attempted: Fabretti's Latin is itself baroque and quirky, and it is certainly not presented in a style that recommends itself for translation into idiomatic English. The result, however, is that his work is not readily accessible to modern scholars, not to mention the much larger group of general readers interested in the history of archaeology and epigraphy.

The present study, a full English translation of the *De aquis*, with commentary focusing primarily on the topographical problems discussed in the text, seeks to make accessible to a much wider audience this important work of a leading scholar of the seventeenth century—and certainly that century's most important aqueduct hunter—and to demonstrate more clearly Fabretti's contributions to our understanding of the Roman aqueduct system.

The *De aquis* is worth much closer attention than it has received, especially since Fabretti frequently cites and describes physical evidence in the Roman Campagna that has long since disappeared or been destroyed. But even more important than Fabretti's text itself is the evidence it provides for assessing the state of topographical research in Rome during his period, a time of intense interest in antiquity. The *De aquis* gives us an extremely valuable picture of a serious scholar at work, writing for an equally serious audience of fellow learners and attempting to put archaeological investigations on a sound scientific basis by integrating them with ancient literary evidence.

As it has well been stated in a recent exemplary study of Joseph Justus Scaliger, "It is not easy to meet the great scholars of the later sixteenth century on terms of intimacy."[9] Much the same can be said for a pioneer epigrapher and aqueduct hunter working a century later, in Rome of the seicento. But given the importance of both Fabretti's topic itself and the scholarly world in which he worked, it will be well worth our effort to do more than simply make his acquaintance.

9. A. Grafton, *Joseph Scaliger: A Study in the History of Classical Scholarship*, vol. 1, *Textual Criticism and Exegesis* (Oxford, 1983), 1.

CHAPTER ONE

Fabretti the Aqueduct Hunter

An informed assessment of a scholarly work, especially one of an era quite different from our own, must begin with an introduction to its author and his scholarly world.[1] Fabretti's *De aquis* is no exception: the three dissertations therein were completed within a two-year period between 1677 and 1679 but represent the fruits of long study and exploration of the Roman Campagna by one of the leading antiquarians and scholars of seventeenth-century Italy.[2]

Raffaello Fabretti (ca. 1619–1700) was among the most learned men of his day. Born the second of six sons within a prominent family of Urbino, he undertook an ecclesiastical career, in accordance with both the long established custom of primogeniture in force at the time and the prestige of his family. Following his education in the classics and law, first at

1. For summary accounts of Fabretti's life and career, see *DBI*, 43:739–42 (M. Ceresa); Sandys, 2:280; *EHCA*, 422. G. Mennella's *Il Museo Lapidario del Palazzo Ducale di Urbino: Saggio storico su documenti inediti* (Genoa, 1973) presents a valuable discussion focusing on Fabretti's epigraphical work (13–32), with a balanced overview of his career (13–18) and a full review of earlier biographical studies (13–14 n. 1). Among the more useful longer biographies, see in particular that of Fabretti's assistant Domenico Riviera, written in Latin soon after his death and later translated into Italian and published by G. M. Crescimbene in *Le vite degli arcadi illustri* (Rome, 1708), 1:89–111; see also G. Marotti, "Raphael Fabrettus," in *Vitae Italorum doctrina excellentium qui saeculis XVII et XVIII floruerunt*, ed. A. Fabroni (Pisa, 1780), 6:174–229.

2. The first dissertation is dated to December 31, 1677; the second to August 12, 1679; the third to October 27, 1679.

Urbino and Cagli and then in Rome, where he went in his late teens to complete his studies, Fabretti entered the priesthood, a calling that was quite congenial to a man of comfortable scholarly interests: he was able to devote to research and scholarship whatever time he had free from ecclesiastical duties.[3]

Fabretti's privileged background and personal ability led to important assignments within the church. He first undertook service for thirteen years as a papal diplomat in Spain (1651–64), where he continued his classical studies during his time not spent in ecclesiastical duties and where he appears to have formed his intense interest in archaeology. Recalled to Rome in 1664, he visited Paris and other European cities on his return trip, made with Cardinal Carlo Bonelli, the papal nuncio in Spain; on this journey, he became acquainted with leading scholars, such as Jean Mabillon and Bernard Montfaucon, with whom he corresponded and maintained a close friendship for the rest of his life.[4]

In Rome, Fabretti next served for six years as an ecclesiastical judge of appeals (1664–70), a position that prompted him to begin his own collection of inscriptions. Because farmers in the Roman Campagna regularly brought archaeological materials and inscriptions found on their property into the city for sale at ridiculously low prices, Fabretti began to purchase antiquities and inscriptions that became the nucleus of a private museum.[5] He was then called to his native city, Urbino, for three years (1670–73), as a member of the episcopal staff of Cardinal Carlo Cerri, but in 1673, he accepted the invitation of his patron Cardinal Gaspare Carpegna, the cardinal vicar of Rome, to return to Rome to oversee the publication of papal edicts. As superintendent of excavations of the catacombs and custodian of relics and antiquities (*custode delle ss. reliquie e dei cimiteri*), a position he held for more than ten years, Fabretti found major focus for the remainder of his life and career. This assignment gave him direct access to primary source materials and opportunities to excavate, as well as ample time for his own investigations and research, opportunities for which, in the introduction to the second dissertation in the *De aquis* (II.1), Fabretti warmly

3. Fabretti himself acknowledges much the same thing in his introduction to the second dissertation (II.1), addressed to Cardinal Carpegna.

4. When Mabillon visited Italy in 1685–86, Fabretti served as his personal guide. See *DBI*, 43:741 (M. Ceresa).

5. See Mennella, *Museo Lapidario*, 18–19.

thanks Carpegna, himself a well-known antiquarian and collector of books and coins.[6]

Fabretti's *De aquis* appeared in 1680, published both in Rome and in Paris. The treatise, the result of long interest in antiquity and extensive research undertaken throughout his life, was his very first publication, but it appeared only when Fabretti was already sixty years old. The three dissertations therein seem to have enjoyed an instant success; they quickly established his reputation as a leading antiquarian among European scholars, and there are frequent reports that students of antiquity consulted him regularly during visits to Rome.[7] Among his Roman associates, Fabretti was well known for making long excursions into the Campagna in search of antiquities and inscriptions, always riding the same horse, which friends jokingly named Marco Polo; the horse was reported to have had a unique ability to detect ancient monuments and inscriptions, as if by smell, and to have stopped near them, indicating their location. Fabretti himself, amused by the name given to the horse by his friends, even wrote a letter to one of them in Marco Polo's name.[8]

Fabretti is described as a tall and slender individual, with lively eyes and disheveled hair, which indicated to his contemporaries an intense seriousness.[9] He was also famous for carrying out his archaeological explorations under all sorts of conditions, regardless of weather and rough roads, not to mention physical danger to himself.[10] An excursion he undertook on the Via Nomentana in late 1699 at an advanced age resulted in a serious illness, to which he succumbed in early January 1700. Cardinal Carpegna, who witnessed Fabretti's will, and Cardinal Gianfrancesco Albani, who later became Pope Clement XI, visited Fabretti regularly at his home during his final sickness. Fabretti's death was mourned by all of Rome, where he had been given the extraordinary honor of citizenship, and he was buried, in accordance with his instructions, at the Church of

6. For Carpegna's career, see *DBI*, 20:589–91 (G. Romeo).

7. See *DBI*, 43:740 (M. Ceresa); Marotti, "Fabrettus," 195.

8. See Crescimbene, *Le vite*, 1:97; Marotti, "Fabrettus," 192–93. The story is repeated in numerous accounts of Fabretti's life and work: see, for example, Sandys, 2:280; *EHCA*, 422.

9. See Marotti, "Fabrettus," 224; Mennella, *Museo Lapidario*, 22–23.

10. His descriptions in the *De aquis* of exploring the conduit of the Aqua Appia (I.4e) and discovering a *cippus* (milestone) of the Via Sublacensis (II.4c) in the bed of the Anio River vividly illustrate this.

S. Maria Sopra Minerva, where his funeral monument, erected by his nephew Gaspare, may be seen today in the north aisle of the basilica.[11]

Fabretti's scholarly energy never seems to have flagged, and in the last twenty years of his life, he produced a number of other important topographical studies, much longer and more ambitious than the three dissertations on the aqueducts. Three years after the publication of the *De aquis*, Fabretti completed a monumental work on Trajan's Column, *De columna Traiani syntagma* (Rome, 1683), in a folio volume that included two additional monographs, the first scholarly discussion of the recently discovered Tabula Iliaca (*Explicatio veteris tabellae anagliphae Homeri Iliadem, atque ex Stesichoro Arctino et Lesche Ilii excidium continentis*) and a study of Claudius's engineering works for draining the Fucine Lake (*Emissarii lacus Fucini descriptio*).[12] Fabretti's study of Trajan's Column, a detailed discussion of the spiral reliefs of the monument, focusing on construction techniques and the military engineering represented on them, was in large part a sharply critical response to the recent major publication of the monument by Pietro Santi Bartoli and Giovanni Pietro Bellori [*Colonna Traiana . . . accresciuta di medaglie, iscrittioni e trofei da Gian Pietro Bellori* (Rome, 1672)] and a more sympathetic correction of an earlier interpretation of the column by the Spaniard Alfonso Chacon (1540–99); Fabretti included in his volume the complete text of Chacon's *Historia utriusque belli Dacici a Traiano Caesare gesti* (Rome, 1576).[13]

Fabretti's ecclesiastical career was far from over at the age of sixty. In the years following the publication of the *De aquis*, Fabretti served as secretary of papal memorials under Pope Alexander VIII (1689–91), who conferred on Fabretti canonries in two churches in Rome, S. Lorenzo in Damaso and S. Pietro. In 1691, Pope Innocent XII appointed Fabretti prefect of the Archivio Apostolico at Castel Sant'Angelo, a highly sensitive assignment that points to papal confidence in his judgment and discretion. Fabretti's archaeological and epigraphical studies continued, and he also carried on extensive correspondence with leading scholars of his day, while finishing his largest and most important work, *Inscriptionum anti-*

11. For the text of Fabretti's funerary inscription, composed by his assistant Domenico Riviera, see Crescimbene, *Le vite*, 1:103; Marotti, "Fabrettus," 228.

12. The work had been in progress for some time, because Fabretti alludes to his study of the column in his second dissertation (II.4d).

13. See *DBI*, 43:740 (M. Ceresa); Marotti, "Fabrettus," 200–201.

quarum quae in aedibus paternis asservantur explicatio et additamentum (Rome, 1699), a monumental compendium of more than four thousand ancient inscriptions, intended to be a supplement to and updating of Janus Gruter's magisterial corpus published early in the seventeenth century.[14] More than four hundred of these inscriptions were in Fabretti's own collection at Urbino; his holdings later became the nucleus of the epigraphical collection of the Palazzo Ducale in that city.[15] A major work long in progress on the topography of ancient Latium (*De veteri Latio*), to which he alludes twice in the *De aquis* (I.4g, III.6), was left unfinished at the time of his death.[16]

Fabretti is worth attention not simply for his own highly significant contributions to scholarship but also because of his position as the leading antiquarian and epigrapher of his day. He was a man in the center of Rome's scholarly world, well connected with figures prominent in the city's learned circles, among them Giovanni Ciampini, founder of the Accademia Fisico-matematica in the city, and many individuals in Queen Christina's Accademia Reale, which later became the nucleus of the famous Accademia degli Arcadi established in 1690. Fabretti himself was a member of the Arcadians, by whom he was called, by pastoral name, "Iasitheus Nafilius," and he appears to have been active in the Academy.[17] He names throughout the *De aquis* many figures prominent in the intellectual circle of Queen Christina.

Rome was a major center of scholarly and intellectual interest at this time, a reputation it had enjoyed since the flourishing of Renaissance humanism.[18] Although the political influence of the papacy had declined in the aftermath of the Thirty Years' War, with the court of Louis XIV in France becoming the leading power in Europe in the last half of the sev-

14. Gruter's *Inscriptiones antiquae totius orbis Romani in corpus absolutum simul redactae* (Heidelberg, 1603), the *CIL* of its day and in many ways the symbol of seventeenth-century scholarship, certainly needed updating in light of the many discoveries since its publication. See Mennella, *Museo Lapidario*, 26–27.

15. See M. Luni and C. Gori, *1756–1986: Il Museo Archeologico di Urbino* (Urbino, 1986), 15–23.

16. See Ashby, 3 n. 2 (21 n. 18). Fabretti's notes for his work left unfinished seem to have been lost after his death; see Mennella, *Museo Lapidario*, 30–31.

17. See *DBI*, 43:741 (M. Ceresa); Marotti, "Fabrettus," 225. For a brief history of the Arcadians, see H. Gross, *Rome in the Age of Enlightenment* (Cambridge, 1990), 286–97.

18. For a presentation of this heritage, see A. Grafton, ed., *Rome Reborn: The Vatican Library and Renaissance Culture* (Washington, D.C., 1993).

enteenth century,[19] the city of Rome was still a most stimulating place. Thanks to its growth during the previous two centuries, the cultural impact of the Counter-Reformation, the rise of baroque architecture, recent advances in scientific research, and the ambitious urban renewal projects of popes from Sixtus V (1585–90) through Alexander VII (1657–69), Rome was effectively transformed into the image of the grand theater of the world.[20] Fabretti played a key role in the scientific and scholarly life of the city, and his house in the Borgo, built according to a design by Baldassare Peruzzi, became the center of conversations, discussions, and meetings among Roman antiquarians.

All of Fabretti's research merits further analysis for understanding how a seventeenth-century scholar worked and communicated with fellow scholars. This book focuses on Fabretti's treatise on the aqueducts. His first publication, the *De aquis*, is a work of manageable scope that shows this aqueduct hunter at his most innovative and original, engaged in a topic to which his own investigations made extremely important contributions.

19. For the political situation in late seventeenth-century Italy, see G. Spini, "Italy after the Thirty Years War," in *The Ascendancy of France, 1648–88: The New Cambridge Modern History*, ed. F. L. Carsten (Cambridge, 1964), 5:458–73.

20. For a general account of intellectual life of the period, see M. Petrocchi, *Roma nel Seicento*, Storia di Roma 14 (Bologna, 1970), 134–52. More specific, although not focused directly on Fabretti's period, is Gross, *Rome*, 15–23, 247–53. On the impact of papal building, see R. Krautheimer, *The Rome of Alexander VII* (Princeton, 1985).

CHAPTER TWO

Fabretti's *De aquis*

Given the huge size and highly ambitious scope of his later works, Fabretti's three dissertations on the aqueducts in the *De aquis* are hardly to be considered the primary achievement of his scholarly career. But this early publication must be considered the most significant of all his writings: as the first serious topographical study focused on the aqueducts of the ancient city, it had enormous influence on all later scholarship on the subject.

Fabretti's dissertations appear to have become a standard topographical reference soon after their publication. Certainly, they were read and quickly became controversial in the scholarly world. In a digression in his third dissertation (III.6), Fabretti severely attacked the Dutch scholar Jacob Gronovius for an emendation of Livy recently published in his *Epistolae in quibus multa Titi Livii loca geographica emendantur et illustrantur* (Amsterdam, 1678). Gronovius published a sharply polemical reply, *Responsio ad cavillationes Raphaelis Fabretti* (Leiden, 1684), to which Fabretti counterreplied under his Arcadian pastoral name, as "Iasitheus Nafilius," in an equally spirited polemical treatise, *Isathei ad Grunnovium apologema, in eiusque Titilivitia sive somnia de Tito Livio animadversiones* (Naples, 1686).[1]

Such scholarly controversy, while in many ways typical of the learned

1. See Ashby, 2–3 (21–22); Marotti, "Fabrettus," 195–96.

world in the seventeenth century, is clear indication that Fabretti's work was being noted—and not simply in Roman circles. All three dissertations were included in Johannes Graevius's massive twelve-volume collection of treatises on Roman civilization, the *Thesaurus antiquitatum romanarum congestus* (Venice, 1732).[2] Just forty years after its initial publication, Fabretti's work was cited frequently by Giovanni Poleni in his famous edition and commentary on Frontinus, *Sex. Iulii Frontini de Aquaeductibus Urbis Romae Commentarius* (Padua, 1722); Poleni also directly reproduces many of Fabretti's figures in his commentary.[3] The *De aquis* was also cited repeatedly by Alberto Cassio in his *Corso dell'acque antiche* (Rome, 1756–57).

Fabretti's dissertations were reissued in a second edition published by Natale Barbiellini in the late eighteenth century (Rome, 1788), with extensive footnotes added by an unidentified commentator.[4] Included in this second edition was a map of the Roman *suburbium* originally published in the *Ad Grunnovium apologema* and intended for Fabretti's study of the Roman *suburbium* still unfinished at the time of his death.[5] The notes of Barbiellini's edition are, however, primarily bibliographical, with virtually no discussion of topographical problems. Because later topographers of Rome and the *suburbium* focused on subjects other than the aqueducts, Fabretti's treatise remained the single most important study of Rome's water system until Rodolfo Lanciani published his monumental commen-

2. Graevius, 4:1677–1778.

3. Poleni (26) writes: "Illud tamen praetermittere non possum, quod pertinet ad Dissertationes de Aquaeductibus Raphaelis Fabretti; quippe quia mihi haud mediocre lumen attulere: etsi enim Fabrettus illustrare Frontini commentarium sibi non praeposuerat; attamen, cum res multas ex reliquiis antiquitatis summa cura vestigatas, incredibili diligentia mensuris definitas, bono in lumine collocaverit, ope earundem facilius deinde fuit etaim quaedam a Frontino tradita exponere atque illustrare. Sane ille suppediavit nonnullas figuras singulari fide delineatas, quas si non habuissemus ab illo, comparare tamen nobis ad institutum nostrum, debuissemus" [I cannot omit the following observation pertinent to Fabretti's dissertations on the aqueducts, since they have brought considerable illumination to me. Although Fabretti had not planned a commentary on Frontinus, nevertheless, since he brought to light many things from ancient remains, researched with the greatest care and measured with incredible diligence, it has been considerably easier to set forth and illustrate certain things handed down by our author. Indeed, Fabretti has furnished us some figures rendered with extraordinary accuracy; if we had not had them as a result of his work, we would have had to prepare them for our own undertaking].

4. The publisher, in his dedication to Pope Pius VI, describes Fabretti's treatise as "longo temporis intervallo perrarum sane, et emunctae naris viris iamdudum optatum" [quite rare and long sought by men of refinement] (Barbiellini, vi), but he does not identify the author of the notes. See Ashby, 2 n. 2 (20 n. 13).

5. The map appears opposite p. 90 in the second edition. See Ashby, 3 n. 3 (21 n. 19).

tary on Frontinus's treatise *De aquaeductu urbis Romae* in the late nineteenth century.[6]

Much of Fabretti's achievement in the *De aquis* is owed to his extensive learning and firsthand knowledge of the Roman Campagna, combined with a highly original approach to his topic. Before he published his *De aquis*, there had been relatively little scholarly interest in the water system of ancient Rome; indeed, the attention the ancient aqueducts had received was largely practical, as a result of ongoing efforts to repair existing conduits to supply water needs of the growing city of Rome in the sixteenth and seventeenth centuries.

Pope Nicholas V had begun the process of repairing the conduits, having first undertaken the reconstruction of an ancient Roman aqueduct in his restoration of the Aqua Virgo in 1453, a project completed only after eighty years, in 1570, under Pius V.[7] Less than twenty years later, Sixtus V tapped the sources of the ancient Aqua Alexandrina to introduce his own Acqua Felice to the hills of Rome, thus opening higher elevations in the eastern regions of the city for residential development.[8] In the early seventeenth century, Paul V reworked the ancient Aqua Traiana as the Acqua Paola to bring water to the Janiculum, with branches to Trastevere and the Vatican.[9]

Scholarly interest in the aqueducts was certainly not lacking, but it was primarily philological in its approach, focusing on the text of Frontinus's *De aquaeductu*.[10] In 1425, the book hunter Poggio Bracciolini had discovered at Monte Cassino the oldest surviving manuscript of Frontinus's commentary on Rome's water system, written in A.D. 97 when Frontinus began his service as *curator aquarum* under Nerva; Poggio's manuscript was a copy made by the twelfth-century scholar and writer Peter the Deacon around

6. R. Lanciani, *I commentarii di Frontino intorno le acque e gli acquedotti*, *MemAccLinc* 3d ser., 4 (1881): 213–616. Ashby, 1–9 (19–25) provides a review of topographical study of the aqueducts up to the time of Lanciani. See also Pace, 13–16.

7. See R. Motta, "L'acquedotto Vergine (sec. XVI–XVIII)," in *Trionfo*, 205–8.

8. See R. Motta, "L'acquedotto Felice," in *Trionfo*, 220–25.

9. The papal inscriptions on the terminal fountain of the Acqua Paola and Arco del Tiradiavoli over the Via Aurelia Antica wrongly identify the water as that of the Aqua Alsietina, an observation made by Fabretti himself twice in the *De aquis* (I.4g, II.4a). On the Acqua Paola, see C. Cancellieri, "L'acquedotto Paolo (sec. XVII–XX)," in *Trionfo*, 225–31.

10. Poleni (25–26) provides a comprehensive but surprisingly short list of writers who had treated the subject up to the eighteenth century ("De scriptoribus rerum ad Aquaeductus pertinentium"), dividing them into three main groups: (1) those interested in legal matters, (2) architectural writers, and (3) students of *opera Romana* (Roman construction).

the year 1130. The editio princeps of Frontinus's treatise, based on a fifteenth-century copy of this manuscript, appeared in Rome in 1489, together with the text of Vitruvius; a second edition, published in 1513 by Giovanni Giocondo in Florence, inspired later publications of it throughout the sixteenth and into the seventeenth centuries, many of them combined with Vitruvius's treatise on architecture.[11]

There was in the fifteenth and sixteenth centuries a renewed general interest in the antiquities and ruins of ancient Rome, which had become a popular destination for collectors and artists in search of classical culture, as well as pilgrims visiting the seat of the church. A well-established tradition of publications about the city existed from late antiquity through the Middle Ages and into the Renaissance. These works made little attempt to be systematic or scientific in their descriptions of the ancient monuments until the fifteenth century, when Poggio Bracciolini's *Historiae de fortunae varietate libri quatuor* (Rome, 1448), in an advance over earlier guidebooks, presented detailed descriptions of many of the city's antiquities, and when Flavio Biondo's *Roma instaurata* (Verona, 1481–82) introduced a truly systematic approach to the topography of ancient Rome; Biondo's work was not superseded until the publication of Bartolomeo Marliani's *Antiquae Romae topographia libri septem* in 1534.[12]

In the sixteenth century, the prominent antiquarian and cartographer Pirro Ligorio (ca. 1513–83) had included notices on water and aqueducts in his encyclopedic volumes on classical antiquity,[13] but the only scholarly treatment focusing on the aqueducts themselves, *De aquis, in urbem Romam olim influentibus* (Venice, 1576) by Aldus Manutius II, was a sketchy account presenting little more than a summary of Frontinus and other literary sources and citations of a few well-known inscriptions.[14] Justus Lipsius had included a chapter on the aqueducts in his *Admiranda sive de magnitudine Romana libri quattuor* (Antwerp, 1617), which was essentially a

11. For a summary of the textual transmission of the *De aquaeductu*, see M. D. Reeve, "Frontinus," in *Texts and Transmission: A Survey of the Latin Classics*, ed. L. D. Reynolds (Oxford, 1983), 166–70. R. H. Rodgers is completing a new edition of Frontinus's treatise, with full commentary (Cambridge, forthcoming).

12. For a review of earlier topographical study, see *EHCA*, 550–55 (C. Sperling); Richardson, xxii–xxiv.

13. For a review of Ligorio's career and publications, see *EHCA*, 680–82 (R. W. Gaston).

14. Aldus Manutius II, *De aquis, in urbem Romam olim influentibus*, in *De quaesitis per epistolam libri III* (Venice, 1576), 26–46 ("Lib. I ep. II ad Io. Cardinalem Moronum"). Fabretti severely criticizes this work in the first dissertation in the *De aquis* (I.3).

collection of quotations from Pliny, Strabo, Cassiodorus, Procopius, and Frontinus.[15] This was the approach currently in vogue when Fabretti tackled the problem.

In the early seventeenth century, several highly influential guidebooks on ancient Rome appeared. *Roma vetus ac recens* (1638) by the Jesuit Alessandro Donati (1584–1640) and *Roma antica*, a large and enormously popular description of the city by Famiano Nardini (d. 1661) published posthumously in 1666, combined extensive quotations of ancient texts with accounts of ancient ruins in their present state, as well as later buildings in the modern city. However, these recently published works said relatively little about the water system of ancient Rome. Donati included in his guidebook a brief chapter on the aqueducts, focusing mainly on a discussion of the Aqua Virgo, remains of which had been discovered during construction of the Church of S. Ignazio; his treatment of the other lines was much more cursory.[16] Nardini's book, comprised of more than five hundred pages, followed an organization based on the fourteen regions of the ancient city, using the regional boundaries as topographical markers. Nardini devoted only a short section to the aqueducts of ancient Rome, citing Frontinus and well-known inscriptions, such as those at the Porta Maggiore.[17]

Lacking at this time was a systematic study of the Roman Campagna. Lucas Holste, the German librarian of the Vatican Library, had corrected and updated the topographical work of his mentor Philip Clüver in *Annotationes*, published posthumously in 1666.[18] Even more recently, the famous Jesuit polymath Athanasius Kircher had completed an illustrated historical and topographical account of Latium, *Latium: Id est, nova et parallela Latii tum veteris tum novi descriptio* (Amsterdam, 1671). Fabretti cites both of these works and criticizes them severely in the *De aquis* and elsewhere.[19] Holste had also collated the Monte Cassino manuscript of Frontinus and had prepared an annotated edition of the *De aquaeductu*, which was never published after his death in 1661.[20]

---

15. Lipsius, 157–61 ("Aquaeductuum mira"), reprinted in *Iusti Lipsi opera omnia* (Antwerp, 1637) 3:444–46.

16. See Donati, 296–302 ("Aqua Virgo, et Aliae Aquae, et Aquaeductus") = Graevius 3:830–40.

17. Nardini, 501–10 ("L'Acque").

18. See Ashby, 1–2 (19–20).

19. See Ashby, 3 (21).

20. See Ashby, 1–2 (19–20); Pace, 13.

Fabretti's three dissertations certainly draw from earlier scholarship. Indeed, the *De aquis* in many ways constitutes a dialogue with other antiquarians and scholars; the author demonstrates his familiarity and dissatisfaction with the works of his predecessors and contemporaries, as well as detailed knowledge of Frontinus and other ancient sources. But Fabretti's work also marks a significant advance in modern topographical study of the aqueducts and the Roman Campagna itself. Intended for a scholarly audience, rather than the casual reader or visitor to the city, each dissertation focuses on a particular topographical problem concerning the aqueduct system. A brief review shows that the three sections of the *De aquis* were planned as individual monographs, perhaps originally presented for informal presentation as they were individually completed, but also intended to complement each other when issued as a finished publication.

DISSERTATION I

The first dissertation in the *De aquis*, the shortest but most important of the three dissertations therein, treats the route and purpose of the ancient Aqua Alexandrina, the last of the ancient conduits introduced by Alexander Severus into Rome in A.D. 226. This aqueduct seems to have been of particular interest to Fabretti because Sixtus V had tapped its sources in the late sixteenth century to supply his Acqua Felice, the introduction of which made possible the urban development of Rome in the seventeenth century. Although Fabretti credits his colleague Adrien Auzout with identification of the Aqua Alexandrina itself, Fabretti's dissertation presents the earliest full discussion of the line (I.2), along with a detailed map of its route through the Roman Campagna (fig. 2), and demonstrates that the Alexandrina was a conduit separate from the nine aqueducts listed by Frontinus and from the other major imperial line, the Aqua Traiana of A.D. 109 (I.3). In his discussion of the sources of the Alexandrina, Fabretti also takes the opportunity to correct the recently published topographical work of Lucas Holste (I.7).

The first dissertation makes several other significant contributions to our understanding of the water system and topography of ancient Rome. First, it establishes the conduit levels at the Porta Maggiore and elsewhere within the city (I.4a, I.5). Fabretti appears to have been the first scholar to examine systematically the individual elevations of the aqueducts entering the city at Spes Vetus, thereby establishing the relative height of each (fig. 7). Although the conduits of the Aqua Claudia and Anio Novus had been identified by name through Claudius's inscriptions on the Porta Maggiore, as well as through references in Frontinus's text, and although the Aqua Marcia, Aqua Tepula, and Aqua Julia were also known through the Augustan and Antonine inscriptions at the Porta S. Lorenzo, no one before Fabretti seems to have noted or documented their respective levels.

Second, Fabretti's first dissertation identifies the Aqua Traiana (I.4g). Fabretti approaches the problem of its identification as a corollary to his main topic, primarily to demonstrate that the Aqua Alexandrina cannot be identified with a Trajanic reworking of the Aqua Marcia. Yet his reading of the evidence here had far-reaching results. In his argument that Trajan's aqueduct was a wholly new conduit, separate from any reworking of an earlier line and certainly not to be associated with the Aqua Alexandrina, he demonstrates that the aqueduct on the Janiculum reworked by Paul V for his Acqua Paola was indeed the Traiana.

Third, this dissertation identifies the "Trophies of Marius" as a *castellum* (or distribution tank) supplied by one of the Claudian aqueducts (I.4c). Before Fabretti's work, the "Trophies of Marius" were generally thought to have been supplied by the Aqua Marcia, specifically through a Trajanic reworking of that aqueduct. Fabretti rejects that theory, arguing that only the Claudian aqueducts were high enough to supply the *castellum*.

Finally, Fabretti's first dissertation gives the location of the Porta Capena, confirmed by observations on the underground course of the Aqua Appia in the immediate area (I.4d). Fabretti rejects the contemporary identification of the Porta Capena with the Porta S. Sebastiano of the Aurelian Wall, thus correcting a topographical error common at the time, and fixes more accurately the location of this gate of the republican "Servian" Wall.

DISSERTATION II

The second dissertation in the *De aquis*, the longest and most wide-ranging of the three, discusses the sources of the Aqua Marcia and Aqua Claudia in the upper Anio Valley east of Tivoli. Although Fabretti begins with a digression on the measurements of the ancient Roman foot and mile

(II.3) and includes a lengthy discussion on a possible representation of the Aqua Claudia, illustrated through comparison of similar reliefs of female divinities (II.4d), his main focus is the topography of the upper Anio. He provides an extremely detailed map of the area (fig. 17) and a description centering on two problems. The first problem concerns the course of the ancient Via Valeria, from which Frontinus reports statistics in the *De aquaeductu* concerning the sources and intakes of individual aqueducts. Fabretti, attacking the topographical work of Lucas Holste, argues that the original route of the Via Valeria ran directly from the Osteria della Ferrata, past Riofreddo, to S. Giorgio (II.4b). The second problem on which Fabretti centers his description concerns the sources of three ancient aqueducts: the Aqua Marcia, the Aqua Claudia, and the Aqua Augusta, the supplement added by Augustus to augment the Marcia's capacity (II.4c, II.4e–f).

In addition, Fabretti's second dissertation introduces other topics not directly germane to the topography of the upper Anio but intended to enlighten readers on points made by Frontinus. Here, Fabretti discusses the location and functioning of a settling tank of the Aqua Marcia on the Via Latina, providing shorter descriptions of settling tanks of the Aqua Claudia/Anio Novus and Aqua Virgo (II.5a–b). He also discusses here Frontinus's measurement of water (II.5c) and the identification of the Roman to be credited with the introduction of the Aqua Marcia (II.6).

DISSERTATION III

The third dissertation in the *De aquis* has a primarily literary, rather than archaeological or topographical, focus, examining the discrepancy between the fourteen aqueducts cited by Procopius and the twenty water sources listed in the fourth-century regionary catalogs. Arguing that ancient Rome must be understood as comprising not only territory within its walls but also outlying settlements on the main roads from the city, Fabretti presents two illustrations: a map giving detailed coverage of the Roman Campagna (already treated in the first dissertation), with a particular focus on settlements near the city (fig. 31); and a plan of the walls and gates of ancient Rome that illustrates the courses of main Roman roads issuing from them (fig. 32), the circuit of which Fabretti discusses in detail (III.3).

Fabretti next reviews Procopius's notice that the city was supplied by fourteen conduits, to identify those not named by Frontinus in the *De aquaeductu* (III.4a–b). The lengthy discussion that follows concerning settlements along the major roads outside the city (III.5) includes a detailed description of the warehouse along the Tiber south of the Aventine commonly identified today as the "Porticus Aemilia" (III.5a[1]) and an attempt (at III.5c) to make sense of a famous topographical puzzle, Pliny's discussion of the size of Rome based on the measurements of the roads from the Golden Milestone in the Roman Forum.

Finally, Fabretti turns to his primary topic, identification of the twenty water sources listed in the regionary catalogs (III.6). He gives a long digression on the topography of the *ager Tusculanus* and attacks Gronovius's emendation of Livy's text, before ending with a brief discussion of ancient drains and sewers in seventeenth-century Rome (III.7).

Together, the three dissertations of the *De aquis* constitute a roughly coherent whole, each treatise complementing the others in subject matter and focus. Although Fabretti was not attempting to present a comprehensive overview or complete account of the Roman aqueduct system, his first and third dissertations together come close to providing just that, supplemented by the second dissertation's discussions of the sources of the Marcia and Claudia in the upper Anio Valley and the functioning of settling tanks and problem of water measurement. Fabretti's overall title for his three dissertations is therefore apt: all three together focus on both the water sources and the conduits of the ancient city.

Fabretti's dissertations are much more than simply treatments on their primary subjects. Each is an impressive presentation in itself, including extensive topographical information, extremely careful sketches and diagrams, and detailed maps. Each draws from literary, epigraphical, and archaeological evidence in the development of its argument. It is time to let the aqueduct hunter's text speak for itself.

CHAPTER THREE

Dissertation I on the Waters and Aqueducts of Ancient Rome

To the distinguished Giovanni Lucio of Trau, Raffaello Fabretti, son of Gaspare, of Urbino, sends greetings.

1. INTRODUCTION

Thanks to your interest in my behalf, the things I wrote to you earlier about the aqueducts I observed around the Anio River do not at all displease me. You have influenced my diligence by your expressions of praise, both in your own name and in the names of your most learned friends (whom you also have in very large number). As a result, I feel that I am much more eager to pursue the investigation set forth on this subject; I would already have completed it had the abundance of waters from heaven not shown itself opposed to my own watery task. But you should not think that I have been completely idle: indeed, although I was not able to approach for a second time the sources of the Marcia and Claudia, at some distance from me, and not able therefore to follow up my ideas by surer reasoning, not uselessly, perhaps, will I show you that I have been engaged in the more immediate neighborhood of that aqueduct introduced by Pope Sixtus and called the Acqua Felice from his own name before his pontificate, a source of water also brought once before to Rome (as you will hear), though by another route.

2. THE AQUA ALEXANDRINA: COURSE, STATE OF REMAINS, DESCRIPTION, AND ROUTE

I have first classified the construction into a certain order, and I have discovered that all of it, both on arcade and substructure above ground, belongs to one and the same aqueduct. This construction, I have also found, occupies many valleys around the Via Labicana and, last of all, that plain between the Lake of Castiglione and Lake Regillus, in the highest part of which, toward the south, the course of this aqueduct emerges, as well as the sources of the Osa stream. Adrien Auzout of Rouen, a man of refined taste and a very shrewd investigator, no less of physical science than of antiquities, impressed this on me through many exchanges of letters while I was at Urbino and still skeptical about the matter; indeed, the similarity of construction everywhere, the width of the conduit, and its height do not allow me to doubt it. Learn, therefore, the line of the entire conduit attached in the topographic map [fig. 1].

You should not think that we have labored needlessly because Fr. Athanasius Kircher has already displayed of that region "the topography of Latium proper, delineated and measured not so much with the pen as his own steps," as he announces in his *Latium*.[1] To confess the truth, Kircher's map has been spoiled in unhappy ways—either because the man, most distinguished and most skilled in geometric matters, "thought it unworthy to look away from things incorporeal and perceived by reasoning alone to those tangible things that require ample leisure spent in common pursuits," as Plutarch [*Marc.* 14.6] hands down from Plato, or because he was betrayed by his lack of interest in physical structures. The result is that all things on his map appear lying this way and that, as if scattered by chance. Among these matters pertaining to our topic (for I pass over the others freely and willingly), in the very place near the town of Colonna from

1. Kircher, opposite p. 70 ("Tusculani territorii topographia"). Footnotes to the translation give bibliographical references to works cited by Fabretti in the text; however, all citations of ancient authors appear in square brackets within the text itself. In addition, all inscriptions cited from Janus Gruter's *Inscriptiones antiquae* (Heidelberg, 1603) are identified in the text by their listings in *CIL* and *ILS* when appropriate.

where this aqueduct is derived is said to be the beginning of the Aqua Virgo,[2] the sources of which are about ten miles distant from here—indeed, on the Via Collatina; at the eighth milestone from the city, according to Frontinus; and near the Anio River, which appears nowhere on Kircher's map.

Before turning to a fuller description, I will go beyond those things that are noted in general terms on my map [fig. 1]. It occurs to me to wonder at the fickleness of human affairs and the oblivion that threatens accomplishments, however distinguished. Besides the damages of time—which someone might believe that this aqueduct, given the strength of its construction, could scarcely have been able to suffer—the very name of the builder has been lost, along with the purpose of the line. As a result, I do not know whether more truthfully or more elegantly, reference should be made to the summation of our native poet:

> So then, if, successfully at one time, famous works
> Make war on Time, with slow step
> Time brings down both the works and names together.

Indeed, to be sure, huge constructions of this sort, built for the public good, did not deserve the same fate as the useless pyramids; in their case, Pliny [*HN* 36.79] rightly declares that "the authors of such great vanity have been extinguished by fortune most equitable." But most unworthily has the fame of this excellent and most useful construction vanished because of the fault of time and the silence of writers. Let us concede that the name of its builder had been surrendered to oblivion for some centuries up to this time. In no way, however, can it be said that this construction has collapsed completely on the ground, for in the very ruins and the tumbledown remains, there is still so much that we may behold as if awestruck. Just as the remains have so long resisted voracious time, they promise a certain everlasting endurance and the admiration of posterity.

The work itself appears everywhere to be of brick-faced concrete, with the outer surface of the sides finished and rows of bricks arranged exactly in a straight line. The squared piers fill a thickness of eight feet. The distance between the piers, or the opening of the arches, is everywhere twelve feet, except in the most distant valley in the direction of the

2. Kircher, opposite p. 70.

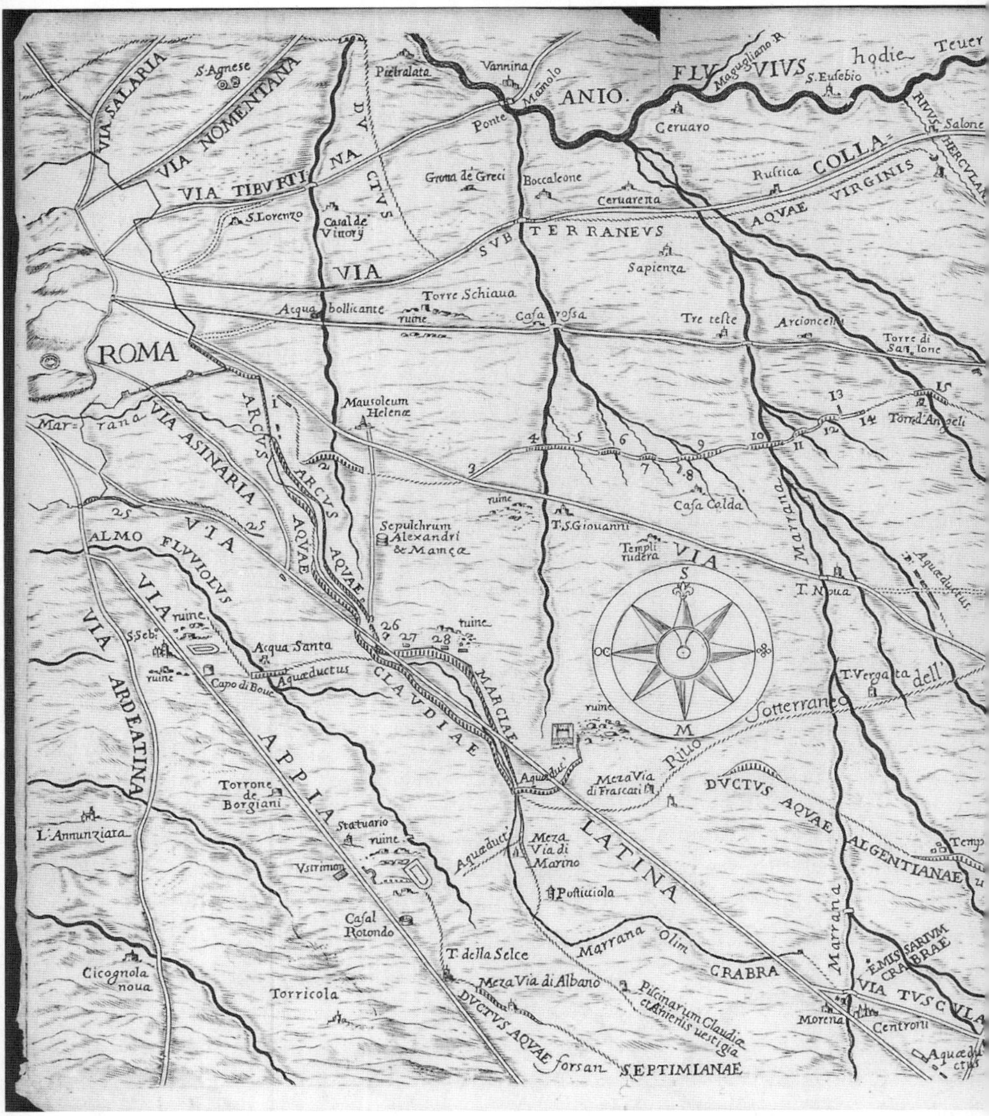

Fig. 1. Topographical map of Latium from Rome to Labicum showing the course of the Aqua Alexandrina and other aqueducts

1. Final substructure of this aqueduct in the direction of Rome in the vineyard of the Chartusians
2. Fifty-two arches in the valley of the Acqua Bollicante
3. Substructure after the intersection of the Via Labicana
4. Ninety-two arches, the highest of all, in the valley of the Casa Rossa
5. Seven arches of moderate height in the small valley following
6. Twenty-three arches in another valley
7. Eighteen arches in another
8. Branch line from the main conduit to the left, below the Casa Calda
9. One hundred two arches, which occupy two successive valleys
10. Twenty-eight arches in the valley of the Marrana
11. Twenty-two arches in the following valley, arranged in a curve
12. Fifty arches, likewise in a curve
13. Branch line from the main conduit to the right

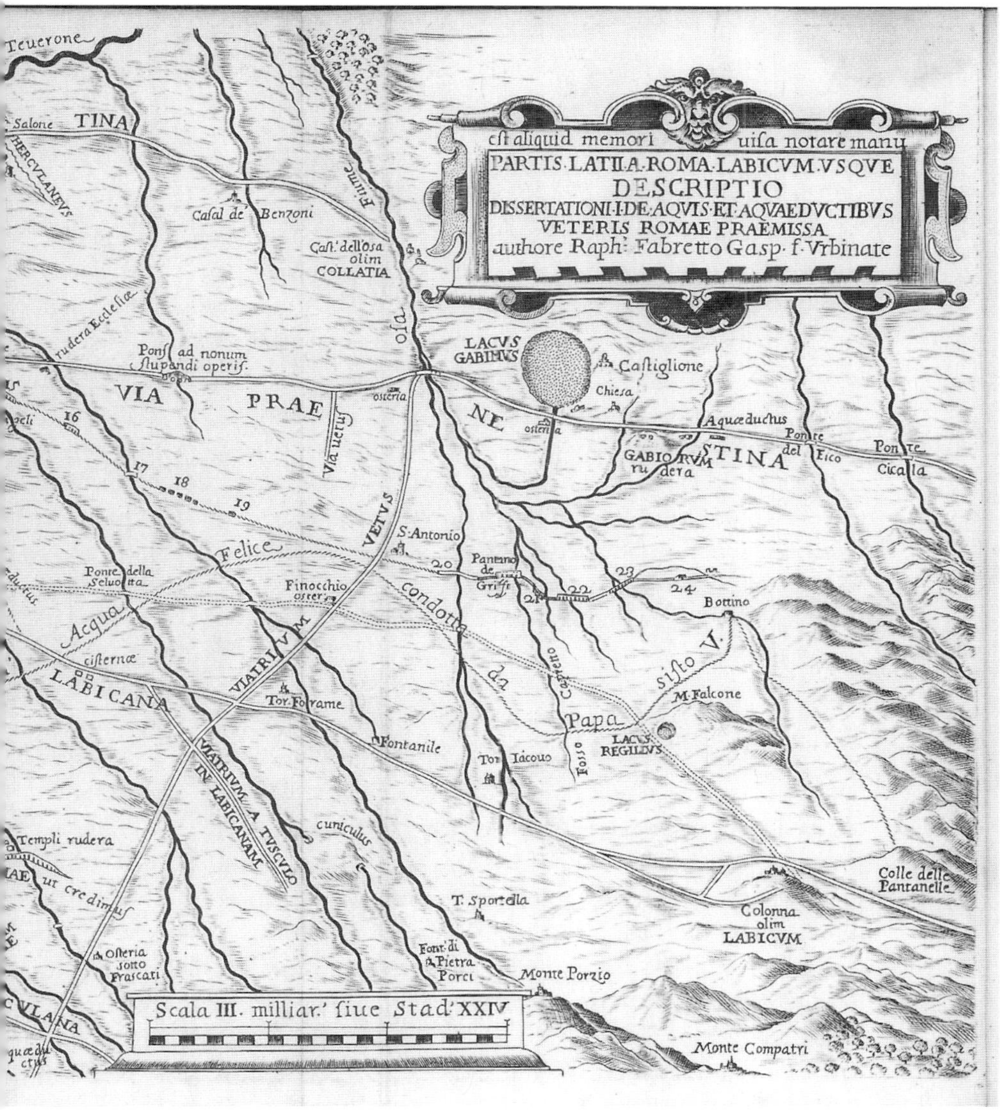

14. Single arch in the middle of a substructure of twelve paces
15. Twenty-eight arches, below the Casale di Tor Angela, closer to the Via Praenestina
16. Four arches, lower and almost buried
17. Single arch of rough Gabine stone, above this streambed
18. Five wellheads and evidence of another three, described separately
19. Another wellhead, like those cited already
20. New appearance of substructure after a long underground conduit, below the Casale di S. Antonio
21. Sixty-eight arches behind a stable, commonly called Pocoio de Pantano
22. Sixty-two arches in the middle of the plain
23. Forty-five arches, the first of this aqueduct
24. Settling tank, described separately
25. Remains of a conduit carrying a portion of the Anio Vetus, we believe
26. Ruins of a temple at the fourth milestone of the Via Latina, perhaps that of Fortuna Muliebris
27. Settling tank of the Marcia, we believe
28. Other settling tanks of the Aqua Julia and Aqua Tepula

sources, where it is ten and a half feet. Where the arches are lower, they are formed from a single series of two-foot bricks, but there is a double series in the higher spans. To add a greater solidity to the construction, the higher of these arches have another curved span below, by which the piers are joined between themselves, so that they are not split apart with a huge gap. The two plans that we add show better the dimensions of the channel, the reworkings, ornamentation, mutules, and other more detailed elements. The first [fig. 2] depicts the line closer to the city; the second [fig. 3], the arches in the furthest and marshy area called the Valle di Pantano.

To them, we have added a third plan [fig. 4], of the settling tank at number 24 [in fig. 1], in which the water was received and cast off its mud, as if the course of its conduit were taking its breath, so to speak—this is the particular function of a settling tank, according to Frontinus [*Aq.* 19.1]. Indeed, from conduit A, the water entering tank C through mouth B and there purified, again unburdening itself through D into E, continued its course toward Rome.

From the same things, the original state of the construction has been set before our eyes, today changed to a very great extent, with arches damaged here and reworked there, not all at one time. In some places, you might see arches propped up with brick-faced reinforcement and piers encased. Elsewhere (and in most places), there are restorations crammed with unworked tufa, from which the ignorance of skill, the limited resources of the restorers, and their purposes betray themselves. That part of the aqueduct that was closer to the city has so felt the damage of attacks and devastation that it appears that the first substructure and first arches running to the right of the Via Labicana [fig. 1, nos. 1–2], have been reconstructed from the ground up, for there is nothing in that remaining structure that knows the elegance or the proportion of the rest. There is also to be noted a recent use, or abuse, of this aqueduct, where, over the same arches [at no. 2], where there is no conduit but a channel dug out in the remains, there is carried today a portion of the Aqua Crabra in the opposite direction, taken off toward the east:

> So much of time can long decay change.
>
> [Verg. *Aen.* 3.415]

Both the variety of the reworkings already cited and the scarcely believable thickness of the incrustation demonstrate that this conduit has

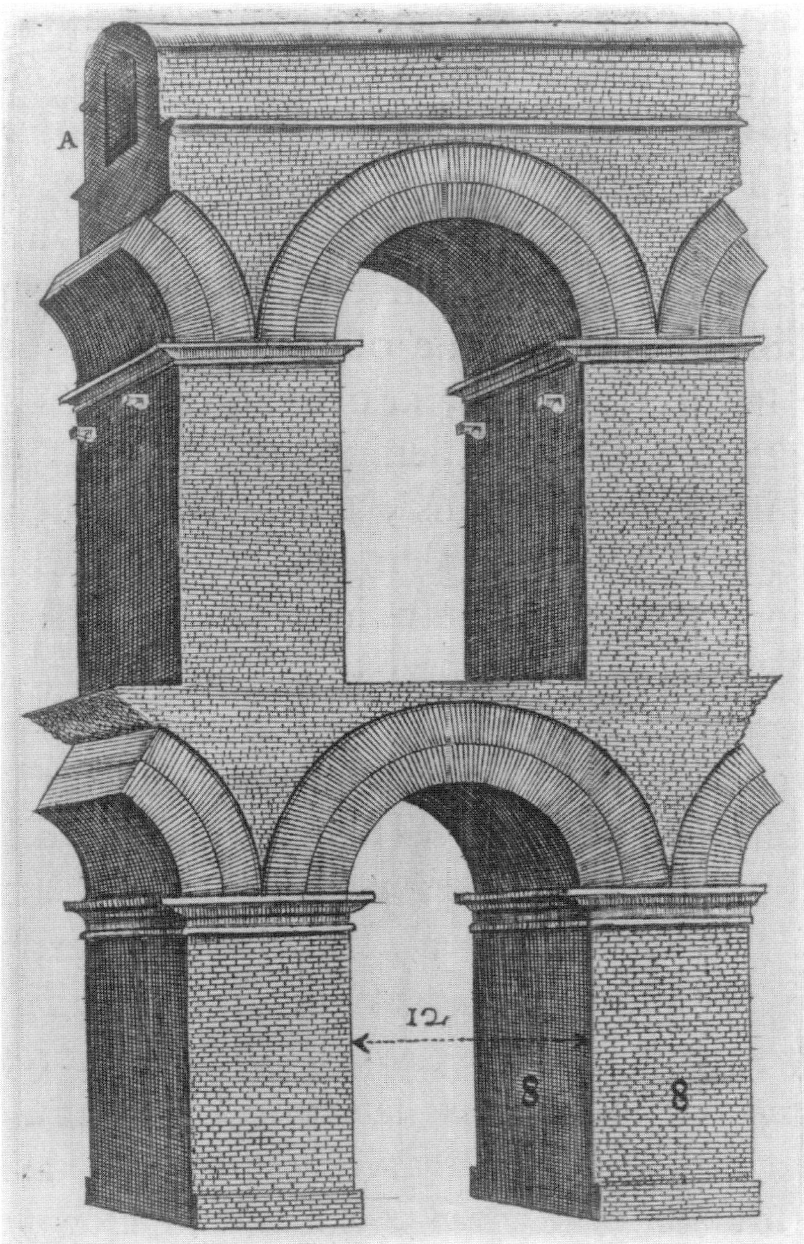

Fig. 2. Aqua Alexandrina: double arcade close to the city

A. Conduit, two and a half feet wide, four and a half feet high up to the curvature of the arch. The arch itself is one foot four inches. The sides are uniformly two feet three inches thick.

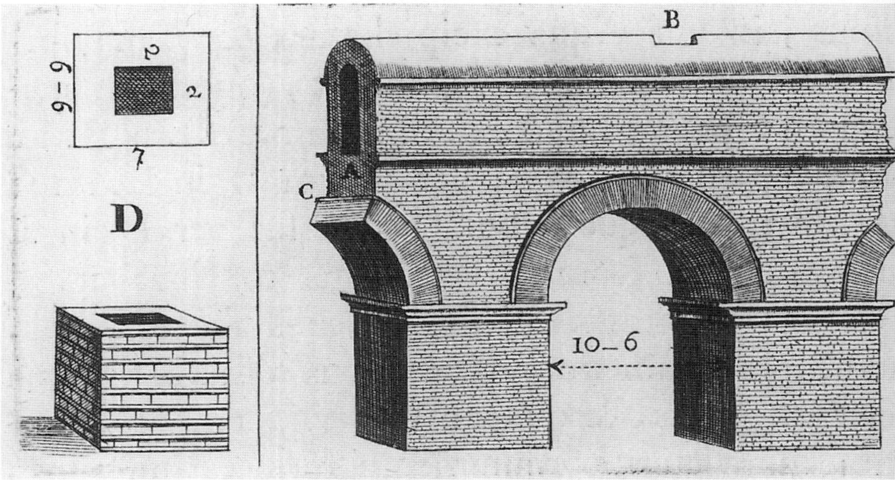

Fig. 3. Aqua Alexandrina: single arcade in the Valle di Pantano

A. Conduit, two feet eight inches wide. The sides are of the same thickness.
B. Airholes of two and half square feet appear at various intervals.
C. Brick cornice around the sides of the arches
D. Representation of wellheads for the excavation of earth and an airhole for the excavators of the aqueduct, with side openings arranged for climbing. Their construction is of alternating tufa and brick.

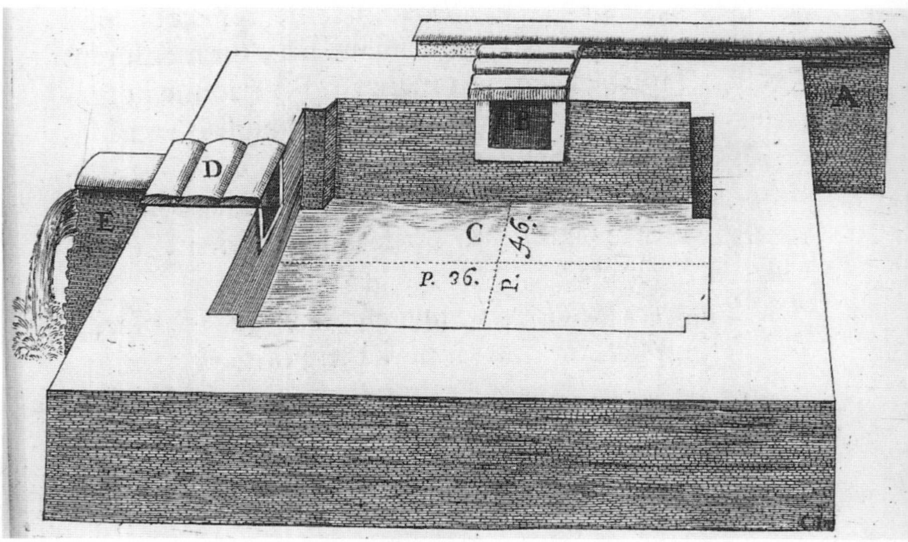

Fig. 4. Aqua Alexandrina: settling tank near the aqueduct intake

survived for a very long time. The deposit is to be found not so much in the channel itself, almost blocked and quite often reduced to three-quarters of its space, but along the exterior walls of the conduit, where we see that moisture issuing from the same place has hardened into huge masses that resemble a wagon loaded with hay. This is a fault the aqueduct has in common with those of the region of Tivoli; even the Marcia itself—

> Which shines so bright, so clear
> That you would suspect no water at all there

and would believe that pure Lygdian marble is gleaming, as Martial [6.42.19–21] has sung about it; "bestowed on the city by a gift of the gods," as Pliny says [*HN* 31.41]—has hardened into a deposit far thicker than this that we are describing and more densely formed. The incrustation of this aqueduct is indeed darker and more muddy but is at the same time thinner and more brittle.

However, it greatly escaped me for a long time what reasoning influenced the architect—after scorning a direct route that, although higher, would have required less construction on arches—to plan the course of the aqueduct through lower and sunken places. As a result, on account of the great curvature of the line from the corner near the Via Praenestina [fig. 1, no. 15], I judged that the aqueduct went straight toward Tivoli from that corner, because it looks back in that direction. Finally, from repeated and more frequent inspection of the area and the advice of my friend [Auzout], as I have said, I thought, when the rest of the line had been found, that I had discovered the cause of the matter and could explain the foresight of those times. Indeed, I believe that building on firm terrain would have been the goal, although at the cost of winding along a longer course, which is not found as equally on a straight path as are the low-lying places. That region, the more closely it approaches the Anio, consists of much more solid earth and is full of quarries of the red Collatine stone celebrated by Strabo [5.3.11]. Not only was it right that the foundations lie in a stable place, but since the aqueduct itself, as excellent as it is very great, is constructed entirely of concrete (if you exclude its brick-faced exterior as a shell), it was important for the long life of the construction to seek fill as solid as possible for the concrete, not at a distance, but collected near the construction area. I would add, too, even though an argument for it might be lacking, that the appearance of magnificence could have also been sought through the display of its very high arches, by which the line is raised in many places to the height of seventy feet.

3. PROOF THAT ALEXANDER SEVERUS BUILT THIS AQUEDUCT

Now that the construction has been described, it remains that we seek the builder of this aqueduct and set forth for the judgment of scholars a matter to this time either untouched or despaired of by students of antiquity. Indeed, Aldus Manutius the Younger seems to have attempted this in his treatise with the impressive and all-inclusive title *On the Ancient Aqueducts Once Flowing into Rome*.[3] But though his introduction stimulated my appetite and held promise of a sumptuous dinner, I found with disgust nothing except the barest summary of Frontinus—full of mistakes besides—and certain other sweepings from Pliny. Therefore, I will pave my own way for my inferences, although no one is assisting. I state the following as a premise: this aqueduct cannot be numbered among the nine that Frontinus described, a point that is obviously proved by review of the lines themselves.

Indeed, this was not the Appia; that aqueduct took its beginning between the seventh and eighth milestone of the Via Praenestina to the left and came to the city entirely underground (with the exception of sixty paces next to the Porta Capena), according to Frontinus [*Aq.* 5.5]. This aqueduct comes out of the ground to the right of the Via Praenestina, quite close to the Via Labicana, in the area of the fourteenth milestone (there are more than three miles beyond, which run from the source and the intakes; their beginning is at the Via Labicana itself, up to the point of emergence cited). There is a good deal of construction carried both above ground and conspicuous on arcade, as we have set forth in particulars on the map.

In this and other citations of Frontinus, we follow the edition of Benedetto Millini, already disseminated through all the books of the city, as the most careful and fullest. (If, by chance, this is not at hand, readers should know that the second book of Frontinus begins at chapter 49, pro-

3. Manutius, *De aquis*, 26.

ceeds through the listing of *quinariae* [pipe capacities] of each individual aqueduct to chapter 74, and takes its end with chapter 105.)

[This is] not the Anio Vetus, since its source, according to Frontinus [*Aq.* 6.5], is above Tivoli. The source of our aqueduct, beneath the town of Colonna—where we, with Holste, believe Labicum to have been[4]—excludes every confusion. There is therefore no need to examine the difference in substructure above ground. In the Anio, it is 702 paces; in this aqueduct, however, 4,250 paces. The Anio Vetus also totally lacks construction on arcade; in this aqueduct, it runs for 2,325 paces. Likewise, its construction is more recent than that of the Anio Vetus. We know from Frontinus [*Aq.* 6.1] that the Anio Vetus was built in the flourishing times of the republic, in the 481st year from the founding of the city [272 B.C.]. A brick found with the pottery stamp "AVG.N." [fig. 5 = *CIL* XV, 211.14], commemorating "Augustum Nostrum," indicates that this line was introduced after the empire; indeed, the brick was built in the original arches, so that we need not be uncertain whether it was introduced on the occasion of repairs.

[It is] not the Virgo, the sources of which are even now conspicuous at the eighth milestone on the Via Collatina, where Frontinus [*Aq.* 10.5] describes them.

[It is] not the Marcia, the Claudia, or the Anio Novus, all aqueducts that have their origin on the Via Valeria or Via Sublacensis, according to the same Frontinus [*Aq.* 7.6, 14.1, 15.1].

[It is] not the Julia or the Tepula, which, tapped to the right of the Via Latina, were received onto the same arches, those of the Aqua Marcia, as is clear from Frontinus [*Aq.* 19.3].

Finally, [it is] not the Alsietina, which was sought from the lake of this name in Etruria, on the Via Claudia at the fourteenth milestone, on a side road to the right at sixty-five hundred paces, according to Frontinus [*Aq.* 11.3]. This aqueduct originated in the middle of Latium.

With all these aqueducts of Frontinus excluded and since we know as a fact that Severus Alexander Augustus introduced an aqueduct, I argue that this aqueduct cannot be attributed to another emperor; no other line can be attributed to this builder. Clearly, it follows that Alexander was the builder of this construction.

Spartianus reveals to us in his *Life* that Alexander introduced an aque-

4. Holste, 194.

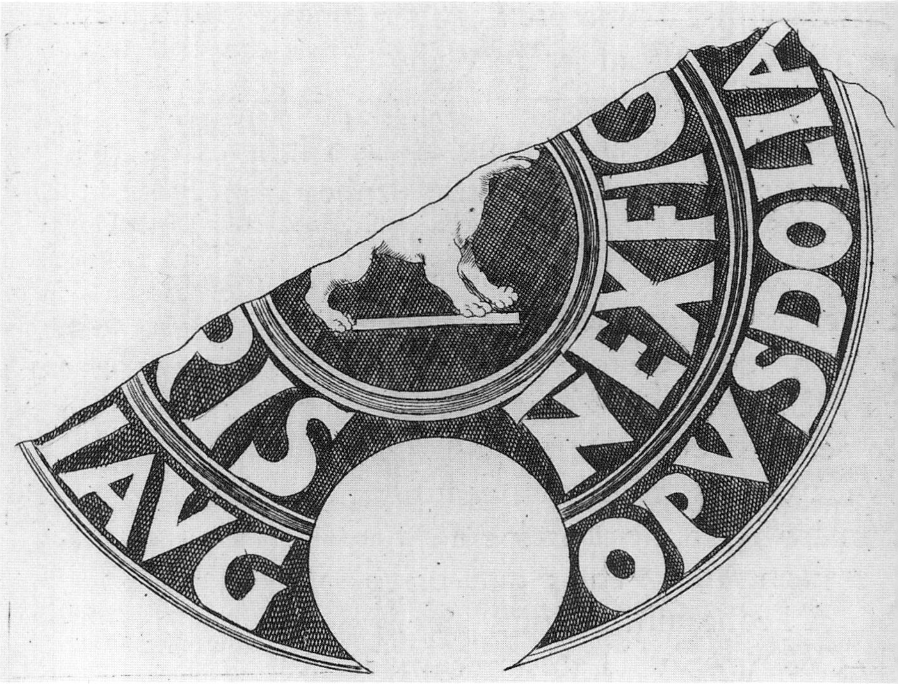

Fig. 5. Brick stamp from Aqua Alexandrina [*CIL* XV, 211.14]

duct: "He restored works of the old emperors and himself established new ones; among them were baths in his name, built next to those that were the Neronian baths, after he introduced an aqueduct that is now called the Alexandrina" [SHA *Alex. Sev.* 25].

A coin of Alexander himself shows this as well, with his baths and a fountain, by which it was customary to indicate an aqueduct; in the collection of Alessandro Rondinini, it is of very fine appearance and notable for its state of preservation (as those knowledgeable in numismatics put it). Angeloni has illustrated an example of it in the coinage of this emperor;[5] Oisel, however, believes that the bay of Baiae and a ship are indicated.[6] Publius Victor[7] and the *Notitia Imperii* [VZ, 1:256; Jordan, 2:570] also cite this Aqua Alexandrina.

5. Angeloni, 276–77, table 282.3.
6. Oisel, 533–34, table cix.7.
7. Publius Victor, *De regionibus Romae libellus aureus* (= VZ, 1:256; Graevius, 3:112E).

Finally, it is evident that neither can another aqueduct be attributed to this emperor nor another builder to this aqueduct (indeed, these two points already follow each other in turn). Aside from the aqueducts cited by Frontinus, later generations have recognized only two in this city: the one that is the subject of our discussion and that of the Aqua Traiana across the Tiber. As we shall prove, the Traiana, brought along the Via Aurelia for almost two miles, enters the city through the gate that is either the Aurelia or the Pancratiana, as it was called from the times of Procopius, and displays itself at the top of the Janiculum.

4. REVIEW OF OTHER AQUEDUCT CONDUITS IN ROME

Our purpose here is to review the other arcades and water conduits that have been scattered throughout the city—on this occasion with the pen, as we have done so many times with our eyes—so that we may rule out the possibility that they received water from this aqueduct or another with the name of Alexander.

a. Aqueducts at Porta Maggiore

Those aqueducts, highest of all from the eastern corner of the city, that come along the city walls down to the Porta Maggiore carried the Aqua Claudia and Aqua Anio Novus, as Frontinus has reported [*Aq*. 18.4] and as the following four-line inscription on each side demonstrates.

> Tiberius Claudius, the son of Drusus, Caesar Augustus Germanicus, pontifex maximus, in the twelfth year of tribunician power, consul five times, imperator twenty-seven times, father of his country, provided for the introduction of the Aqua Claudia from sources that were called the Caeruleus and Curtius from the twenty-fifth milestone and likewise of the Anio Novus from the sixty-second milestone, at his own expense. [*CIL* VI, 1256]

Next to these, on the right for those coming to Rome, were other arcades that carried the conduits of the Aqua Marcia, Aqua Tepula, and Aqua Julia; scarcely a trace of them is left, since their remains yielded to the arcade of Sixtus V. Where they still remain intact, at a distance from the city in a farmhouse of the property called Settebassi, at the fifth milestone of the Via Latina, I found them built to most exact form and greatest solidity [fig. 6]. As a result, it was rightly said by Propertius [3.22.24], "eternal work, the water of the Marcia"; and truly Plutarch [*Cor.* 1.1] called this "the greatest and most beautiful aqueduct," since its elegance was also sought from the different hues of its stone construction.

There is no doubt that these aqueducts came into the city, joined and in the ascending order set forth already, as evidenced by Frontinus: "from the settling tanks they are received onto the same arches, the highest of them that of the Julia; below, that of the Tepula; and then that of the Marcia" [*Aq.* 19.3]; "The highest is the Anio Novus, next is the Claudia, the Julia holds third place, the fourth the Tepula, then the Marcia" [18.4]; "The Anio Vetus holds sixth place in elevation," and so on [18.6]; "The Virgo follows the elevation of this line, then the Appia" [18.7]. All of these statements correspond excellently to the cross section of each conduit that we set forth, the Claudia and the Anio Novus at the Porta Maggiore [fig. 7]. Outside of this, for those entering on the right, the channels of the Aqua Julia, Aqua Tepula, and Aqua Marcia (not, indeed, that of the Claudia, as Fabricius has wrongly surmised)[8] present themselves, as you see here [fig. 7].

These conduits then reappear at the modern Porta S. Lorenzo, with the name of the Marcia inscribed on the following inscriptions of Titus and Caracalla (so that no hesitation remains about them).

> The emperor Titus Caesar Augustus, son of a god, Vespasianus Augustus, pontifex maximus, in the ninth year of tribunician power, nine times imperator, fifteen times censor and consul, eight times consul-designate, rebuilt the channel of the Aqua Marcia, collapsed from old age, and brought back its water, which had ceased to be in use. [*CIL* VI, 1246]

> The emperor Caesar M. Aurelius Antoninus Pius Felix Augustus Parthicus Maximus Britannicus Maximus, pontifex maximus, saw to the introduction into his sacred city of the Aqua Marcia, blocked by various mishaps, with its source cleaned out and mountains tunneled and cut through, the conduit having been restored, and a new source, the Antoninian, having been acquired. [*CIL* VI, 1245]

8. Fabricius, 201 (= Graevius, 3:527B).

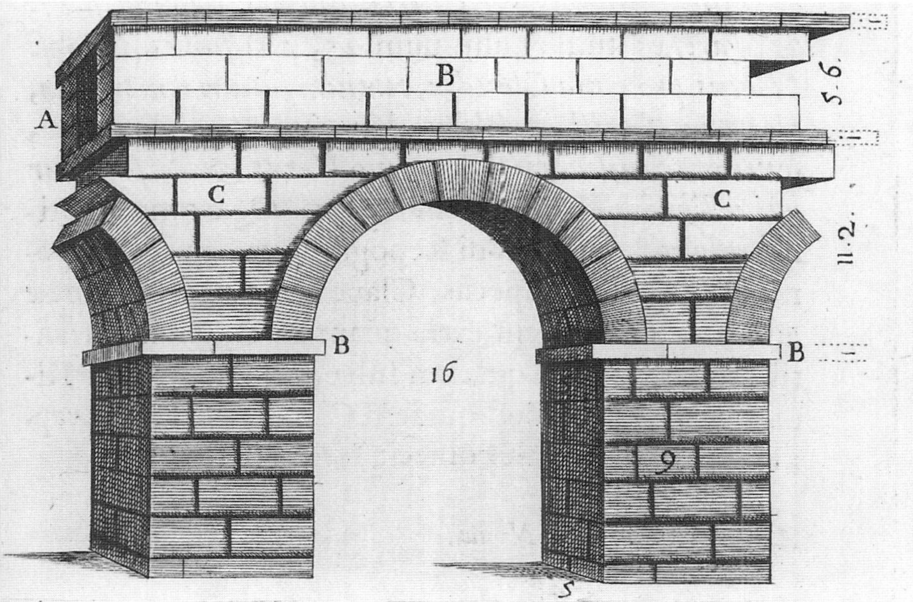

Fig. 6. Aqua Marcia: arcade at Settebassi

A. Conduit, two and a half feet wide, five and a half feet high. Its walls are one foot three inches thick.

B. The whole height of the conduit and the band below the arches is constructed of red Collatine stone.

C. Triple rows of stones below the conduit, from Tusculan stone of yellowish color, called *sperone*. The remaining work is in dark Gabine stone.

b. The Arcus Caelimontani

The arches that run from the Porta Maggiore to the left, over the Caelian, were called Neronian and carried part of the Aqua Claudia, as we are taught by Frontinus [*Aq.* 20.2–4]: "Their arches [those of the Claudia and Anio Novus] end behind the estate of Pallas, from where they are distributed in pipes to supply the city's needs. The Claudia, however, transfers part of itself beforehand at Spes Vetus to the so-called Neronian Arches. This arcade runs over the Caelian Hill and terminates next to the Temple of the Divine Claudius."

From these words, it is obvious that the distribution of the Aqua Claudia that Fabricius (cited earlier) has set forth, as if "one branch were

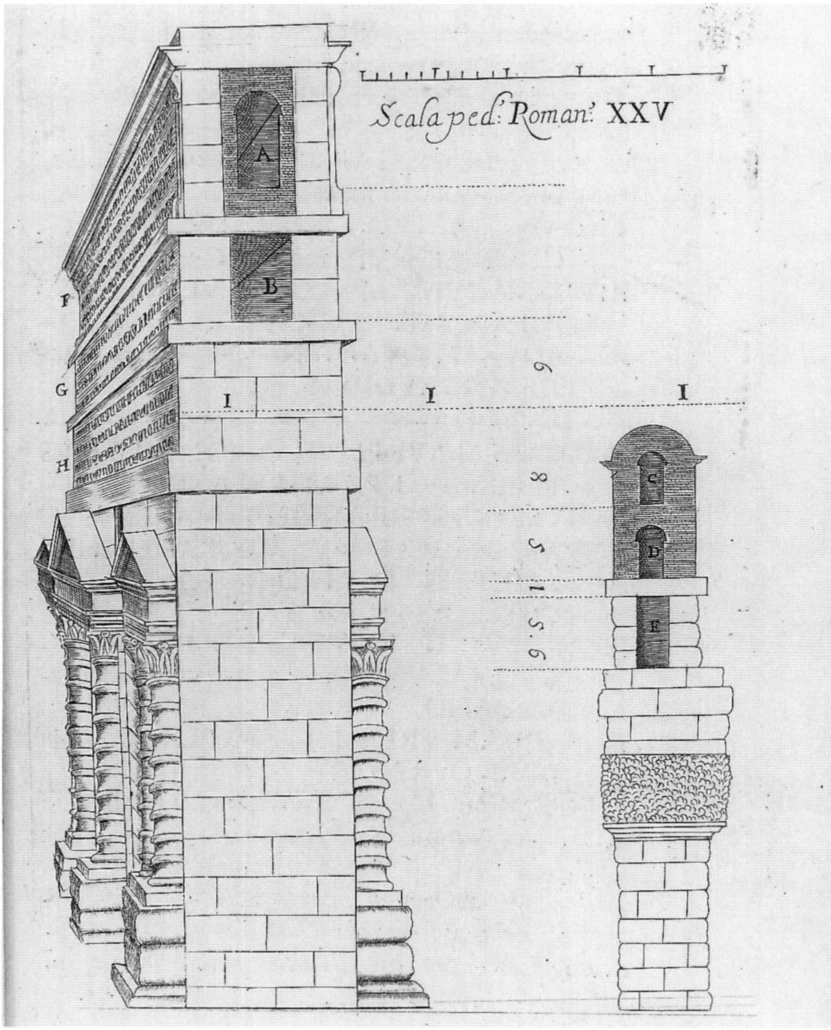

Fig. 7. Cross section of aqueduct arcades entering the city at Spes Vetus: Aqua Anio Novus, Aqua Claudia, Aqua Julia, Aqua Tepula, Aqua Marcia

A. Conduit of the Anio Novus
B. Conduit of the Claudia
C. Conduit of the Julia
D. Conduit of the Tepula
E. Conduit of the Marcia
F. Location of the inscription of Tiberius Claudius Caesar, set forth already
G. Location of the inscription of Vespasian Caesar, set up on the reconstruction of these aqueducts
H. Location of the inscription of Titus Caesar, set up on the same occasion
I. Height of the convex arch of the Acqua Felice on the outside, six feet below the base of the Claudia, eight feet above the base of the Julia

extending from the Caelian onto the Capitoline, the other onto the Aventine,"[9] is incorrect and reduced. In contradiction to Frontinus, Fabricius announced his subdistribution of the line as the first and highest distribution of it. There still remained the water of that portion that was transferred over the Neronian Arches; and the other portion, which, as we have seen, "was distributed in pipes behind the estate of Pallas to supply the city's needs," was disregarded.

The following inscription shows that the arches were also named Caelimontani.

> The emperor Caesar, son of the Divine Marcus Antoninus Pius Germanicus Sarmaticus, brother of the Divine Commodus, grandson of the Divine Antoninus Pius, great-grandson of the Divine Hadrian, great-great-grandson of the Divine Trajan, great-great-great-grandson of the Divine Nerva, Lucius Septimius Severus Pius Pertinax Augustus Arabicus Adiabenicus Parthicus Maximus, pontifex maximus, in the ninth year of tribunician power, eleven times imperator, twice consul, father of his country, proconsul, and the emperor Caesar, son of Lucius Septimius Severus Pius Pertinax Augustus Arabicus Adiabenicus Parthicus Maximus, grandson of the Divine Marcus Antoninus Pius Germanicus Sarmaticus, great-grandson of the Divine Antoninus Pius, great-great-grandson of the Divine Hadrian, great-great-great-grandson of the Divine Trajan Parthicus, and great-great-great-great-grandson of the Divine Nerva, Marcus Aurelius Antoninus Pius Felix Augustus, in the fourth year of tribunician power, proconsul, with their own money restored from the ground up the Arcus Caelimontani, which had collapsed from age in various ways and been damaged. [*CIL* VI, 1259]

Although this inscription is said to exist in the region of the Arenula, it was "once affixed on this aqueduct in front of the hospital of S. Giovanni," as stated in an old manuscript that the most distinguished Marcello Severoli keeps in his most well appointed library, himself even more well appointed in every kind of literature.

I do not understand what Nardini means by the following observation: "The Neronian Arches appear today of brick construction, but the other arches constructed of large stones, which are outside the walls, give indi-

9. Fabricius, 201 (= Graevius, 3:527B).

cation that these arches were of no less magnificence, where that which is seen today was a facing given them perhaps in less happy times," and so on.[10] Indeed, I do not know why Nardini demands the same stone structure in this part of the Aqua Claudia—constructed as a branch by Nero and running in a different direction from the main conduit—as in the main aqueduct as a whole, with the result that he has doubts about the reworking of these arches and the brick facing alone. Where is there even a slight trace of stone construction on this whole course? How would Nardini find an excuse for those brick piers, which he has judged faulty from their excessive thinness and the width of their arches, unless the reason was to overshadow as little as possible the urban structures adjacent to the line? I ask, could a greater and almost double construction of other stones have been supported on those piers, only eight feet long and seven feet wide? We may understand this from a comparison of each conduit, which we show on this occasion [fig. 8].

All these arches, however, show reconstruction by Severus and Caracalla, about which the inscription speaks. They fall short of the simplicity and gracefulness of the arches that we have described, and we believe that Nardini observed only these reworked arches when it pleased him to call a most careful work "a facing given . . . in less happy times." Indeed, just because Nero used brick to construct his arches (in other respects, quite elegant and most worthy of their builder), you would not judge them less splendid or useful. Vitruvius [2.8.16] especially approves this material, when he recounts the considerable number of works, public or regal, built in brick throughout the whole world, finally ending with these words: "Since, therefore, kings with such great power, for whom it would have been permitted through taxes and booty quite often to have buildings not only in concrete or squared stone but also in marble, have not scorned structures of brick walls, I do not think it right to disapprove of buildings constructed with brick facing, provided they have been brought to completion correctly."

c. The Claudia Branch to the "Trophies of Marius"

Other arches, which from the same Porta Maggiore to the right terminate at the *castellum*, or *emissarium*, as Gruter calls it,[11] near the Arch of Gal-

10. Nardini, 507.
11. Gruter, cclxxiv.5.

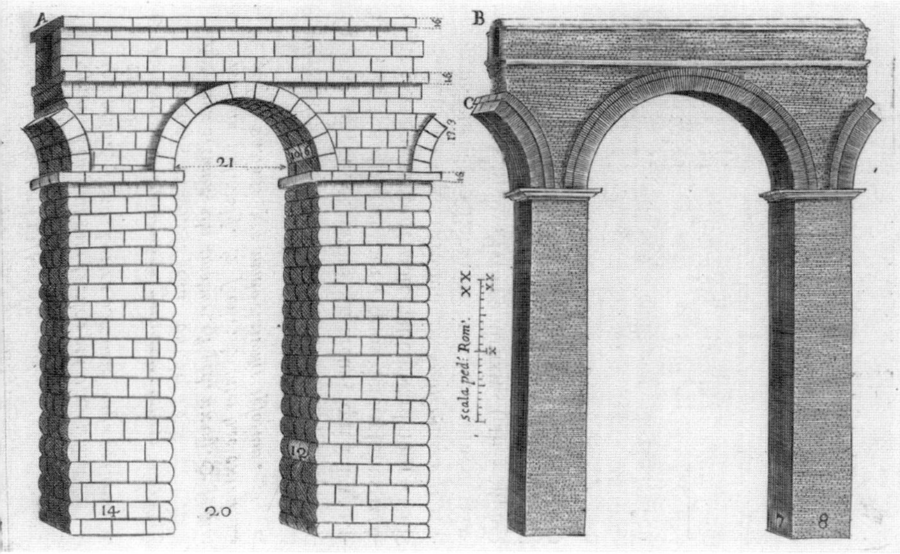

Fig. 8. Arcades of the Aqua Claudia and the Arcus Caelimontani branch

A. Conduit of the Claudia, four feet three inches wide, six feet high. The walls are three feet thick.

B. Conduit of the Neronian Arches, two feet nine inches wide, five and a half feet high to the curving of the arch. The arch itself is one and a half feet high. The walls are uniformly one foot ten and a half inches thick.

C. Bricks of the arches of different size: the interior, two feet square; the exterior, one foot four inches. The opening of the arches, where it is wider, is twenty-seven and a half feet; where smallest, eighteen and a half feet; and between these two extremes, varying everywhere, in proportion to the height of the arches themselves.

lienus, are indeed at the same elevation as the Neronian Arches already discussed. From this, it is certain that they carried the remaining part of the Aqua Claudia (as Gruter correctly observes at that place) or also the Anio Novus superimposed on it. These two waters were mixed before the time of Frontinus, as he himself states [*Aq*. 72.6], and this is the same place, behind the estate of Pallas, from where, according to a passage just cited [*Aq*. 20.2], when the arcades of the conduits had ended, the immense supply of water, displayed through five openings (as we shall state), used to be distributed in pipes for the use of the city. Indeed, no other aqueduct (since these were the highest of all) could climb to so great an elevation.

The theory of a modern writer therefore remains colder than the Anio itself or the chilly Simbruvium: from this "*castellum* of the Aqua Marcia" (as he labels it), supposedly erected by Trajan when that emperor restored that line and increased its capacity or extended it onto the Aventine—this, however, I remember having read as never cited from Frontinus himself but only in Angeloni, more than once and in an especially pompous manner with respect to one of his Trajanic coins[12]—he assigns the "Trophies" once standing there to Trajan. The Aqua Marcia, the exact and very nearby elevation of which we have in the channel of the same aqueduct already discussed at the Porta Maggiore, remains at a level twenty-five feet and more below the elevation of the Claudia, as it appears in the cross section set forth earlier, and the elevation of those five aqueducts [fig. 7]. It could not, therefore, have reached the height of this *castellum*.

This established fact, tested by mathematical demonstrations, destroys Nardini's altogether shaky conjecture "And perhaps in the *castellum*, which still remains under the Trophies of Marius, near S. Eusebio, the three openings that are seen there at equal level distributed water of the same three aqueducts individually (understand the Julia, the Tepula, and the Marcia)"[13] and, indeed, his confident assertion "There can be no doubt that a *castellum* of the Aqua Marcia stands there, seeing the clear division into three intakes."[14] I say that this is shaky because to those three openings seen by Nardini, another two would have to be added at its sides (that *emissarium*, indeed, shows so many). All five (so that Nardini not have the task to bring in still two more other aqueducts) were supplied by one conduit, distributed as you see here [fig. 9].

This distribution is made through a delivery system that no one (as far as I know) has noticed up to this time. Fabricius was about half wrong; he states that the same conduit he recognized between the Porta Maggiore and the "Trophies of Marius" was that of the Aqua Julia, higher than the other two.[15] He also is too little concerned about elevation but, like others, belches forth what comes into his mouth.

d. The "Arco di Druso" Aqueduct and Specus Octavianus

It will be more important (as a result of the common delusion of antiquarians, as you yourself will judge) to determine to which of the ancient aque-

12. Angeloni, 114, table 123.27.
13. Nardini, 505.
14. Nardini, 161.
15. Fabricius, 202 (= Graevius, 3:527C).

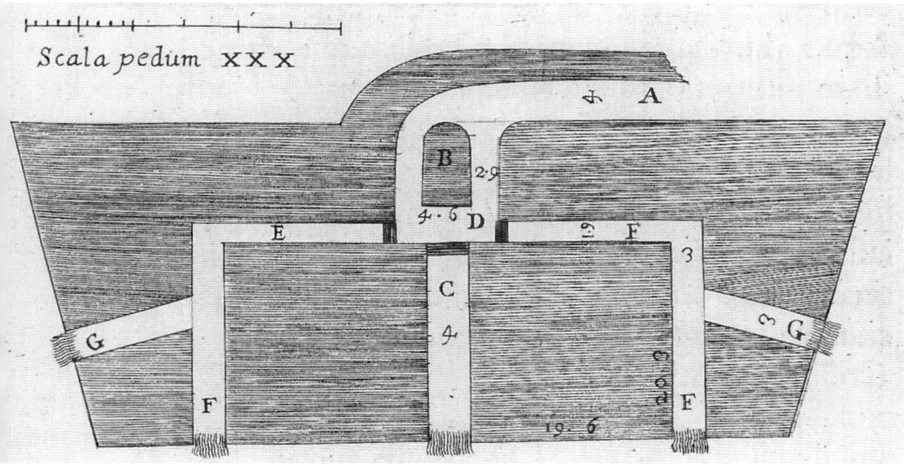

Fig. 9. "Trophies of Marius" *castellum*: plan

A. Principal channel of the conduit, through which the aqueduct was brought to the *emissarium*

B. Blocking wall, where the aqueduct, on contact, divided itself into two channels

C. Middle opening, through which the aqueduct was sent forth from level *D*, descending somewhat

E. Side channels, through which a part of the aqueduct directed itself down again into *F* and *G*, with double openings on both sides

ducts the following conduit pertains—the one that, proceeding almost three stades along the Via Latina on the left (for those coming toward Rome), enters the city in a corner of the walls further to the south, crosses the Via Appia above an arch next to the Porta S. Sebastiano, and continues toward the areas of Region XII and the Piscina Publica. Indeed, up to this point, some have asserted fearlessly that that very arch is the Porta Capena, damp (from the Aqua Appia falling above it) and dripping, as the poets describe it. First, we must first fight and overcome that common mistake and demonstrate that the Aqua Appia did not reach this point and that the Porta Capena stood not in this place but, rather, far back from it, toward the city.

Does the aqueduct we are discussing here have anything to do with the Aqua Appia? The matter is so absurd that it is a wonder that first Luca Peto, in his treatise *On the Reconstruction of the Aqua Virgo*,[16] and then

16. L. Peto, *De restitutione ductus Aquae Virginis* (Rome, 1570), 2.

other noted and distinguished men have been so unmindful of Frontinus. They thought that the Appia, which was the lowest of all the aqueducts (the Alsietina across the Tiber excluded), according to his ranking already cited [*Aq.* 18], climbed to the top of this hill. The Virgo itself is found to be far below it, as measurement demonstrates. No less an argument against it is the shortness of the Appia's substructure and the sixty paces in total of its arcade above ground, which Frontinus [*Aq.* 5.5] attributed to the Appia "next to the Porta Capena." Approximately 450 paces of substructure by which the conduit in question comes down from the Via Latina to the Via Appia are evident.

Concerning the location of the ancient Porta Capena, the matter is even more clear, by the authority of Strabo [5.3.9], who in these words demonstrates that the Via Latina branched from the Appia outside the city and therefore beyond the Porta Capena: "The Via Latina begins from the Appia, deviating from it to the left near Rome; passes over the Tusculan Mount, between the town of Tusculum and the Alban Mount; and descends to the town of Algidus and Pictae Tabernae." Strabo's words are so clear that they do not permit different readings of critics and forced interpretations. It should also be added that since Pliny [*HN* 3.67] and Dionysius [*Ant. Rom.* 9.68.2–4] bounded the circumference of the city by the slopes of the hills, the *agger* [rampart] of Tarquinius, and the river, the location of the Porta Capena cannot more conveniently be conceived than in a line drawn from the rock of the Caelian Hill, below the Orti Mattei, toward the spur of the Aventine, beneath the Church of S. Maria in Navicella and the fortification of Pope Paul III.

Since this is the special claim of truth, that all things agree with it, the result is that all the statements of authors speaking on this subject, which have drawn others in different directions, present themselves, in our opinion, altogether obvious and easy. What is more appropriate—indeed, more necessary—than that at this spot, we should allow a passage for the Aqua Appia, the lowest of all the aqueducts this side of the Tiber, "next to the Porta Capena," and bring it back from high onto level ground? Where could the substructure and arcaded construction of the Aqua Appia, not greater than sixty paces, more fittingly have shown itself than in the spot indicated by us—indeed, through the jaws of the Caelian (if I put it this way) and its slope turned toward the Aventine from the south, on which the Church of S. Balbina was built?

What overcomes every argument and agrees with this opinion in a marvelous way is, in actuality, the site where we most recently, while writing this study, discovered the channel of the Aqua Appia, unrecognized over several centuries (concerning which I shall speak shortly). Moreover, Sextus Pompeius Festus [p. 102L] spoke in this fashion; according to him, "The Lemonian Tribe has been named from the Pagus Lemonius, which is beyond the Porta Capena on the Via Latina" (written as it befitted a famous grammarian). Festus will also not submit to correction by Holste, who, in his note, thus moves that *pagus* and, with it, Ortelius, arguing, "Therefore the Pagus Lemonius was between the Via Appia and the Latina, for the Via Appia begins from the Porta Capena, but the Via Latina from the gate of the same name."[17] Rather, I say, with Festus, that the Pagus Lemonius could be approached on the Via Latina itself and not in any other way except from the Porta Capena. Certainly, it was not approached from the Porta Latina, of which there is neither physical evidence nor a name before the extension of the city by Aurelian's walls.

Finally, the position of our Porta Capena is exactly consistent with the spot where the column of the first milestone of the Via Appia [*CIL* X, 6812] was found and from which it was afterward transferred to the Capitolium: precisely, in the Naro vineyard, which is the first on the right for those leaving from the modern Porta S. Sebastiano. Going back from here to the Porta Capena, a thousand paces are completed in a line drawn by us. Someone will say that the column, perhaps set up somewhere else, could have been moved to the spot in later centuries; however, he would use natural possibility (as the schools say) for attacking according to whim. We will be content to sustain our position by moral probability also supported by other arguments.

What blindness, then, has seized minds so that Strabo, speaking so clearly [5.3.9], has escaped the keen sight of so many philologists? From just the ancient testimony of the scholiast on Juvenal's line [*Sat.* 3.11], "He stood at the ancient arches and the damp Capena," we might seem to be able to draw the conclusion that at Juvenal's time, the name of the Porta Capena had been transferred elsewhere, but the original place was still known. The scholiast writes as follows: "Damp, for the reason that an aqueduct is above it, which they now call the dripping arch [*arcus stillans*]; indeed, first, there had been also gates in the very same place, which used to be called the Porta Capena." There existed, therefore, that "dripping

17. Holste, 104.

arch," the conduit of the Aqua Marcia, which constituted the "damp Capena" (by fault of its great antiquity at the time of Juvenal) or, as Martial [3.47.1] describes it, was "dripping with great drops."

Why, indeed, do scholars distort the passage of Frontinus [*Aq.* 5.5] that indicates the substructure of the Aqua Appia and its arcade as "next to the Porta Capena"? They do violence to his very words and, at the same time, to the nature of the place when they bring the Aqua Appia not "next to," as Frontinus intended, but "on top of" the Porta Capena. They have overlooked and ignored the account of the same Frontinus about the Aqua Marcia: "The Marcia, however, behind the estate of Pallas, casts off part of its volume onto the Caelian into a channel that is called the Rivus Herculaneus. The (portion? a word seems to be lacking here) of the conduit itself, supplying no needs of the hill because of its low level, is fixed above the Porta Capena."

These statements of Frontinus [*Aq.* 19.8–9] and, in addition to his authority, the very nature of the spot, as I have said, and the most authentic conduit of the Aqua Appia, recognized by me, as I shall show later, make it impossible that the Appia ran atop the Porta Capena. Although its level would indeed permit the height of approximately eight paces needed for mounting the arch (no higher than the arch at the modern Porta S. Sebastiano), the remaining part of the construction would have occupied the entire width of the valley, and neither four times nor five times sixty paces, which Frontinus assigns to the crossing of the Appia, would have sufficed.

We will therefore permit the Aqua Appia (after we have installed the Marcia atop the Porta Capena) to be carried on the "old arches" at the bottom, crossing the valley on that short substructure, below the gate that rose on the side of the Caelian at a fairly high level, as we believe. The result is that there are two different structures, although quite close in their location: the "old arches" of the Aqua Appia (for none preceded them in age) near the Porta Capena and the Porta Capena itself, with the conduit of the Marcia atop it. After these arguments in support of this (in my opinion) new identification, I shall not omit another to answer those opposing me (although it can already be regarded as superfluous, given the great evidence of the matter); the modern arch at the Porta S. Sebastiano, which they have foolishly substituted, shows itself even today impenetrable to moisture, by its own mass and the very strong incrustation of the conduit that it supports.

A second inscription excavated in the same vineyard and carrying the statement

> The Senate and Roman people saw to the reworking of the Clivus of Mars into level ground at public expense [*CIL* VI, 1270]

shows that that *clivus* [slope], reworked into level ground (not precisely level, to be sure, but lightly sloping, as we see) by the work of excavation that is seen at the Porta S. Sebastiano, was once called the Clivus of Mars; indeed, it was so named from the Temple of Mars next to the Porta Capena. The temple was built near the gate, as Servius testifies [at *Aen.* 1.292]: "There are two temples of this god in the city: one of Quirinus within the city, as it were, of Mars the Guardian, and of Mars the Peaceful; the second on the Via Appia outside the city near the gate, as it were, of Mars the Warrior or Gradivus." It was also in sight of the gate, on the right, as Ovid [*Fast.* 6.191–92] has written: "The same day is a festival for Mars, whom the Porta Capena looks upon, set outside next to the road on the right." We therefore believe that the temple was situated at the bottom of the valley, before the *clivus* had risen significantly, from the evidence of a letter of Cicero. Writing to his brother and reviewing the lowest places of the city damaged by flooding, Cicero [*Q Fr.* 3.5.8] says that the temple itself or the level ground beneath it was washed by an extraordinary flood.

We cite all ancient authorities as witnesses of our earlier argument that absolutely no mention of the Porta Latina is found before the times of Aurelian. Indeed, whatever modern writers treating ancient lore have made mention of this gate at a time when it did not yet exist have spoken through prolepsis. I say they have spoken through prolepsis since the appearance of the site has been changed from that time when our ancient sources had written to explain more clearly its actual location, at that time on the Via Latina outside the Porta Capena. We must especially understand in this way the account of the immersion of S. Giovanni into a caldron of boiling oil in front of the Porta Latina included in the ancient martyrology on May 6th[18] and any other witness of this miraculous occurrence—even more for the reason that generally executions were carried out outside the gates and along the more heavily traveled roads. Indeed,

18. *Martyrologium Romanum*, 4th ed. (Vatican City, 1956), 108.

Prochorus, a contemporary of S. Giovanni, by whose authority (with Fr. Aringhi making approving reference) is told "the passion of S. Giovanni before the gate and the dedication of the church by the faithful who were there in that place,"[19] was one of seven deacons who collected the deeds of the martyrs into a whole. As is noted in the alphabetical index of authors of the *Maxima bibliotheca veterum patrum* [*Most Comprehensive Library of the Church Fathers*], according to the judgment of Giovanni Brisichella, the head of the holy apostolic palace, "That history, however, and the *Life* of S. Giovanni are falsely ascribed to Prochorus, for the *Life* is indeed apocryphal and the stuff of fable."[20] Cardinal Bellarmine (whom Lipsius appropriately calls the champion and teacher of the church) has branded similar censure on this book. Bellarmine was especially disturbed since, he says, "it seems incredible that the church would have been built in his [Giovanni's] honor by the faithful who were there in that place; for who would dare that under the eyes of the raging Domitian?"[21] As a result, I am amazed that Aringhi cites the authority of an apocryphal writer to posit the existence of a Porta Latina at the time of S. Giovanni, whom he believes contemporary with the emperor Domitian, and corrupts through anachronism what others, using anticipation of chronology, have handed down correctly.

Now that the Porta Capena has been pushed back inside the intersection of the Via Latina, along with both the substructure of the Aqua Appia and its course on arcade, a greater task still awaits us: that of establishing the name of the aqueduct and the arch at the modern Porta S. Sebastiano. In matters of this sort, uncertain and obscure because of their very antiquity, it has always seemed easier to attack other people's positions than to prove one's own. That conduit was, by more than one indication, I think, that same one that is called Octavianus by Frontinus [*Aq.* 21] and that received part of the Anio Vetus, as is understood by these words: "The Anio Vetus this side of the fourth milestone . . . of the . . . Via, where one passes from the Via Latina to the Via Labicana, crosses on arches and has itself a settling tank. From here, this side of the second milestone, it delivers part of its volume into a conduit that is called the Octavianus and

19. Aringhi, 2.16.

20. La Bigne, Marguerin de, ed., *Maxima bibliotheca veterum patrum et antiquorum scriptorum ecclesiasticorum* (Lyons, 1677), vol. 2, title page ("series authorum et opusculorum quae in hac bibliothecae maximae ss. patrum tomi II parte continentur").

21. La Bigne, *Maxima bibliotheca,* 2:46.

reaches the region of the Via Nova and the Asinian Gardens, from where its volume is distributed throughout that district. The main conduit, however, coming behind Spes Vetus, is distributed inside the Porta Esquilina into deep channels throughout the city."

There are indeed many facts in agreement to confirm this opinion. The first and most important evidence is the ranking of elevation, especially appropriate to the Anio Vetus, sixth place in height, between the Marcia and the Virgo, as Frontinus ([*Aq*. 18.6] cited earlier) has revealed. The second, not insignificant, is furnished by the branch line and the course of this conduit, similar to that described in Frontinus. This Specus Octavianus was indeed found around the second milestone from the city, to the left of the Via Latina for those leaving Rome, in the place indicated [fig. 1, no. 25]. In its deviation from this spot again to the left, toward the Via Labicana, on which it was looking back, and toward the Monument of Alexander—near it there exist traces of the side road from the Via Latina to the Via Labicana, which Frontinus cites—the branch was able to show straight and intact the channel from which it had diverged. The higher aqueducts, the Claudia and the Anio Novus placed on top of it, which were brought on a downward course to this first region over the Arcus Caelimontani, according to Frontinus [*Aq*. 86], would not indeed have required so greatly circuitous a route. To this region, the Marcia, or that part of it which Frontinus [*Aq*. 19.8–9] calls the Rivus Herculaneus, also "used to distribute itself over the Caelian," as we just saw. Likewise, the presence of the Via Nova and the Asinian Gardens in the twelfth region, where are located the Thermae Antoninianae, which the conduit seems to seek as its destination, shows that this was the Specus Octavianus, from the same terminus that Frontinus assigns to the distribution of the Aqua Anio Vetus carried by it.

Finally, we compared the deposit from it (so that we might leave no type of proof unexplored) with another that was taken from the conduit of the Anio Novus. As the waters of each Anio line were tapped from the very same river, according to Frontinus [*Aq*. 73.6], so, in proportion to the distance from which they were carried, we found this one like that, and we brought the matter for testing to the Experimental Academy (so they call it) that the most honorable Giovanni Ciampini has opened in his house in a courageous and certainly praiseworthy enterprise, so that our conjectures might also be supported by proofs of the alchemist's art. It was certainly surprising and at the same time very pleasing to have found in this comparison a variety of incrustation in one and the same conduit of the Anio Novus. That older deposit attached to the *opus signinum* [waterproof concrete lining] from the time when the water used to be drawn unmodified from the river itself, was of turbid color, "as the river itself, although flowing from a very clear lake, nevertheless because of its crumbling banks carries away something by which it is muddied" (so Frontinus [*Aq.* 90]). However, the water that was added last of all to the conduit, "after Nerva Caesar saw that the shortcomings of the Anio Novus could be eliminated and ordered that its water be drawn from the lake, where it is very clear, with the river bypassed" [Frontin. *Aq.* 93.1–2], has produced a newer and cleaner deposit and a whiter one. As a result, the truth of Frontinus's account is shown to be confirmed.

The reckoning of the distance, as I touched on it in passing, of twenty-two miles between the beginning of each conduit—the Anio Vetus was brought from the twenty-second milestone, according to Frontinus [*Aq.* 6.8]; the Anio Novus from the forty-second milestone, according to the same author [*Aq.* 15.1]—was the reason that the deposit of the one seems the same as the other but indeed darker, inasmuch as it was necessary for the water tapped far below to have arrived more muddy "because of the greater number of cultivated areas lying beside it."

Certainly, that conduit that we saw recently uncovered by diggers in the stone quarry of the vineyard of Benedetto Santori could not have been anything except an extension of this Specus Octavianus. The conduit lies at the corner of the road that runs from the seating of the Circus to the Porta Ostiensis, with another thoroughfare by which there is a route to the left to the Church of S. Balbina the martyr. In addition, the course of the Marcia, which we have shown from the evidence of Frontinus to have crossed above the Porta Capena, intersects here with the Specus Octavianus, at a level of about six feet above it [fig. 10].

e. The Aqua Appia

Under these two conduits—as here a meeting of aqueducts occurred in the way as at Spes Vetus or in the vicinity of the modern Porta Maggiore, where we saw that the Marcia, before "it cast itself down over the Caelian," met with four other higher lines—a certain excavator most recently showed me another conduit lying very deeply hidden and, with difficulty, brought me crawling to it along the ground. When I had seen it,

A B. Ductus Anionis Veteris, ex parte A Arcum Druſi per Eurum grad. xij. versùs Auſtrum; ex B latus orientale Auentini è diametro reſpiciens

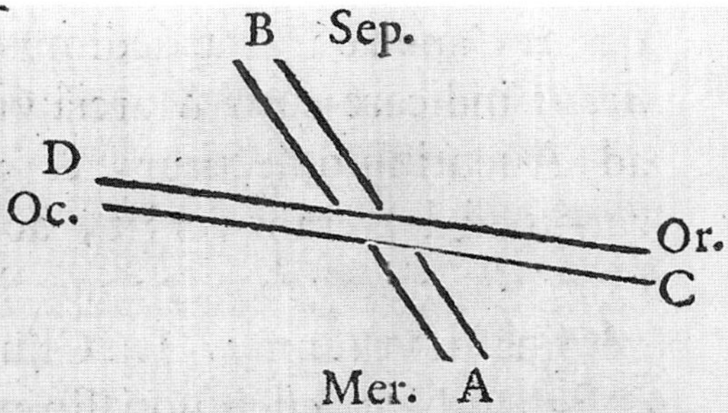

C D. Ductus Marciæ, ex parte C rupem meridionalem Cœlij, & locum, quem Portæ Capenæ aſſignauimus, per Solanum grad. viij. verſus Eurum; ex D medium Auentinum (vbi rurſus è regione Templi Sanctæ Priſcæ ſubterraneus viſitur) è diametro ſpectans.

Fig. 10. Aqua Anio Vetus and Aqua Marcia: plan of conduits underground near Porta Capena

AB. Conduit of the Anio Vetus, from direction A (the Arch of Drusus) through the east twelve paces toward the west; from direction B, the east side of the Aventine looking crossways

CD. The conduit of the Marcia, from direction C, the southern cliff of the Caelian, and the place that we have assigned to the Porta Capena, through the south eight paces toward the east; from *D*, the middle of the Aventine (where again, from the region of the Church of S. Prisca, it is seen underground) looking crossways

the freshness of an object long sought drove all weariness and fear of the place from my body and mind. That this is the conduit of the Aqua Appia, the first fruit of Roman foresight and greatness, I certainly convinced myself and will be eager to prove to you.

Its level and depth, as Frontinus calls it [*Aq.* 65.7], do not at all rule out the identification, since of all the aqueducts this side of the Tiber, I found this one to be the lowest (as Frontinus [*Aq.* 18.7] describes it), from a comparison with the conduits of the Marcia and the Anio very close by. It was running twenty-eight whole feet below the Anio Vetus, to the extent that I could measure it because of the narrowness of the place, through levels taken several times in the tunnel. Likewise, by its downward course to this spot, the place we have assigned to the Porta Capena, it, along with the Marcia, seems to indicate "ancient arches" near the gate. In cavelike fashion, however, it turns toward the Aventine, and after about fifty paces, it bends again to the right; from its appearance, it indicates the place where it emerged, according to Frontinus, namely, the Salinae and the Porta Trigemina [*Aq.* 5.9] or the Clivus Publicii [*Aq.* 22.3].

Not lacking is the most significant evidence, the same dimension of width and height Frontinus describes for the Appia [*Aq.* 65]. Although that conduit is for the most part different in form from the others, since the walls widen themselves gradually—I have noted this in no other aqueduct, and I think it happened when Agrippa was restoring this and the others that had "almost collapsed," as Frontinus indicates [*Aq.* 9.9]—when, nevertheless, you advance about seventy paces further, you will find walls straight and parallel, a foot and three-quarters distant between themselves, according to Frontinus, and a height of five and a half feet, excluding the layer of *opus signinum*. This is the distance of the very five feet of Frontinus above the *signinum*, which is commonly accustomed to have a thickness of half a foot at the bottom of conduits. Besides all these things, the majesty of the work, built with simplicity and, at the same time, combining graceful and natural strength, demonstrates clearly the age of Appius and the already increasing splendor of public construction and magnificence. To judge the construction as quite old from its very appearance, there is, in addition, the very thick accumulation of blackish deposit everywhere, stony and solid within, but on the surface, somewhat porous and smoothed. It is generally of the appearance of that stone brought from Tiburtine territory to Rome, which, most frequently for the sake of ornamentation, more sophisticated people are accustomed to use in urban gardens to imitate rustic fountains. But learn the thing better through drawing and explanation [fig. 11].

f. The "Arco di Druso" Itself

Our ignorance of the builder of our Specus Octavianus, which we almost seemed to have forgotten, and of the arch below it at the Via Appia and the Clivus of Mars detracts not at all from the discussion we have presented. Nevertheless, the more we search, the more it seems to favor the opinion set forth, if we argue that this is the arch that Augustus by decree of the Senate set up in honor of his stepson (or rather his son?) Drusus, according to Suetonius [*Claud.* 1.3]. Ligorio certainly, in his own manner, protests that this is not a triumphal arch but a distribution tank of an aqueduct; with that license, in the same treatise, he called the arch at the Porta

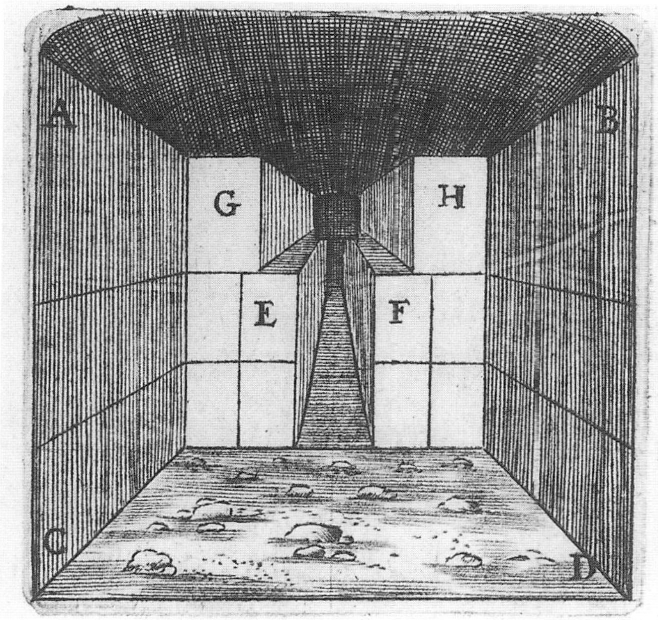

Fig. 11. Aqua Appia: cross section of underground conduit

ABCD. Size and shape of the gallery excavated for building the aqueduct in it, within the rock itself. The height from C to A is five and a half feet. The breadth from C to *D* is equal to the height. The arching is very low and is scarcely raised by a half foot. It does not rest on the walls of the conduit but stands on its own.

EFGH. Walls of Alban stone, by which the conduit was narrowed. The breadth from *E* to *F* is one foot four inches. The breadth from G to *H* is two feet ten inches.

S. Lorenzo a distribution tank (which, according to Roman law [*Dig.* 43.20.38], is the same as a reservoir).[22] At the Porta S. Lorenzo, there are in like manner single conduits of three aqueducts and no more than here the ghost of a distribution tank. Nevertheless, we scorn Ligorio's credibility as well as his judgment and, though he barks in the way, proceed to an investigation of the evidence.

The very great similarity between the coin of this Arch of Drusus and the remains of the arch itself furnishes the first support for our identification, as you can compare from the diagram of each set forth [fig. 12].

---

22. Ligorio, *Paradosse*, 39.

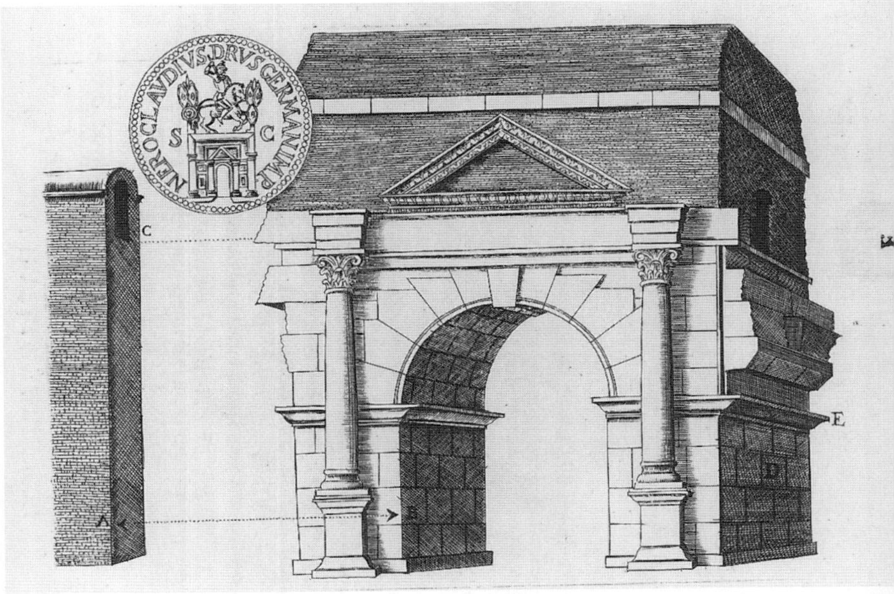

Fig. 12. "Arco di Druso" and coin representation of the Arcus Drusi

C. Channel of the aqueduct at the modern Porta S. Sebastiano, which we believe to be the Octavianus, two and a half feet wide, five feet high to the curvature of the arch. The arch itself is one foot three inches. The walls are uniformly two feet ten inches thick.

Moreover, does not the name *Specus Octavianus,* by which we identified the conduit placed above this arch, seem derived from Octavius Augustus? That emperor, when he "rebuilt the channels of all the aqueducts" (as expressed in the inscription at the arch of the Aqua Marcia, or the modern Porta S. Lorenzo, which we will show shortly later) or, as in his *Res Gestae* [20.2], "rebuilt the channels of the aqueducts collapsing from old age in many places," could have rebuilt this one also, broken off in the valley of the Clivus of Mars, and could have left his own name on it. The words *clivos* and *planitia* smell of this time. Vitruvius, twice in the same chapter [9.7.2], uses this form of the second word to his contemporary Augustus, and often on the coins of Augustus, you will read the inscription "DIVOS IVLIVS," with similar license. It is indeed more than probable that where the conduit was blocking the way, the slope had been smoothed so that the triumphal arch could be erected under the channel itself, and of the three arches along the Via Appia—those of Drusus, Trajan, and Lucius Verus, as many as Victor and the *Notitia* list—Augustus, who preceded the others, chose for his the higher position.

Indeed, even though its parts are half destroyed and buried by earth, this arch shows some special characteristics through which it may be best distinguished from the others. First, its tympanum and pediment are extended a bit beyond the breadth of the gate; in this, it indicates the times of Augustus and the same man as its builder. Augustus affixed architectural features by no means different to his own arch at Rimini, with the same narrowness of pediment, which a later age, or Vitruvius coming on the scene, corrected.

The appearance of the Arch of Augustus at Rimini is seen here [fig. 13] just as today, although it is buried in great part. This arch was erected in the rebuilding of the Via Flaminia, "which Augustus undertook while he instructed that the remaining roads be rebuilt by certain senators, at their own expense"—by a most certain conjecture (as the fragment of its inscription shows [*CIL* XI, 365]), at the time of his seventh consulship and the designation of his eighth. Dio [53.22.2] reports that this was done at this very time. In addition, in this year, he was honored with the name *Augustus* by the Senate and Roman people; we lack this name, perhaps originally displayed on this monument, and an intact inscription ("What has ruinous time not diminished?" [Hor. *Carm.* 3.6.45]).

I would have argued that neither of these arches was designed by Vitruvius, since each falls short of his prescribed standards. On the arch at Rome, indeed, the cornice rests immediately above the epistyle, although Vitruvius [3.5.10] establishes a carved frieze in the middle between these two parts. The arch at Rimini, however, has dentils under its mutules, not only on its cornice, but also in its pediment, whereas Vitruvius [4.2.5–6] decided that the cornice ought to be pure, by example of the Greeks and reasons sought from nature itself. Last of all, the pediment of each, especially that of the arch at Rimini, is raised far above the measure of the arch itself. Indeed, Vitruvius [3.5.12] assigns to a tympanum such a height that would be the ninth part of the length of the cornice from the farthest cymatia; here, the tympanum has in its height a length of the whole fourth part of the cornice. Although this proportion of the tympanum seemed too compressed to Philandrier (in his splendid commentary on Vitruvius, in a place noted by us, where he corrected it very easily),[23] it is certainly not

23. G. Philandrier, *Annotationes in decem libros M. Vitruvii Pollionis de architectura* (Lyons, 1552), 122–23. See F. Lemerle, *Les "Annotations" de Guillaume Philandrier sur le "De architectura" de Vitruve: livres I a IV* (Paris, 2000), 122–23.

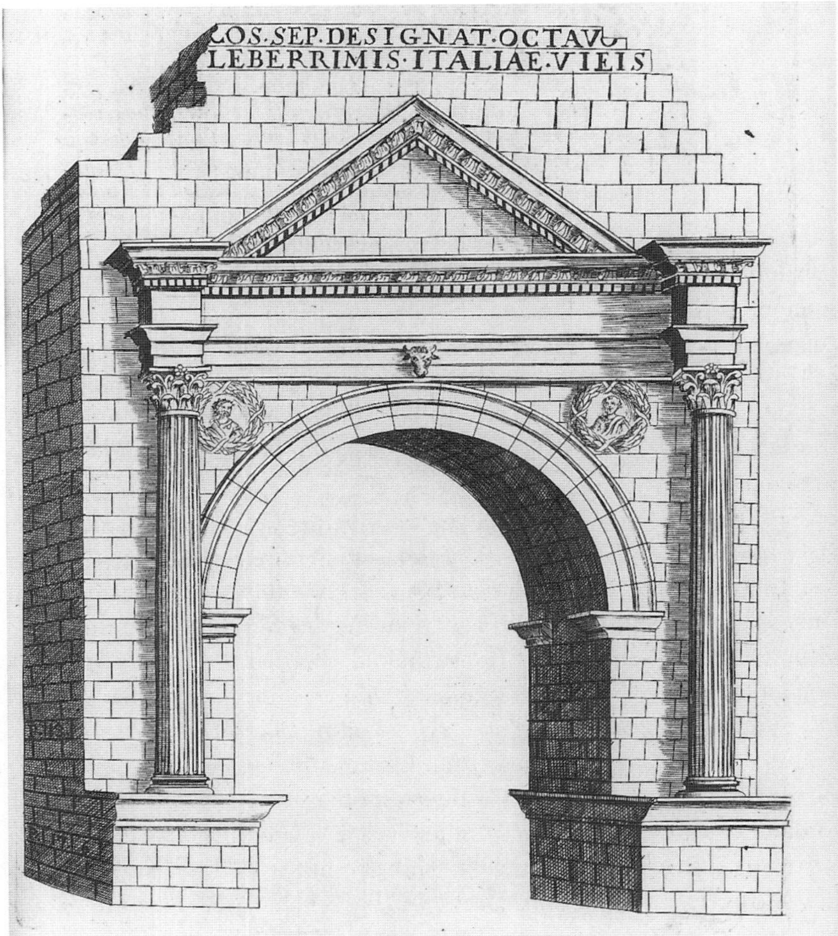

Fig. 13. Arch of Augustus at Rimini

disregarded in a very good example cited earlier, the arch of the modern Porta S. Lorenzo, once situated on the Via Collatina. To demonstrate this in particular, we present it here [fig. 14], just as it appears above the ground.

We point out, however, that the pediment and cornice have indeed been smoothed, so that they might open a space for the later inscription of Caracalla, yet in such a way that there have remained very sure traces of

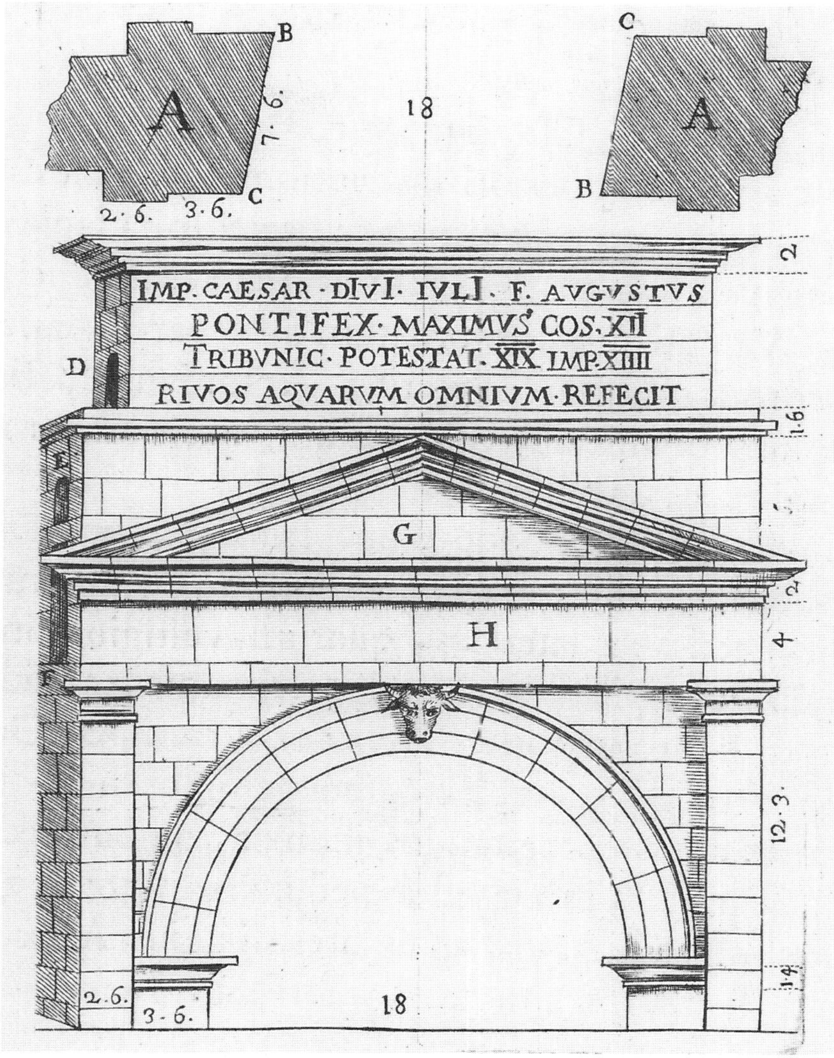

Fig. 14. Porta Tiburtina (Porta S. Lorenzo)

A. Cross sections of the piers of this arch
B. Very similar corners, less than a right angle, precisely of seventy-seven degrees
C. Other corners, greater than a right angle, of 103 degrees, conforming to the slanting course of the Via Collatina
D. Channel of the Aqua Julia
E. Channel of the Tepula, lower than elsewhere, but its capacity supplemented by widening along its breadth
F. Channel of the Marcia
G. Location of the inscription of Antoninus Caracalla, inscribed on smoothed blocks, set forth earlier [I.4a]
H. Location of the inscription of Titus Caesar, set forth above in the same place [I.4a]

the smoothed blocks. Indeed, although Titus Vespasianus earlier inscribed the bands under the cornice, to avoid disturbing the ornamentation, Caracalla occupied and inserted himself in the remaining space in the middle, although resulting in damage to the arch.

Side windows, traces of which are seen on side D, with the same crown E running back around it [fig. 12], also distinguish this triumphal Arch of Drusus from others and make it similar to our coin. This fact perhaps deceived Enea Vico and Ottavio de Strada, leading them to posit twin smaller arches from this point.[24] There cannot be space for them in a narrow area of twenty-four and a half feet, the distance between B and A, where the brick-faced arches of this aqueduct come to an end from this direction. The sketches of the distinguished Francesco Gottifredi, that great and most trustworthy exponent of numismatics, confirm the credibility of our coin; our Francesco Cameli, deprived of physical sight but a Lynceus of the mind, whom I name for the sake of honor and friendship, has kindly suggested the parallel.

It is pleasing now to digress somewhat and to supplement this coin and, at the same time, the history of Nero Drusus from another different coin completely intact, which the most distinguished Count Francesco Mezzabarba Birago has indicated exists in the Septalian collection. It shows a small vessel within the tympanum, two augural staffs at its side from each direction, and, toward the ends but at the same height, a patera in relief on the right and an urn on the left, arranged generally in that order in which we see them presented (somewhat obscurely) in Oisel.[25] As a result, it appears from the staff (which Cicero [*Div.* 1.17.30] calls the most famous emblem of the augurate) that Drusus was an augur, although history says very little about Drusus's augurate. Dio (in whose book 54 there are quite a few things about Drusus) and Suetonius [*Claud.* 1.3] make reference only to his quaestorship, consulate, and proconsular imperium.

For this reason, nevertheless, since the coin that we have compared to the Arch of Drusus, is seen to have been struck by his son Tiberius Claudius Caesar, I do not agree with Angeloni, who asserts that this arch was constructed and adorned by Claudius himself.[26] In addition to our own

24. E. Vico, *Le imaggini con tutti i reversi trovati e le vite degli imperatori* ([Venice?], 1548). Cf. E. Vico, *Discorsi sopra le medaglie de gli antichi* (Venice, 1558), 105; O. de Strada à Rosberg, *Vitae imperatorum caesarumque Rom. orient. et occid. uxorum et liberorum eorundem* (Frankfurt, 1629), 31.

25. Oisel, 475, table xcvi.1.

26. Angeloni, 43–44, 46.10.

arguments about Augustus, which we have set forth already, Dio [55.2.3] seems to speak about the matter, not only decreed but also brought to completion at the time of Augustus: "Drusus himself was carried out into the Campus Martius by the knights, those holding the equestrian rank officially and those who were of senatorial rank. Having been cremated here, he was buried in the mausoleum of Augustus. He was also given the name Germanicus, with his sons, and he received the honors of both a statue and an arch, as well as a cenotaph next to the Rhine." Augustus "made honors, magnificent ones, for Drusus," as Tacitus has entrusted to his history [*Ann.* 3.5], and, to cite the words of Suetonius [*Claud.* 1.5], "loved him so much while alive that he made him coheir with his sons; praised him in public assembly after his death; and, not content to have inscribed an epitaph on his tomb in verses composed by himself, composed a memorial of his life in prose." It is not believable that he would have neglected the honor of an arch decreed immediately after Drusus's death for the twenty-three years in which he survived his nephew and left it uncompleted for so long, especially since his wife, Livia, was urging him. Livia would scarcely have permitted this last honor for her son and solace for herself to have been postponed, given her most bitter and very great grief for him, with which she almost consumed her own life, as Albinovanus Pedo bears witness in these verses:

> She would have been able to survive for scarcely a few hours
> when Caesar brought aid to her unwilling;
> He applied prayers, mixed with them a sense of duty,
> and wetted her dry throat with water poured forth.
>
> [*Consolatio ad Liviam* 419–22]

Moreover, the emperor's coins show openly that this arch on the Via Appia has very little to do with Trajan (as Andrea Fulvio believed);[27] they display a triple arch [built by Trajan], not a single one, such as this one.

To exclude Lucius Verus, whom we know to have had an arch in the same region, as we said, we do not have such ready arguments as in Trajan's case, since no coins of his with his arch are extant. Nonetheless, as a result of so many arguments through which we have established that Augustus built this arch, we will remove from it another builder, after we

27. Fulvio, 246.

have sufficiently declared that the Arch of Drusus, unless I am mistaken, is situated under this aqueduct—indeed, at the highest point of this Clivus of Mars.

Nevertheless, all these matters, perhaps not useless to know in other respects, will seem to have been argued more than enough, given the limits of our investigation. Whatever doubt there can be about the name of this Specus Octavianus, let the following (which is sufficient for us) be beyond every hazard of a doubt: its water could in no way have been brought back at so great a distance to the Baths of Alexander in an opposite flow, against both the nature of the ground and its downward fall, with its *nymphaeum* [terminal fountain] left so far behind, as we shall soon argue.

g. The Aqua Traiana

As a result, the whole investigation falls back on this: whether Alexander brought into the city the aqueduct on the Janiculum, which we will identify as the Traiana, or this aqueduct, which we will call the Labicana, from the territory in which it arises. In this, I would wish that you remember that we have more than once reviewed, not without disgust, each of the inscriptions—the one on top of the impressive fountain on the Janiculum, the other on the arch on the Via Aurelia below the Horti Pamphilj—in which this aqueduct is called the Aqua Alsietina and its conduits are said to have been built by Augustus Caesar. How, except through the greatest stupidity, could the Alsietina be confused with this aqueduct? According to Frontinus [*Aq.* 18.8], it was lower than all the aqueducts, had no arcade except for 257 paces and, moreover, no substructure above ground [*Aq.* 11.4], and was allotted a different source, from the name of its own lake. The Janiculum aqueduct holds a level either higher or equal to the Marcia itself, acknowledges its source in various springs near Lake Sabatini, and is carried around the city (to summarize) on two thousand paces of arcade and substructure. Therefore, if we establish that this aqueduct cannot admit Alexander as its builder, this Labican aqueduct will remain the only one that must be said to have been related to him.

Probability alone almost forces us to believe this, since it is far from every probability that anyone ever thought to bring a Transtiberine aqueduct to this side of the river for the Baths of Alexander. We learn in particular from Victor [VZ, 1:213] that the Nymphaeum of Alexander stood in Region V, the Esquiliae—indeed, on the very direct course of this aqueduct from its arcade to the baths. Ligorio writes that he observed traces of this line and its settling tank near the Church of S. Croce in Jerusalem, where he found this inscription, perhaps, as Ariosto jokes about Turpin, "truthfully in this circumstance" [*Orlando furioso* 30.49]. The inscription appears not at all spurious, as are most things connected with that man:

> To Hercules Protector, unconquered companion of our lord Severus Alexander Pius Victor, always Augustus and most excellent emperor, the most distinguished Marcus Aurelius Priscillianus, curator of the *nymphaeum*, vowed from his own money . . . [*CIL* VI, 333*]

Ligorio, in keeping with his habit, was able to add on his own a settling tank to the total, although we are of the opinion and show that the settling tank was elsewhere, and scarcely do we believe that there were two of them.

I will add—and not at my peril, since I have a sufficiently suitable champion of this opinion—that the Janiculum aqueduct could not have been called the Alexandrina (this from the testimony of one of the ancient sources); more accurately, it was one and the same as the Traiana. So recognizes Fr. Aringhi from the Acta, especially those concerning the martyrdom of S. Antonio: "In the same hour, Vitellius ordered Antoninus to undergo the sentence of execution; he, having been led along the Via Aurelia next to the arcade of the Aqua Traiana, was beheaded on August 22nd."[28] The statements of the very ancient manuscript of the Acta of S. Giulio Senatore, which appear in Nardini and elsewhere, also confirm this; in them is cited "the arcade of the Traiana next to the Via Aurelia." In the Instrument of Donation of Charlemagne, "the arcade of the Traiana up to the Porta Aurelia" is given as the boundary of a certain property.[29] The same fact is also dug out from the following citation, although inaccurate, of the librarian Anastasius in the *Life* of Honorius: "At the same time, he built the Basilica of S. Pancrazio at the second milestone of the Via Aurelia from its foundation, . . . and he established there mills up against the wall of the city and an arcade that brings water down into Lake Sabatini."[30]

The authority of Ligorio is not of such great value for us that, as a result

28. Aringhi, 2.346.
29. See Nardini, 509.
30. Anastasius bibliothecarius, *De vitis pontificum romanorum* (Rome, 1718), 121.

of his opinion, we should remove the name *Aurelia* from that road that, by sure traces from the Porta S. Pancrazio (indeed, from the Pons Sublicius), is joined by direct path with another road that Ligorio does not deny is the real Via Aurelia.[31] Ligorio himself puts before us the single relevant passage of Procopius [*Goth.* 1.19.4]: "As a result, it happened that two other gates of the city had been occupied by the enemy, the Aurelia (which is now called the Gate of Peter, after the chief of the apostles of Christ, as it is located near his grave) and another, across the Tiber River."

From this, there is proof that there was a Porta Aurelia located near Castel Sant'Angelo; therefore, Ligorio leads his road with this name from this point and firmly denies that it was elsewhere. He does not refute that the gate at the Janiculum is also called the Aurelia, since so he speaks and ought to be understood. To distinguish between these two gates with the same name, one, as it were, was accustomed to be called the Gate of S. Pietro, the second that of S. Pancrazio, from their proximity to each church. Ligorio's opinion must also be accepted concerning a Via Aurelia Nova. Since we recognize both a Via Aurelia Nova and a Via Aurelia Vetus from an inscription cited by Gruter [*CIL* XIV, 3610], we should not erase this old one, the name of which so many famous documents retain before the occurrence of the second.

The sacred memoirs of the early Christians place the Basilica of S. Pancrazio very close to the aqueduct, in the vicinity of the city on the Via Aurelia, which cannot be Ligorio's Via Aurelia. Moreover, it cannot be doubted that in ancient times and before the Aurelian Wall, there had been no other course of this most ancient road than that already mentioned, from the Pons Sublicius over the top of the Janiculum, where the gate is now seen (whether the Aurelia or Porta S. Pancrazio, as it was called afterward). Even then, no way was open from the city to the Janiculum except over the Pons Sublicius. From here, we see a straight and well-worn route into Etruria Maritima, as excavation, the continuous line of funeral monuments between the villa of Marchio de Nobili and the monastery of S. Pietro in Montorio, and the most recent discovery of basalt pavement running from this point toward the Church of S. Cosimato and to the Pons Sublicius itself openly teach us. What, I ask, is more absurd than to conceive of an ancient and original Via Aurelia over a bridge (which did not yet exist) and through the territory of the Vatican,

31. Ligorio, *Paradosse*, 42.

over windy circuits like the Meander, and to deny that it was there, where true reason and, I might add, the authority of ancient writers put it?

Livy [5.40.8–10] shows that the road that runs from the Pons Sublicius to the Janiculum and then to the city Caere (which is our Aurelia) was trodden by the Vestals in flight, after the city had been captured by the Gauls, when he cites Lucius Albinius, a man of the plebs, who put the Vestals and the holy relics of Rome on a wagon, after his wife and children had been ordered to get off. I would certainly have thought that from this event must be understood the following fragment of precious stone inserted afterward in the floor of the upper and open-air walkway around the dome of the Pantheon, which so indicates the circumstances and the time (when, indeed, the Gauls "were attacking the Capitolium," after Rome had been captured):

> were attacking the Capitolium . . .
> . . . led the Vestals to Caere
> . . . and the solemn rituals so that
> . . . considered it his task
> . . . the sacred objects and the maidens
> . . . led
>
> [*CIL* VI, 1272 = 31583 = *ILS* 51]

As a result, there is very little doubt that it was inscribed for a memorial and praise of such great devotion.

So certain do I consider these things that I am not pleased by the compromise of Aringhi,[32] by which he has also deferred to the opinion of Ligorio[33] and removed the Via Aurelia from the Porta S. Pancrazio. Aringhi leads another street, which Ligorio reported as taking its beginning from the Porta Aurelia of Procopius, through the nearby countryside of the Fornaci, beneath the walls of the city, and joins it to our ancient Via Aurelia at the crossroad of S. Pancrazio. Indeed, by this unlikely curving course, a cure worse than the disease has been brought forward, and Ligorio himself would disagree even more strongly. Ligorio, I suspect, thought of the course of the road neither so close to the city nor so violently twisted back; but more reasonably (so we might support so hardworking a man in some-

32. Aringhi, 1.343.
33. Ligorio, *Paradosse*, 37–38.

thing), he led his Via Aurelia straight to the Taverna del Pidocchio and from here to the left through a side street, a part of which is seen paved with basalt between the Vigna Carpegna and the distribution tank of the Acqua Paola in the Vigna Burana. Even still more consistently, he connected it there to the ancient Via Aurelia through traces of an old road, which, after leaving Cornelia on the right, runs through the Porcareccino property to the Bottachia property. But we shall make these things clear more fully, God willing, in our work *On the Suburban Territory*.

As a result, therefore, now that the ancient Via Aurelia has been restored, contrary to Ligorio, as has the more ancient Via Aurelia Vetus alongside the Trajanic aqueduct, to which it was been joined for almost two miles, to each emperor, both Alexander as well as Trajan, has his own aqueduct been returned. Indeed, it is well known, besides the authority of modern writers already cited, that Trajan introduced an aqueduct, from the evidence of a coin struck by decree of the Senate in honor of the accomplishment and through its representation of the divinity of the source as water drawn from a spring, not taken from other conduits. See the coin, carefully reproduced but very badly interpreted by Angeloni; he says abruptly—and as though Frontinus were confirming it—that "Trajan brought the Aqua Marcia onto the Aventine from the Via Valeria, over a course of twenty-three miles,"[34] and he concludes that this is shown by the coin. We see that this argument has seduced his other followers, who tend to believe more what "the man himself said" than Frontinus and the truth, as we noted earlier [I.4c]. Most recently, indeed, Angeloni has drawn to his side Jacques Oisel in his *Thesaurus* published this year.[35]

5. LEVELS OF THE AQUA TRAIANA AND AQUA ALEXANDRINA

It is by evidence not to be scorned that this Aqua Alexandrina was introduced for the sake of the baths already mentioned. If someone else had built it for the common and accustomed purpose of aqueducts, he clearly could have raised it far more usefully to the same level as did Pope Sixtus

34. Angeloni, 114, 123.27.
35. Oisel, 529, table cviii.8.

afterward with his aqueduct, that is, twelve and a half feet above the Marcia (the elevation by which Sixtus's aqueduct is detected as surpassing the nearby conduit of the Marcia at the Porta Maggiore), and further, to another forty-one and a half feet. The Marcia surpasses the conduit of the Alexandrina by this height, as the measurement demonstrates, taken from the first arches noted [fig. 1, no. 2], near the arches of the Aqua Marcia, very close by at that point. Since the site of Alexander's baths was indeed lower, the emperor, considering only them, neglected the advantage of a higher elevation and lowered his aqueduct by fifty-four feet. Everyone will agree that such a great difference in elevation had its origin not from lack of expertise but from deliberate plan.

However, how, in his book on the water system, Frontinus, Trajan's contemporary and superintendent of the water supply under him, could be silent about this most outstanding work of his emperor is clearly understood from the preface of this same book. Frontinus wrote his treatise at the very beginning of the duty enjoined on him—indeed, when Nerva, to whom he dedicated his work, was still alive. Moreover, the introduction of the Aqua Traiana, from the coinage already mentioned, took place when Trajan was consul for the fifth time, or five times consul, that is, at least after the fifth year of his principate.

But if you should object that I am reducing the number of aqueducts against the expressed opinion of ancient authorities—Victor names twenty (or even twenty-four, according to some manuscripts) [VZ 1:255–56]; the *Notitia Imperii* names nineteen [Jordan, 2:569–70; VZ, 1:154–56]; and Procopius, although he names fewer, clearly names fourteen [*Goth.* 1.19.13]—I will respond that Victor and the *Notitia* do not overly concern me, since they make reference not to conduits but to sources of water. More than one of them could have been channeled into a single conduit, as Frontinus [*Aq.* 12, 14.2, 15.4, 73.8] testifies in the case of the Augusta, Albudine, and Herculean springs and as the inscription of the modern Porta Tiburtina [*CIL* VI, 1245] confirms in the case of the Aqua Antoniniana. I shall prove at some other time that Procopius, who speaks about the aqueducts, is in agreement with me, and I shall bring forward certain new things for the clarification of this point, if I may only be permitted to interpret the words *Rome* and *city* in a different way. I put off this subject for the present time, so as not to go entirely beyond the limits of a letter.

6. EVIDENCE OF BRICK STAMPS AND THE TOMB OF ALEXANDER SEVERUS

If someone desires a still more certain proof of my opinion concerning Alexander as builder of the work, he should take care (as we shall do) to investigate the brick stamps from other monuments of Alexander. Two of them are undisputed: namely, the baths in the Palazzo Medici, for the purpose of which, as we have just said, this aqueduct we are discussing was introduced; and the tomb between the Via Latina and the Via Labicana in what is now called Monte del Grano. Perhaps there might occur either the brick stamp illustrated earlier [fig. 5] or another [fig. 15], with the inscription "SEX.CAEC.PRO . . ." [*CIL* XV, 2313.b.3], which is inscribed in a way similar to the brickwork even now on the arches at number 7 of the topographical map [fig. 1].

Indeed, the same characters will certainly demonstrate that the two constructions were contemporaneous. It is plausible that the bricks, as they had first been fired, served the very great number of projects that were then being built and that they were not accustomed to being kept for a long time in the brickyards, as is the practice even today.

Since, as Flaminio Vacca says, it is the mark of natural modesty to confess by whom you have profited, I report to you news I have learned. The funeral monument at Monte del Grano covered the ashes of Alexander and his mother, Mamaea. I repeat Vacca's account, which you indicated to me, investigated and confirmed: "I remember a mile outside the Porta S. Giovanni, having passed the aqueducts, where the place is called Monte del Grano, there was a large ancient construction, made of cement; the impulse was sufficient for a digger to break into it, enter inside, and then take himself down, with the result that he found a large sarcophagus decorated with the rape of the Sabines, and on top of the lid, there were two figures stretched out, with the portrait of Alexander Severus and his mother, Julia Mamaea. Inside were found some ashes; the sarcophagus mentioned now is found on the Campidoglio, in the middle of the courtyard of the Palazzo dei Conservatori."[36]

Vacca writes these things to Anastasio Simonetti in a letter of November 1, 1594, in a simple style indeed, but with greater truthfulness than Fr. Aringhi, who says that the sarcophagus was found on the Via Labicana,[37]

36. F. Vacca, *Memorie di varie antichità trovate di diversi luoghi dell'alma città di Roma nell'anno 1594* (Rome, 1704), memoria 36.

37. Aringhi, 2.34.

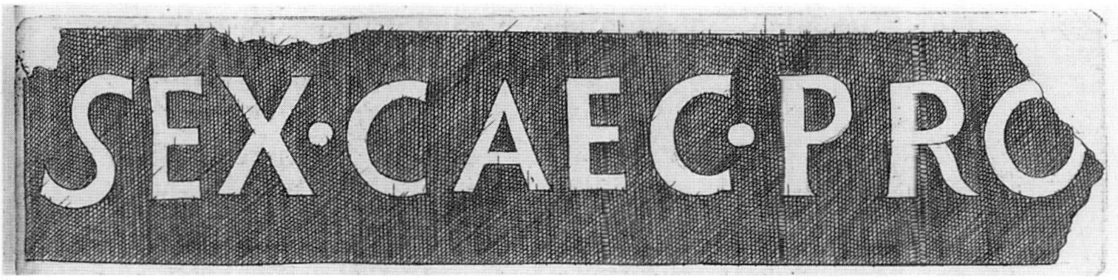

Fig. 15. Brick stamp from Monte del Grano tomb [*CIL* XV, 2313b.3]

although the place is somewhat closer to the Via Latina and is distant from the Labicana by about seven stades. However, in a modern inscription on the base, atop which the sarcophagus has been installed, the provenance is said to be outside some Porta Labiena, of which there is scant trace among authors. Although a contemporary writer from France denies that the images of the sarcophagus just mentioned reflect those of Alexander and Mamaea,[38] the common agreement of scholars and the clear resemblance with their coins give more than enough support to our assertion, and they do not permit an imposition on us. As a result, perhaps, there may be added an important argument from the aqueduct built by Alexander, if we consider that the Senate and Roman people revived the eternal memory of its builder in erecting the mausoleum at a place not at all far from it, where there is a full view of the tomb from the source of the aqueduct itself. The construction, although ugly and stripped of all ornamentation, shows that the size of the tomb was not modest. Lampridius exaggerates this when he says, in his *Life*, "He earned a cenotaph in Gaul and a most magnificent tomb at Rome. [SHA *Alex. Sev.* 63.3]. We add a plan of it here [fig. 16].

7. SOURCES OF THE AQUA ALEXANDRINA AND ACQUA FELICE

The modern aqueduct today is drawn from various sources, especially those that issue beneath the Colle delle Pantanelle, from marshy and moist soil. I am doubtful whether Alexander or his engineers sought these same sources at one time; the opposite, rather, persuades me, the nature of

38. Bernard de Montfaucon, O.S.B., *Diarium italicum sive Monumentorum veterum, bibliothecarum, musaeorum, etc. notitiae singulares in itinerario italico collectae* (Paris, 1702), 138.

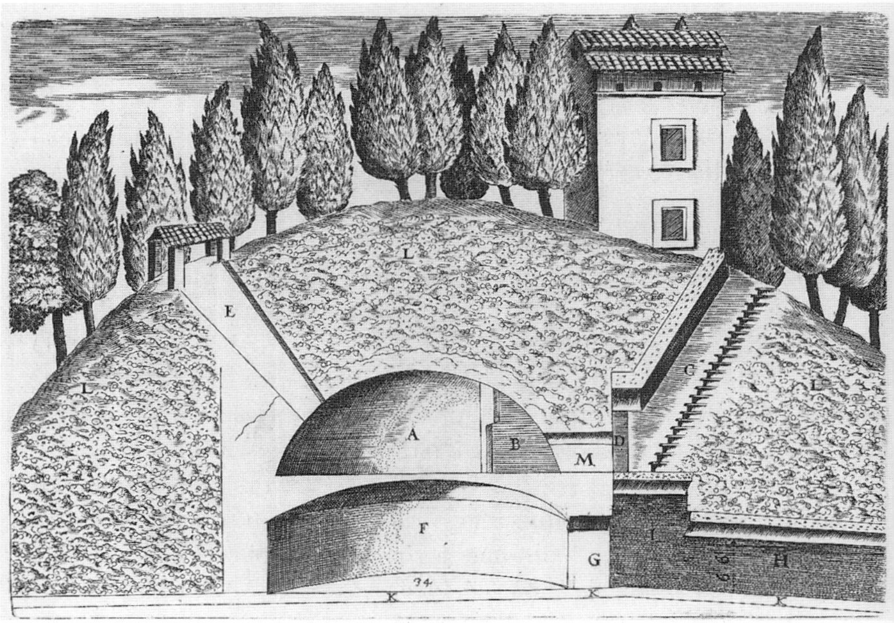

Fig. 16. Monte del Grano tomb: cross section

A. Round chambered cella, in which was found the sarcophagus with the images of Alexander and Mamaea, today on the Campidoglio
B. Small chamber, through which there opened an entrance into the burial chamber already mentioned, through a door in M
C. Steep stair, through which the sarcophagus was carried through the door in M
D. Wall, behind the sarcophagus, in front of the raised door of the burial chamber, so that all approach might be blocked
E. Modern opening, from which the following was discovered
F. Round hypogeum, with a sunken chamber, under the burial chamber
G. Door to the same, five feet three inches wide, having a five foot thickness in the wall
H. Cryptoportico, five feet three inches wide, eighty feet long, inclined six and a half feet along that entire length
I. Space at the head of the cryptoportico, with a raised arch, for supporting the upper stair
K. Pavement laid under the entire structure, of travertine blocks, one foot three inches thick.
L. Ruins of a building, an uncertain structure, but, as can be judged from the huge heap of them, a magnificent one, as according to Spartianus

the modern aqueduct being different from the ancient one. The Alexandrina produced a huge deposit in its conduit, as we have said; the Felice, after a course of nearly a hundred years, shows its conduit still pure and gleaming. As a result, I would determine that the source of the Aqua Alexandrina was in those very springs of the Osa stream that flow in a scattered way here and there at the sides of that valley.

Yet with the greatest certainty, I report that a modern geographer is shamefully mistaken: in his topographic map published in 1674, he shows the source of the Acqua Felice as the lake (today the Lake of Monte Falcone) that Holste calls Lake Regillus,[39] quite correctly in my opinion. Shafts above the underground channel, which are seen near this lake, perhaps furnished the cause of the error. The writer, clearly having taken too little care, noticed neither that they continued up to the source of the aqueduct or to the first distribution tank nor even the tank itself, set apart an entire mile from here. The lake, however, contains water that is stagnant and altogether unsuitable for drinking, and it would not suffice for distribution of a single day besides. In the summer season, moreover, it would supply water that is indeed smelly and polluted, on account of the very great supply of flax and hemp that the inhabitants of the neighboring towns bring here for grinding.

There are those who have said that the Acqua Felice arises to the left of the Via Praenestina, and in this way, in an exemplary inscription and with more than exemplary error, it is described at its principal fountain near the Baths of Diocletian. But they have been mistaken; today, because the Via Praenestina itself is interrupted and impassable, one cannot approach Palestrina except by the Via Labicana, to which, therefore, they assign the name of the Via Praenestina everywhere. Holste recognized this and corrected it with reference to Clüver, but Holste himself was likewise mistaken when, enticed by some old fragments of inscriptions scattered in the structure of the Osteria del Finocchio, he accepted the modern Via del Finocchio, diverging from the Labicana after Torrenova, as the ancient Via Labicana and looked for nothing besides—supported by only censure of his Clüver, who, I say, had championed and glorified him (in this, with the greatest ingratitude and at the same time bitterness of speaking, he makes mistakes too frequently).[40] To the right are certain traces of the ancient Via Labicana, better surely for the mind than for the feet. We have, however, observed an ancient branch road crossing this way from the Taberna dell'Osa, on the Via Praenestina, past the Casale of S. Antonio, which from there, a little above the Tor Forame, crosses the Labicana and runs in the direction of the Via Tusculana and Via Latina, and we have marked it in its place on our map [fig. 1]. To such an extent, as very frequently, "those who do not know the path," to quote Cicero [*Div.* 1.58.132], "show the way to the other."

39. Holste, 191–92.
40. Holste, 191, 194.

I could show you these things face-to-face and on the ground itself more easily and perhaps more usefully, but since you now reject labor, using your age as an excuse, take advantage of the privilege of old age—which I dearly wish you to enjoy for a much longer time, for the common benefit of literature—and present your opinion while sitting down. Farewell. Written from our museum, on the last day of the year of salvation 1677.

Commentary

The first dissertation of the *De aquis* was originally written without section headings, but the edition published in 1680 includes marginal numbering for subjects cited in its index (omitted here because all such topics are included in the general index). The dissertation is here organized as follows:

1. Introduction
2. The Aqua Alexandrina: Course, State of Remains, Description, and Route
3. Proof that Alexander Severus Built This Aqueduct
4. Review of Other Aqueduct Conduits in Rome
    a. Aqueducts at Porta Maggiore
    b. The Arcus Caelimontani
    c. The Claudia Branch to the "Trophies of Marius"
    d. The "Arco di Druso" Aqueduct and Specus Octavianus
    e. The Aqua Appia
    f. The "Arco di Druso" Itself
    g. The Aqua Traiana
5. Levels of the Aqua Traiana and Aqua Alexandrina
6. Evidence of Brick Stamps and the Tomb of Alexander Severus
7. Sources of the Aqua Alexandrina and Acqua Felice

The scope of the dissertation is much broader than simply a discussion of the Aqua Alexandrina. Fabretti also reviews all the aqueduct conduits brought into the ancient city, both those cited by Frontinus in the *De aquaeductu* and the Aqua Traiana of the early second century A.D.

OPENING

Giovanni Lucio of Trau (Ivan Lučić, 1604–79). At one time a mathematical tutor to Fabretti's nephew, Lucio was the author of *De regno Dalmatiae et Croatiae libri sex* (Amsterdam, 1668) and other studies on Trau and Dalmatia. Fabretti cites him in the preface to his corpus of inscriptions (*Inscriptionum*, 2) as the friend whose example had inspired him to undertake his own epigraphical collection. See Ashby, 2 n. 4 (20 n. 9); Mennella, *Museo Lapidario*, 19 n. 2.

1. INTRODUCTION

Aqueducts around the Anio River. This is a possible reference to the second dissertation of the *De aquis*, which treats the sources of the Marcia and Claudia in the upper Anio Valley. Fabretti states that bad weather had prevented a second trip to the area, implying that he had already explored the upper Anio Valley. He seems to have informed Lucio about the work in progress and may have sent him a draft of the second dissertation.

Acqua Felice. This aqueduct was introduced by Pope Sixtus V (1585–90) and named from its builder, who was Cardinal Felice Peretti before his elevation to the papacy. Designed by Domenico Fontana, the Acqua Felice takes a higher and largely underground course to the watershed between the Anio and Tiber Rivers, then parallels the route of the Marcia/Tepula/Julia (the arches of which Fontana used to facilitate construction) and the Claudia/Anio Novus into the city at Porta Maggiore. For the history of its construction and route, see Lanciani, 389–93; Motta, "L'acquedotto Felice," 220–25.

2. THE AQUA ALEXANDRINA: COURSE, STATE OF REMAINS, DESCRIPTION, AND ROUTE

Adrien Auzout of Rouen (1622–91). Auzout was an antiquarian and mathematician particularly interested in problems of ancient engineering and measurement. One of the first members of the Académie des Sciences at Paris, he is credited with the invention of the movable thread micrometer used by astronomers to observe the dimensions of heavenly bodies. He was also the first to decipher the canceled line on the Arch of Septimius Severus (*CIL* VI, 1033 = 31230) and served on the committee to tap the water of Lake Bracciano for the Acqua Paola. Paraphrasing Apuleius (*Met.* 2.23), Fabretti describes Auzout elsewhere as "perspicacior ipso Lynceo, vel Argo, et oculus totus" [more keen-sighted than Lynceus himself, or Argus, every bit of him an eye] (*De columna Traiani syntagma* [Rome, 1683], 118). See *DSB*, 1:341–42 (R. M. McKeon); Ashby, 308 n. 3 (365 n. 3).

Topographic map. Each of Fabretti's dissertations begins with a detailed map depicting the topography of the subject to be covered, based on extensive autoptic investigation of the region. The Latin pentameter that appears above the title in the upper right-hand corner of the map in fig. 1, "est aliquid memori visa notare manu" [It is no small accomplishment to note with mindful hand the things one has seen], a direct quotation of Ovid (*Pont.* 3.4.18), is more than decorative, clearly indicating to the reader Fabretti's pride in his cartography and the firsthand knowledge of the region it represents. He repeats this quotation on the topographical map of the second dissertation (fig. 17).

The map in fig. 1, entitled "Plan of the Area of Latium from Rome to Labicum," presents, according to Ashby (308 [365]), a "very accurate" rendering of the course of the Aqua Alexandrina from its source to the area of Spes Vetus, as well as other topography to be discussed. J. B. Ward-Perkins ("The Via Gabina," *BSR* 40 [1972]: 122) notes, "[Fabretti's] account, though supplemented in detail by modern writers, has proved to be in all respects remarkably reliable." For the orientation of Fabretti's maps with modern topography, see fig. 38.

Fabretti's listing of details on the map in his text is selective, focusing almost entirely on sections of the Alexandrina and other aqueducts cited in this dissertation. The numbered citations in his legend correspond to the following identifications and descriptions by modern topographers:

1. Substructure closest to Rome: Ashby, 315 (374); Van Deman, 356.
2. Fifty-two arches in the valley of the Acqua Bollicante: Ashby, 315 (374); Van Deman, 355–56; Aicher, 106–7.
3. Substructure near the Via Labicana: Ashby, 315 (374); Van Deman, 355.
4. Centocelle arcade: Ashby, 314–15 (373–74, figs. 106–8); Van Deman, 352; Aicher, 106–8. For an account of recent restorations in this area,

see G. Caruso and P. Giusberti, "Acquedotto Alessandrino: Restauro del tratto tra via del Fosso di Centocelle e via dei Fioppi," *BC* 95, no. 2 (1993): 117–21.

5. Seven arches in the next valley: Ashby, 314 (373); Van Deman, 352; FI 1.10, x, 541–42 (no. 439).
6. Twenty-three arches: Ashby, 314 (371); Van Deman, 361; FI 1.10, 535–41 (no. 438); Aicher, 109.
7. Eighteen arches: Ashby, 314 (371); Van Deman, 351; FI 1.10, 530–35 (no. 436); Aicher. 109.
8. Branch line of aqueduct: FI 1.10, 556–57 (no. 448); Van Deman, 351.
9. 102 arches: FI 1.10, 546–56 (no. 447); Van Deman, 349.
10. Twenty-eight arches: Ashby, 312–13 (370); Van Deman, 348; FI 1.10, 504–12 (no. 417). For recent excavations in the area, see C. Calci, "L'acquedotto alessandrino alla tenuta della Mistica," *BC* 89 (1984): 105–6.
11. Twenty-two arches in a curve: Ashby, 312 (370); Van Deman, 347; FI 1.10, 497–503 (no. 415); Aicher, 111.
12. Fifty arches in a curve: Ashby, 312 (369); Van Deman, 346.
13. Branch line to the north: Ashby, 312 (369); Van Deman, 346; FI 1.10, 490–97 (no. 414).
14. Single arch in substructure: Ashby, 312 (369); FI 1.10, 484 (no. 402); Van Deman, 346.
15. Twenty-eight arches: Ashby, 311–12 (369); Van Deman, 346; FI 1.10, 487–83 (no. 399); Aicher, 111.
16. Four lower arches: Ashby, 311 (369); Van Deman, 345–46; FI 1.10, 474–45 (no. 388)
17. Single arch: Ashby, 311 (368–69); Van Deman, 345; FI 1.10, 462–63 (no. 370). Cf. Ward-Perkins, "Via Gabina," 99–100: "There is nothing to be seen at the crossing today, but there is no reason to question Fabretti's record of a single-arched bridge of Gabine stone (which he thought carried the aqueduct); and since the aqueduct is elsewhere faced with brick and of relatively slight proportions, one may confidently accept Ashby's suggestion that the bridge carried, not the aqueduct, but the road beside it."

18–19. Well shafts: Ashby, 311 (368).

20. Substructure: Ashby, 311 (368); Van Deman, 344.
21. Sixty-seven arches: Ashby, 310 (367); Van Deman, 343–44.
22. Sixty-two arches: Ashby, 310 (366); Van Deman, 343–44
23. Forty-five arches: Ashby, 309 (366); Van Deman, 343

24. Settling tank: Ashby, 309 (366); Van Deman, 343; Lanciani, 382.
25. Branch of the Anio Vetus (?): Ashby, 87 (107).
26. Temple of Fortuna Muliebris: Tomassetti, 4:81–82; L. Quilici, *La Via Latina da Roma a Castel Savelli,* Passeggiate nel Lazio 4 (Roma 1978), 62–65.
27–28. Settling tanks of the Marcia/Tepula/Julia: Ashby, 133–35 (166–67); G. Pisani Sartorio, "Punto di derivazione dell'acqua Marcia alla cisterna della c.d. villa delle Vignacce sulla via Latina; strada di manutenzione e cippo terminale delle acque Marcia, Tepula, e Iulia," in *Trionfo,* 55–56; Aicher, 93–95.

Topographical features indicated on the map but not noted in Fabretti's legend include the major Roman roads emanating from the city, with prominent topographical features along them, as well as the major tributaries of the Anio River in the Roman Campagna. Fabretti shows the underground course of the Acqua Felice from its source to the area of "Roma Vecchia" (Old Rome), where it emerges to ride on arches into the city at the Porta Maggiore, as well as the courses of the Aqua Virgo, the Aqua Marcia/Tepula/Julia, and the Aqua Claudia/Anio Novus.

The map also indicates landmarks cited later in the dissertation: the Monte del Grano (identified as the "Sepulchrum Alexandri et Mamaeae"), discussed in I.6; Colonna (identified as the "site of ancient Labicum"), discussed in I.2; and Lake Regillus and topographical features of the *ager Labicanus*, discussed in I.7. The map focuses only on the eastern *suburbium* of the city, giving no information about the Aqua Traiana and topography of the area outside the Porta S. Pancrazio, even though Fabretti discusses this region and its roads at some length (I.4g).

Athanasius Kircher, S.J. (1601–80). A German polymath and antiquarian, Kircher converted to Catholicism, joined the Jesuit order, and became a professor at the Collegio Romano and a leading scholar of his day, particularly in Egyptology and Syriac and Coptic texts. He also established a collection of antiquities in Rome known as the Museum Kircherianum; for assessments of his scholarship and career, see E. Lo Sardo, ed., *Athanasius Kircher, S.J.: Il Museo del Mondo*, exhibition catalog (Rome, 2001); A. Grafton, *The Footnote: A Curious History* (Cambridge, Mass., 1997), 149–54; Sandys, 2:365; *ECHA*, 636–38.

Despite his fame as a polyhistor, Kircher's work on Roman topography was primitive by modern standards and certainly lacking in Fabretti's eyes. Ashby (3 n. 5 [21 n. 21]) writes, "Kircher's inaccuracies in regard to the representation of the aqueducts in his maps are flagrant: and his text is equally full of errors." The map cited by Fabretti (Kircher, opposite p. 70) is a particularly striking example of such faulty cartography: as Fabretti observes, Kircher wrongly indicates the source of the Aqua Virgo near the town of Colonna and omits any representation of the Anio River.

Arcade and settling tank. Fabretti's illustrations of the two sections of arcade and settling tank of the Alexandrina (figs. 2–5) are generally accurate. Ashby (314–15 [366]) and Van Deman (352–55) present without comment or correction Fabretti's depiction of the double arches in fig. 2 (part of the long arcade indicated at no. 4 on the topographic map, in the area of the modern Centocelle).

Fig. 2 presents a section of the arcade in the Valle di Pantano (no. 21 on the map): see Ashby, 309–10 (367); Van Deman, 344. For the settling tank, see Ashby, 309 (366); Ashby notes that while the dimensions as given by Fabretti can be confirmed, Fabretti's representation of the settling tank in fig. 3 is incorrect in its placement of the openings. Van Deman (343) also observes that the outlet "was not exactly in the middle, as represented on Fabretti's plan," and she gives measurements of 12.5 meters in length and 10.64 meters in width for the tank, citing Fabretti's dimensions as "somewhat larger than those here given, including, it is probable by some error, the outer walls of the structure." Ashby (309 n. 5 [366 n. 13]) further notes that Lanciani (382), who follows the description given by Fabretti here, reverses Fabretti's measurements in his description of the settling tank.

This settling tank itself is small and relatively simple in its construction and functioning, as Fabretti illustrates and explains it; we may contrast the much more ambitious settling tanks closer to the city that are discussed and illustrated in the second dissertation (II.5a–b, figs. 28–30). The location of the tank, quite close to the source of the aqueduct, might have been a factor: no provision is indicated for local distribution of the water from the tank, and its sole function seems to have been to clear the water having been tapped from the springs.

Later reworking. Fabretti's description of the conduit in the valley of the Acqua Bollicante (fig. 1) is cited without correction by Ashby (315 [374]). Cf. Van Deman, 356: "Of the original structure but little is now traceable, since the greater part of it is buried in later encasing walls, with reinforcing arches under the older arches. . . . The construction of the later encasing walls and reinforcing arches, of which some fragmentary remains are still left, is of concrete of an inferior grade faced with late brickwork the exact date of which is not clear."

Aqua Crabra. This ancient water source in Tusculan territory is cited by Frontinus (*Aq.* 9.4–9) as intended for local delivery independent of the Roman water system but tapped at one time to supplement the Aqua Julia. Fabretti presents a much longer treatment of it in the third dissertation (III.4c). For further discussion, see Hodge, 249–50, 448 n. 17.

As noted by Lanciani (325–27) and Ashby (315 n. 3 [374 n. 35]), Fabretti is incorrect here in his identification of the water carried by the reworked arcade of the Alexandrina as that of the Aqua Crabra, a mistake repeated in the third dissertation. Fabretti's error was a common one: the papal Acqua Marrana Mariana, introduced by Calixtus II in A.D. 1122 to supply the area of the Lateran, had been identified with the Crabra in many maps of Rome throughout the sixteenth and seventeenth centuries, most recently in Giovanni Battista Falda's 1676 map (see Frutaz, 3: table 360). Nardini (505) makes the same error of identification.

This later channel installed in the conduit of the Alexandrina described by Fabretti carried a branch of the Marrana. Lanciani (327) cites a chapter of the city statutes published by Pope Paul II in 1519 referring to maintenance of a branch of the Marrana: "'vadit ad aquam Bollicantem usque ad formam ruptam quae vadit ad pedem turris sancti Joannis (Tor tre teste) ad turrim Quatraris (Quadraro).' La forma rupta è quella dell'alessandrina. Infatti anche oggi un ramo della marrana traversa la valle d'acqua Bollicante, correndo nel nucleo dei cinquantadue archi del vetusto aquedotto" ["It goes to the Acqua Bollicante up to the broken conduit that goes at the foot of Tor tre teste to the Tor Quadraro." The broken conduit is that of the Alexandrina. In fact, even today a branch of the Marrana crosses the valley of the Acqua Bollicante, running in the center of the fifty-two arches of the ancient aqueduct]. John Henry Parker also described this reuse of the Alexandrina in *The Aqueducts of Ancient Rome* (Oxford, 1876), 133: "This arcade had been originally double, and the lower one only now remains, with a modern *specus* [channel] made upon it . . . , but the water now flows *from* the Marrana, at the foot of the arcade of the great aqueducts . . . and *runs down* upon this arcade to the garden and small monastery of S. Peter and Marcellinus, at the Mausoleum of S. Helena." On the Marrana, see R. Motta, "La decadenza degli acquedotti antichi e la conduzione dell'acqua Mariana," in *Trionfo*, 203–5.

Incrustation. Cf. Ashby, 310 (367): "There are considerable overflows of heavy deposit from small leaks, noted by Fabretti."

Route. Ward-Perkins ("Via Gabina," 125) comments on the course of the aqueduct, "The Aqua Alexandrina was able to follow an unusually direct route to Rome for two reasons: because its objective was the low-lying quarters of Rome and because Severus Alexander was prepared to spend what even so must have been a large sum of money on carrying the conduit above ground for an unusually high proportion of its course."

Ashby (311 [369]) repeats Fabretti's comments concerning the route of the Alexandrina and also notes that the winding course may well have been chosen to achieve a gradual fall for the conduit: the entire Alexandrina had a fall of not more than .438 percent, "far lower than that of the great aqueducts" (Ashby 309 [366]). Van Deman (342–43) also notes that its course is unusual, in contrast to the route chosen by Sixtus V for the Acqua Felice, and cites reconstruction to raise the channel floor in the section at Procoio del Pantano (fig. 1, no. 21) "to correct some error of construction or in the rate of fall of the water" (Van Deman, 344).

Neither Ashby nor Van Deman comments on the third reason for the route cited by Fabretti in this passage: Alexander Severus planned his aqueduct as a line to be carried on extensive arcade above ground, despite its low level, and the conduit itself is most impressive. The aqueduct bridge at Centocelle, surrounded by modern apartment buildings, is striking even today.

It is surprising that Fabretti says nothing about suburban distribution of the Alexandrina, although he notes branch lines of the aqueduct in the legend to his map (fig. 1, nos. 8, 13). Supply to villas and property in the *ager Labicanus* was no doubt a factor in fixing the route of the line. For such distribution along the lines of the Aqua Claudia and Anio Novus, see F. Coarelli, "L'*urbs* e il *suburbium*," in *Società romana e impero, tardoantico*, ed. A. Giardina (Rome, 1986), 2:43–45; Blackman and Hodge, 86. For another parallel in the *ager Tiburtinus*, see H. B. Evans, "*In Tiburtium usum:* Special Arrangements in the Roman Water System (Frontinus, *Aq.* 6.5)," *AJA* 97 (1993): 447–55. The imperial "Villa of the Gordiani" on the Via Praenestina (indicated as "Torre Schiava" on fig. 1) might well have been supplied by a branch of the Aqua Alexandrina outside the city. See Bruun, 128–29.

3. PROOF THAT ALEXANDER SEVERUS BUILT THIS AQUEDUCT

Aldus Manutius the Younger (Aldus Manutius Pauli f. Aldi n., 1547–97). Son of Paulus Manutius and grandson of Aldus Manutius, Aldus

Manutius the Younger was a professor in Venice, Bologna, and Rome. Aldus published learned treatises by the age of eleven but never achieved the standing of his father. Joseph Justus Scaliger's comment on this last member of the Manutius family (*Scaligerana* 149) is quoted at length by Sandys (2:101 n. 4): "Aldus filius miserum ingenium, lentum; quae dedit valde sunt vulgaria: utrumque novi; Patrem imitabatur, solas epistolas bonas habet: sed trivit Ciceronem diu. Insignis est Manucii commentarius in Epistolas ad Atticum et Familiares. Manucius non poterat tria verba Latine dicere" [Aldus the son was a man of pathetic and slow talent. His works were really quite common. I knew each man. He imitated his father and alone wrote good letters, but he worked on Cicero for too long a time. There is a notable commentary of Manutius on the *Letters to Atticus* and *Letters to Friends*. Manutius could not speak three words of Latin].

Manutius's twenty-page treatise *De aquis, in urbem Romam olim influentibus* presents a summary of Roman aqueducts listed chronologically, based primarily on Frontinus, whom, surprisingly, Manutius mentions only once in his discussion, citing the *sedulitas* (attentiveness) of the *curator aquarum* in instituting administrative reforms (*De aquis*, 43–44). Manutius also includes citations from Pliny and Suetonius, as well as the inscriptions on the Porta Maggiore, Porta S. Lorenzo, and "Arch of Dolabella," but his short work hardly fulfills its stated goal "quae antiquorum monumenta sparsim apud scriptores scripta leguntur, in unum collecta ob oculos posita essent" [to have collected in one place the written notices of ancient monuments scattered among the writers] (26). The treatise also contains some striking errors: Manutius associates the Aqua Marcia, for example, with Ancus Marcius and locates its sources "in ultimis montibus Paelignorum" [in the furthest mountains of the Paelignians], east of the Fucine Lake and Marsian territory (31). Fabretti's sharp criticism of the tract is therefore justified. Poleni (26), commenting on Fabretti's censure in this passage, writes, "Si quicumque oculis non lippis illud legat, per se id certe sit percepturus" [If anyone will read Manutius's work with clear vision, he will certainly perceive its shortcomings].

Review of other lines. By process of elimination, Fabretti demonstrates that the conduit described in section 2 cannot be identified with any of the nine aqueducts listed by Frontinus. His arguments in this section are based largely on literary evidence, which explains his concern to identify the edition of Frontinus he cites.

Benedetto Millini. Librarian to Queen Christina of Sweden during the period 1661–65, Millini published (under the pseudonym "Modello Tientibene") the play *La costanza delle donne* (Rome, 1647), as well as scholarly and religious treatises in his own name. He also contributed a preface to Famiano Nardini's 1666 guidebook, *Roma antiqua*.

Millini's edition of Frontinus is a mystery; it was surprisingly also unknown to Poleni, who states in his preface (only forty years after the publication of Fabretti's work), "Is tamen neque Romae in bibiothecis codicum refertissimis neque inter Millini libros, neque inter libros Fabretti ab eis, qui ut gratum mihi facerent, conquisivere inventus est" [It was not found, either at Rome in the libraries with the largest collection of editions, or among Millini's books, or among those of Fabretti, which those seeking to assist me searched] (30). Millini's edition was also unknown to the commentator of the Barbiellini edition (12 n. b): "Nos quoque in eiusdem Milliniani Indicis inquisitione versati sumus, eamdemque fere diligentiam adhibuimus; nec tamen, ut reperiremus, contigit" [We, too, engaged in the same search for the Millini edition and applied almost the same diligence, but it was not our lot to find it].

Lucas Holste (Holstein, Holstenius, 1596–1661). A native of Hamburg, Holste studied at Leiden, converted to Catholicism, and from 1627 until his death, lived in Rome, where he became librarian of the Barberini Palace and the Vatican. A protégé of the geographer Philip Clüver (1580–1622), whom he accompanied on extensive travels throughout Italy, Holste had planned to produce editions both of ancient geographers and of Frontinus, but these projects were never realized before his death. His annotations on Clüver's *Italia antiqua* were published posthumously by Cardinal Carlo Barberini as *Annotationes in Geographiam sacram Caroli a S. Paolo, Italiam antiquam Cluverii, et Thesaurum geographicum Ortelii* (Rome, 1666) and are cited frequently, with criticism, by Fabretti. For accounts of Holste's life and geographical work, see R. Almagià, *L'opera geografica di Luca Holstenio*, Studi e Testi 102 (Vatican City, 1942), 1–24; Sandys, 2:364–65; Pökel, 125.

Holste's identification of Labicum with the modern town of Colonna has been accepted by some modern topographers, most recently by L. Quilici and S. Quilici Gigli (see *Barrington Atlas*, map 43 and Directory, 1:628). Other topographers place Labicum at Montecompatri: see S. Barbetta, *Via Labicana* (Rome, 1995), 53; T. Ashby, "The Classical Topography of the Roman Campagna-I," *BSR* 1 (1902): 256–63. Fabretti returns to the site of Labicum in his third dissertation (III.6); the issue is not of critical importance here.

Brick stamp "AVG.N." (fig. 5). Ashby (309 n. 6 [366 n. 14]) corrects

Fabretti in attributing to the original construction the brick stamp found in the seventh arch at Acqua Bollicante (*CIL* XV, 211.14); Ashby describes it as "reused material."

Fabretti is frequently cited as a pioneer in epigraphical studies for his attention to brick stamps, especially in his monumental *Inscriptionum antiquarum;* see, for example, H. Bloch, *I bolli laterizi di Roma e la storia edilizia romana, Studi e materiali del Museo dell'Impero Romano* 4 (1947): 3–4. As a pioneer, however, Fabretti was not aware that the Roman brick industry collapsed after the first quarter of the third century A.D. up until the principate of Diocletian; as a result, virtually no brick stamps are attested for the period of Alexander Severus. Fabretti's argument here from the abbreviation "AVG.N." is therefore not compelling. See Bloch, *Bolli laterizi,* 315, 339; E. M. Steinby, "L'industria laterizia di Roma nel tardo impero," in *Società romana e impero tardoantico,* ed. A. Giardina (Rome, 1986), 2:99–163, especially 99–109.

Citation of the SHA. While Spartianus's *Life,* like most works of the SHA, is frequently discounted by scholars as a historical source for building activities, the list of constructions it attributes to Alexander Severus can be largely confirmed by archaeological and topographical evidence. For a recent assessment, see F. Coarelli, "La situazione edilizia di Roma sotto Severo Alessandro," in *L'urbs*, 429–56, especially 439–42. Cf. Ward-Perkins, "Via Gabina," 122: "Severus Alexander, it should be noted, is also credited with an interest in the bathing facilities of the city as a whole. We are told of his construction, or restoration, of lesser bath buildings (*balnea*) in those of the Regions which were ill-supplied; of his completion of the *thermae* of Caracalla; of his imposition of a tax on a curious assessment of craftsmen, mainly in the clothing and metal trades, together with donations of forest land and oil, for the upkeep of the city's baths."

Rondinini collection. Angeloni (276–77) cites as part of the collection of Alessandro Rondinini the coin here discussed by Fabretti. The seventeenth-century numismatist Exechial Spanheim, in his *Dissertationes de praestantia et usu numismatum antiquorum* (Amsterdam, 1671), also cites coins in the collection of Felicia Zacchia Rondinini, whom he describes as a "praestantissima matrona" [most outstanding matron] (42), "illustre matronarum decus" [noted glory of matrons] (584), and "illustris matrona" [noted matron] (612).

Francesco Angeloni (1559–1652). An antiquarian, Angeloni was secretary to Cardinal Ippolito Aldobrandini and author of the *L'istoria augusta* that Fabretti cites frequently. See *DBI*, 3:241–42 (A. Buiatti). Angeloni was also the mentor of Giovanni Pietro Bellori, whose publication of Trajan's Column Fabretti criticizes sharply in this dissertation; the close association between the two men may have prompted some of Fabretti's criticism of Angeloni's work here. However, Fabretti appears to accept without critical comment Angeloni's reading of this coin of Alexander Severus. For the type, see *RIC*², 4.2:65–66.

Jacques Oisel (1631–86). A Dutch legal scholar, Oisel studied under Salmasius, Gosius, and Daniel Heinsius and became professor of public law in Groningen. His *Thesaurus selectorum numismatum* was published in 1677, the same year this dissertation was written by Fabretti. See J. C. F. Hoefer, ed. *Nouvelle biographie générale* (Paris, 1863), 21:567.

Publius Victor. This name refers to an amplification of the original regionary catalogs that was first published in 1503–4 by Aulo Giano Parrasio under the title *De regionibus urbis Romae libellus aureus* and that was attributed to one Publius Victor; the name seems to have been a concoction inspired by the presence of works by Aurelius Victor in the same codex. Its text may be found in VZ, 1:207–58. This listing was reprinted many times during the sixteenth century and, indeed, well into the nineteenth. However, L. Preller (*Die Regionen der Stadt Rom* [Jena, 1846], 38–46) proved clearly that Victor's listing was not authentic. For a detailed discussion, see VZ, 1:201–6; Jordan, 1.2:302–12. Fabretti considers Publius Victor an authentic source, citing it separately from the regionary catalogs throughout the three dissertations.

Notitia Imperii. This name is commonly applied to mid-fourth-century catalogs of the fourteen regions of imperial Rome. For their text, see Jordan, 1.2:539–74; VZ, 1:89–192.

4. REVIEW OF OTHER AQUEDUCT CONDUITS IN ROME

Moving from literary to archaeological evidence, Fabretti reviews the extant remains of other Roman aqueducts to demonstrate that the Alexandrina is a line separate from them. His survey begins with the most impressive display of aqueducts in the city, that at Porta Maggiore.

a. Aqueducts at Porta Maggiore

Porta Maggiore inscription (*CIL* VI, 1256). For a full discussion of this famous inscription, see Ashby, 191–92 (220–21); R. H. Rodgers, "The Mystery of 62 Miles: *CIL* VI 1256," *ZPE* 63 (1986): 157–60.

Aqua Marcia at Porta Maggiore. Fabretti's illustration is confirmed by Lanciani (299), Ashby (130–31 [161–62]), and Van Deman (105). For recent excavations in the area, see R. Volpe, *Aqua Marcia: Lo scavo di un tratto urbano* (Florence, 1996), 63–72. However, the identification of the three conduits of the Marcia, Tepula, and Julia within the Aurelian Wall at Porta Maggiore presented here is not original; Nardini (505) had made the same observation.

Georg Fabricius (Georg von Goldschmied, 1516–71). A pupil of Rivius in Saxony, Fabricius studied at Padua and Bologna and explored monuments and inscriptions in Rome. His *Descriptio urbis Romae* appeared in 1550. See Sandys, 2:268; Pökel, 75.

Fabricius's description of the aqueduct lines at Spes Vetus is vague, citing distribution of only the Claudia, to which he assigns the three superimposed conduits preserved in the Aurelian Wall north of the Porta Maggiore: "ruinas eius in Aventino prope S. Priscam et prope Portam Naeviam in muro vidimus, canales scilicet saxeos tres, alterum alteri impositum, tantae amplitudinis, ut unus a me perreptari posset" [we saw its remains on the Aventine, near S. Prisca and near the Porta Maggiore in the wall, three channels, one atop the other, of such a width that I could crawl through one of them].

Although Fabretti's criticism of Fabricius is accurate, he fails to note in this passage Nardini's earlier identification (505) of the conduits of the Marcia, Tepula, and Julia; the discussion presented here therefore gives the incorrect impression that Fabretti's observations and arguments are entirely new.

b. The Arcus Caelimontani

Fabretti's discussion of this high-level branch of the Aqua Claudia is brief, presenting a correction of Fabricius on the Claudia's distribution in Rome and an inscription attesting to Severan repairs to the conduit in the early third century A.D. On the Arcus Caelimontani, see H. B. Evans, *Water Distribution in Ancient Rome: The Evidence of Frontinus* (Ann Arbor, 1994), 118–24.

Marcello Severoli (1644–1707). Severoli was a friend of Fabretti and a member of the Accademia degli Arcadi. For his career, see Crescimbene, *Le vite*, 2:275–96. Fabretti also praises Severoli in his *De columna Traiani syntagma* (346, 383).

c. The Claudia Branch to the "Trophies of Marius"

Janus Gruter (Jan Gruytère, 1560–1627). A Dutch philologist and epigrapher, Gruter lived in England, was educated at Cambridge, studied at Leiden, held professorships at Rostock and Wittenberg, and ended his career at Heidelberg, where he spent thirty-five years. He was appointed librarian there in 1602, the same year that he published his corpus of ancient inscriptions, begun at the instigation of Joseph Justus Scaliger. Renowned for his tireless erudition, Gruter also produced editions of at least seventeen Latin authors, including Tacitus (1607), Livy (1608), and Cicero (1618). See Sandys, 2:359–62; Pökel, 104; *EHCA*, 545–46.

"Trophies of Marius" *castellum*. In his discussion, Fabretti challenges two theories common at the time: that the "Trophies of Marius" were a *castellum* of the Aqua Marcia and that the *castellum* was built by Trajan in connection with that emperor's reworking of the Marcia.

Scholars who have credited Trajan with improvement of the Aqua Marcia and its extension to the Aventine cite as evidence Frontinus's *De aquaeductu* 87.3 and 93.4, passages that discuss reworking and improvements of the city's aqueducts, including the Marcia and Anio Novus. For example, Poleni (13) argued that Frontinus completed his *commentarius* under Trajan and therefore paid tribute to that emperor's restoration in progress. The text of *De aquaeductu* 93.4, however, is not without difficulties; the manuscripts name Trajan as the *auctor* of the refurbished Anio Novus ("novum auctorem imperatorem Nervam Traianum Augustum praescribente titulo"), but the citation of Trajan himself in them has recently been questioned as a later gloss: Frontinus wrote his *commentarius* when he assumed responsibility for the *cura aquarum* under Nerva (*Aq.* 1–2.1, 93.1), and an inscription commemorating improvements would have named Nerva, not Trajan, as their *auctor*. See the discussion in R. H. Rodgers's forthcoming commentary.

Fabretti does not identify the contemporary writer whose attribution of the "Trophies of Marius" *castellum* to Trajan he attacks in this passage, but he is without a doubt Giovanni Pietro Bellori (1613–96), the prominent antiquarian, critic, and librarian to Queen Christina of Sweden in Rome. Alberto Cassio (1:121), working in the mid–eighteenth century, certainly understood Fabretti's reference and cited Bellori by name as the object of Fabretti's attack. It is surprising, however, that the commentator of the Barbiellini edition in 1788 did not know his identity (see Barbiellini, 19). For Bellori's career and publications, see *EHCA*, 140–41 (E. Cropper).

Bellori had twice assigned the "Trophies of Marius" to Trajan. He did so first in his notes to Pietro Santi Bartoli's study of Trajan's Column (*Colonna Traiana . . . accresciuta di medaglie, iscrittioni e trofei da Gian Pietro Bellori* [Rome, 1672], pl. 118), basing his argument on similarities between the "Trophies" themselves and representations of trophies on the reliefs on Trajan's Column. He did so a second time, in the following year, in his study of the Severan Marble Plan (*Fragmenta vestigii veteris Romae xx. tabulis comprehensa* [Rome, 1673], table IX, "Monumenta Comitis Herculis, seu Monumenta Mariana" [= Graevius, 4:1955]).

Scholars today discount Bellori's arguments, citing differences in detail between the "Trophies of Marius" and reliefs on the column and arguing that any similarities between the two are the result of stylistic conventions rather than any direct connection between the two monuments. In addition, archaeological evidence does not support arguments for an earlier Trajanic version of the *nymphaeum;* excavations carried out in connection with restoration of the "Trophies" in 1982–88 have revealed that Alexander's *nymphaeum* was built over structures of *opus reticulatum* dating from the Augustan period. For a summary of the evidence, see *LTUR*, 3:351–52 (G. Tedeschi Grisanti).

Fabretti seems to have had intense dislike for Bellori's scholarship, since he attacks his work on Trajan's Column frequently and vigorously in his own *De columna Traiani syntagma* (102–5), where Fabretti identifies Bellori only by the scornful reference "Neotericus" and repeats the same arguments made here about the level of the aqueduct supplying the "Trophies of Marius" *castellum*, reproducing figure 7 from this dissertation. It is therefore not surprising that Fabretti does not think it necessary to identify Bellori by name here, although he cites Angeloni, Nardini, and Fabricius (all dead in 1677), attacking them vigorously in this section. Cassio (1:121–22) comments critically on Fabretti's censure here, "E nel tempo stesso tratta da vano il collettor di numismi Angeloni: da inavveduto Nardini, e da ignorante Fabricio: quali aggravi non si vogliano qui rapportare potendo da ciascuno vedersi nella di lui Dissertazione" [At the same time, he treats the numismatist Angeloni as vain, Nardini as rash, and Fabricius as ignorant; such charges need not be repeated here, since they can be read by anyone in Fabretti's own dissertation].

In attacking Bellori's view that Trajan built the "Trophies of Marius" *castellum*, Fabretti rejects completely the interpretation by Angeloni of a *sestertius* of Trajan's fifth consulship (= *RIC*², 2:248 [Trajan nos. 463, 607–9]; see Ashby, 299 [357]). Indeed, Fabretti returns to this same point later in this dissertation (I.4g), when he attacks Angeloni again, as well as Oisel, for reading the coin as a representation of a Trajanic *nymphaeum*. In his third dissertation (III.6), Fabretti revisits the issue, this time focusing not on numismatic evidence but on the text of Frontinus (*Aq*. 87.3) that credits Nerva with improvement and extension of the Marcia to the Aventine.

Assigning construction of the "Trophies of Marius" *castellum* to Trajan is the first issue Fabretti addresses here; assignment of the *castellum* to the Aqua Marcia is the second. This longstanding identification had been advanced since the sixteenth century: the *castellum* was attributed to the Marcia by Leonardo Bufalini in his map of 1551 (Frutaz, 2: tables 193–94, 198), an identification followed in the maps of Pirro Ligorio in 1553 (Frutaz, 2: table 225), Giovanni Antonio Dosio in 1561 (Frutaz, 2: table 229), and Alo Giovannoli in 1616 (Frutaz, 2: table 284). It had been supported as well by Nardini in his *Roma antica*, in the passages quoted by Fabretti in this section. For a history of this identification with the Marcia, see G. Tedeschi Grisanti, "Primo contributo ad una livellazione urbana sistematica degli antichi acquedotti di Roma," in *Trionfo* II, 59–69.

Nardini cited three outlets in the *castellum* and suggested that they might have served to distribute the Marcia, Tepula, and Julia separately. Fabretti rejects Nardini's hypothesis on two grounds: (1) there are actually five outlets in the *castellum*, not three; (2) none of the three lines Nardini cites was high enough to supply the *castellum*, and only the Claudian aqueducts could have reached the "Trophies of Marius." Fabretti directs this same argument against Georg Fabricius but cites him only in passing at the end of this section; Fabricius (201 [= Graevius, 3:527C]) had assigned the branch line to the "Trophies" to the Aqua Julia: "ductus in Esquilino rudera apparent, inter portam et Marii tropaea" [remains of the conduit appear on the Esquiline, between the gate [the Porta S. Lorenzo] and the "Trophies of Marius"].

Fabretti is incorrect here in his identification of the "Trophies of Marius" *castellum* as the principal *castellum* of the Aqua Claudia and Aqua Anio Novus. According to Frontinus (*Aq*. 20.2), the main point of distribution for the two lines was actually situated "post hortos Pallantianos" [behind the estate of Pallas], where the waters of the two conduits were mixed for delivery throughout the city. The structure, destroyed by fire in

1880, has been securely located not far west of the Porta Maggiore (see Ashby, 243–44 [295–96]; Van Deman, 251, textcut 33; Nash, 1:37:, fig. 29).

In addition, Fabretti's argument based on conduit levels is vague: although he cites his own "certitudo, mathematicis demonstrationibus explorata," he does not provide explicit statistics about levels at the "Trophies of Marius," nor does he clearly state that he actually took levels there, in contrast to his detailed statistics for the levels cited for the five aqueducts at Spes Vetus. Later, in his *De columna Traiani syntagma* (102–5), Fabretti still gives no details about conduit levels at the "Trophies."

Fabretti's failure to cite detailed evidence on levels at the "Trophies of Marius" had far-reaching consequences, leading to misidentification of the *castellum* for more than three centuries. In the early eighteenth century, Poleni (72) accepted Fabretti's assignment of the Claudian aqueducts to the "Trophies of Marius" *castellum*. Forty years later, however, Giovanni Battista Piranesi published new levels in his monumental study of the "Trophies" (*Del castello dell'Acqua Giulia* [Rome, 1761], 1–6), rejecting both Nardini's and Fabretti's arguments and identifying the structure as a *castellum* of a branch of the Aqua Julia. Piranesi's reading of the evidence and identification were supported in 1821 by the architect A. M. Garnaud ("Memoire explicatif de la restauration de château de l'eau Jules (Aqua Julia) à Rome," reprinted in G. Tedeschi Grisanti, *I "trofei" di Mario: Il ninfeo dell'Acqua Giulia sull'Esquilino* [Rome, 1977], 75–77) and in 1879 by Lanciani (383–85).

Lanciani's assignment of the "Trophies of Marius" to the Aqua Julia also won wide acceptance, with only Ashby (297–98 [355–56]) questioning the identification because of the differences in elevation. Pace (160–61) publishes a letter of Ashby on the subject. Only in 1986, through new levels taken during restoration of the *castellum*, has it been demonstrated that all the leveling previously reported was incorrect: the height of the aqueduct conduit supplying the "Trophies of Marius" *castellum* has now been established as 62.28 meters above sea level (masl), considerably higher than that of the Julia at Spes Vetus (59.37 masl). Only the two Claudian aqueducts, the Aqua Anio Novus and Aqua Claudia (respectively, 65.99 masl and 63.85 masl at Spes Vetus), would have been high enough to supply the *castellum* and the branch aqueduct leading to it (see Tedeschi Grisanti, "Primo contributo"). Fabretti's assignment of the *castellum* to the Claudian aqueducts was therefore correct, even if he was wrong in identifying the "Trophies" with the main distribution tank of the two Claudian lines and did not definitively document his argument about levels here. For a summary of the history of identification of the "Trophies," see Tedeschi Grisanti, *Trofei*, 39–40.

d. The "Arco di Druso" Aqueduct and Specus Octavianus

Porta Capena. In an attempt to identify the aqueduct carried by the "Arco di Druso" just inside the Porta S. Sebastiano of the Aurelian Wall, Fabretti first confronts the erroneous identification of that gate with the Porta Capena of the republican "Servian" Wall. He cites no specific authority for this identification, but it appears to have been widespread at the time, attested in fifteenth-century itineraries and so identified by Poggio Bracciolini in 1518; see Tomassetti, 2:50–51. For similar seventeenth-century identifications of the gate with the Porta Capena, see Donati, 69–70 (= Graevius, 3:593D); Nardini, 42; Giovanni Battista Falda's 1676 map in Frutaz, 3: table 360.

To demonstrate that the Porta S. Sebastiano cannot be identified as the Porta Capena, Fabretti notes that according to Frontinus (*Aq.* 5.5), the low-level Aqua Appia ran next to the Porta Capena, where that aqueduct consisted of a substructure and arcade of only sixty paces, much shorter than that of the conduit carried by the "Arco di Druso." All the topographical evidence, Fabretti concludes, puts the Porta Capena in the valley between the Caelian and Aventine, the location accepted by modern topographers; see G. Säflund, *Le mura di Roma repubblicana* (Uppsala, 1932), 146–48. Fabretti cites as topographical landmarks for its location the Villa Mattei (Nolli, no. 944), the Church of S. Maria in Navicella (Nolli, no. 942), and nearby fortifications constructed by Pope Paul III; this last citation, however, is too vague to be useful. On the fortifications of Paul III in the area, see R. Lanciani, *Storia degli scavi di Roma e notizie intorno le collezioni romane di antichità* (Rome, 1902–12), 2:98–99. Fabretti's citation from Strabo does not add significant weight to his argument here but rather serves to introduce his discussion of the Porta Latina in the following paragraphs.

Luca Peto (Lucas Paetus, 1512–81). A Roman jurisconsult and administrator, Peto wrote an account of Pope Pius IV's restoration of the Virgo, begun in 1559 and completed by Pius V in 1570. See L. Quilici, "Sull'acquedotto Vergine dal Monte Pincio alle sorgenti," *QITA* 5 (1968): 126; Pökel, 198; N. del Re, "Luca Peto Giuresconsulto e magistrato Capitolino (1512–1581)," in *Scritti in onore di Filippo Caraffa,* ed. G. Giammaria (Rome, 1986), 309–37. Fabretti cites Peto here primarily to illustrate ignorance of levels.

Porta Latina. Fabretti introduces this argument to counter possible objections to his location of the Porta Capena, which would have been based on the proximity of the Porta Latina of the Aurelian Wall to the Porta S. Sebastiano. He does not identify the proponents of such arguments, but the reasoning is implied by the discussion presented by Donati (69–70 [= Graevius, 3:593D]) in his treatment of the gates of Rome.

First milestone of the Via Appia (*CIL* X, 6812). Fabretti cites this inscription, which was moved to the Capitoline and installed on the balustrade of the Piazza del Campidoglio in 1588, as sure proof of his location for the Porta Capena, although he appears to acknowledge some uncertainty about whether the inscription was found in situ. In actuality, the milestone, part of a private collection when it was moved to the Capitoline, does not strengthen Fabretti's argument. On its uncertain provenance, see *CIL* X, 6812; *CIL* X.2, p. 991; Helbig[4], 2:16–17.

Arcus stillans. With reference to the scholiast on Juvenal, Fabretti addresses what remains a longstanding topographical problem, identification of the *arcus stillans.* Jordan (2:19) interpreted the *arcus* as the Porta Capena itself, wet because of leaks from a nearby aqueduct. The evidence of the twelfth-century *Mirabilia Urbis Romae* ("intus portam arcus stillae," 10.2 [= Jordan, 2:615]), which Fabretti does not cite, suggests that the *arcus* was a separate structure inside the gate, not the gate itself.

Fabretti's identification of the *arcus stillans* with the Rivus Herculaneus of the Marcia has been repeated by modern topographers: see Ashby, 155 (188); Van Deman, 141. Poleni (71 n. 12) observes that Fabretti is following here the manuscript reading "initur" at *De aquaeductu* 19.9 accepted by Giovanni Giocondo in his 1513 edition of the *De aquaeductu;* Poleni himself adopts Rubenius's emendation "finitur," accepted by most later editors. Poleni also questions Fabretti's proposed addition of "modus" to the text: "bona venia manium tanti viri dixerim: etiamsi verbum modus adderetur, nondum tamen sententia perspicua fieret" [by leave of the spirits of so great a man, even if the word *modus* were added, the meaning still would not be made clear].

Clivus of Mars inscription (*CIL* VI, 1270). Fabretti's notice is correct about the provenance of the inscription cited, but the passages of Servius and Ovid that he introduces do little more than confirm that the area of the Temple of Mars mentioned was situated outside the Porta Capena; the evidence does not significantly strengthen his arguments about the location of the Porta Capena itself.

S. Giovanni's martyrdom. Fabretti's citation of Prochorus's account of the martyrdom of S. Giovanni at the Porta Latina now permits Fabretti to introduce an attack on Paolo Aringhi (1600–76), an Oratorian who translated into Latin and expanded Antonio Bosio's systematic study of the Roman catacombs, *Roma sotterranea* (Rome, 1651).

Maxima bibliotheca veterum patrum. A collection of writings of the church fathers first published by Marguerin de la Bigne in 1575 and augmented in a series of editions. Fabretti cites the most recent edition of 1677 (in twenty-seven volumes), paraphrasing here the notice of Giovanni Brisichella in its alphabetical index and inserting Bellarmine's censure of Prochorus's account (also appearing in the *Maxima bibliotheca* [2:46]).

Robert Cardinal Bellarmine, S.J. (1542–1621). A famous scholar and teacher, Bellarmine became rector of the Collegio Romano and a leading champion of the church against Protestant criticism. This citation seems gratuitous, to strengthen Fabretti's argument against Aringhi.

Justus Lipsius (Joest Lips, 1547–1606). Born in Belgium, Lipsius studied Roman law at Louvain. He spent two years in Italy, exploring libraries and examining inscriptions, and became one of leading textual critics of his day, as well as a striking example of the religious fervor of the times, leaving the Roman Catholic Church to convert to Protestantism, then reconverting to Catholicism. See Pfeiffer, 124–26; Sandys, 2:301–5; Pökel, 158; *EHCA*, 686–87 (A. Grafton). Fabretti's citation here seems gratuitous; later, in the third dissertation (III.5), he attacks Lipsius's views on the size of ancient Rome.

Specus Octavianus. The course and purpose of this branch of the Anio Vetus remain problematic because of the scanty evidence available. For modern readings of the evidence, see Lanciani, 264–67; Ashby, 86–87 (107); Van Deman, 66.

Fabretti is certainly incorrect in his assignment of the aqueduct carried by the "Arco di Druso" to the Specus Octavianus branch of the Anio Vetus. Yet his attempt here to link pertinent passages of Frontinus to the physical evidence at hand was in many ways farsighted; he was on the right track, even if he drew the wrong conclusions. Fabretti's identification is based on mistaken assumptions concerning levels, the route of the branch, and incrustation.

Lanciani (316–17), citing the level of Fabretti's *specus* (fig. 1, no. 25) as midway between that of the conduit of the Marcia and that of the aqueduct carried by the "Arco di Druso," identified this specific section of conduit as part of Caracalla's branch of the Marcia known as the Aqua Antoniniana. In contrast to his earlier detailed discussion about levels of the Claudian aqueducts and the Marcia/Tepula/Julia at Spes Vetus, Fabretti presents here only general remarks (based on Frontin. *Aq.* 18.6) about the relative level of the Anio Vetus with respect to the other ancient aqueducts.

De aquaeductu 21, the passage Fabretti cites for the divergence of the branch from the main conduit of the Anio Vetus, is corrupt—a fact he appears to recognize in quoting it. Textual difficulties within the passage make it impossible to determine how Frontinus measured the distance of two miles he cites: was it along the Via Latina itself or along the conduit? If it was along the conduit, as Lanciani (264–65) and Ashby (86 [107]), argue, the *specus* depicted by Fabretti on his map cannot be in the correct position to be identified with the Specus Octavianus.

Fabretti makes reference to chemical analysis of the incrustation of the channel at the "Arco di Druso" to demonstrate its similarity to the water of the Anio River. For a modern analysis of the deposit of the "Arco di Druso" aqueduct, see G. Garbrecht and H. Manderscheid, "*Etiam fonte novo Antoniniano*," *ArchClass* 44 (1992): 193–234, especially 211–26. In the nineteenth century, Lanciani (318) cited his own chemical analysis, demonstrating that the incrustation was similar to that of the Marcia, tapped by Caracalla for the Aqua Antoniniana, although Lanciani seems to acknowledge elsewhere (*Acquedotti*, 264) the scientific validity of Fabretti's analysis in recognizing similarities in Anio River water. By modern standards any chemical analysis done in the seventeenth century must have been primitive; for the difficulties in analyzing incrustation in aqueduct conduits, see C. Puliti, A. Bergioli, and C. Terzano, "Studio chimico-fisico su formazioni calcaree prelevate da antichi acquedotti romani," in *Trionfo*, 195–98. On incrustation in general, see Hodge, 227–32; Blackman and Hodge, 109–15; and A.T. Hodge, "Purity of Water," in Wikander, *Handbook*, 98–99.

Giovanni Ciampini (1633–98). Fabretti makes warm reference to this prominent humanist and scientist of seventeenth-century Rome, who abandoned a legal career to devote himself to research, while working in the Cancelleria Apostolica. Ciampini founded an academy for church history in 1671 and a second for natural sciences, physics, and mathematics in 1679, under the patronage of Queen Christina. As Fabretti remarks here, Ciampini converted his home into a museum of antiquities and became the author of many scientific works in Italian and Latin. He died as a result of asphyxiation from mercury vapors in a laboratory experiment. See *DBI*, 25:136–43 (S. Grassi Fiorentino).

Specus Octavianus extension. In this discussion, Fabretti combines two separate observations: the first concerning a conduit that he identifies as a continuation of the Specus Octavianus under the Aventine: the second, that of the Aqua Appia itself in the same area. Fabretti's attribution of the channel to the Specus Octavianus, intersecting with another channel identified as that of the Marcia, was initially rejected by Lanciani (265–66), who identified both structures as drains; however, Lanciani (*FUR*, 41) later listed them as the Specus Octavianus, with a query. Ashby (87 [107]) noted that in his day, it was impossible to examine these remains and therefore impossible to assess Fabretti's reading of them.

e. The Aqua Appia

In contrast to Fabretti's mistaken reading of the channels just cited, modern topographers have accepted his identification of the conduit of the Aqua Appia in the Vigna Santori. Lanciani (249–50) describes his own descent into the same quarries in 1876. Ashby (52–53 [68]) repeats Fabretti's description in full. See also Van Deman, 27. The site was no longer accessible in Ashby's day.

f. The "Arco di Druso" Itself

Fabretti now introduces a lengthy digression on his erroneous identification of the aqueduct conduit atop the "Arco di Druso," arguing that the arch itself is the work of Augustus, the honorary arch erected by the Senate after Drusus's death in 9 B.C., according to Suetonius (*Claud.* 1.3).

Fabretti's identification is incorrect; the arch over the Via Appia Antica inside the Porta S. Sebastiano is actually contemporary with and a part of Caracalla's Aqua Antoniniana, the branch of the Marcia supplying the Thermae Caracallae. See Garbrecht and Manderscheid, "*Etiam fonte*," 193–206; Nash, 1:79, with bibliography cited; Richmond, 138–39. For a full discussion of the evidence for Augustus's Arch of Drusus itself, see F. S. Kleiner, *The Arch of Nero in Rome* (Rome, 1985), 33–35; Kleiner (28–30) also discusses the Augustan arch at Rimini (cited by Fabretti in this section).

Pirro Ligorio (ca. 1510–83). A Neapolitan architect and painter, Ligorio came to Rome and devoted himself to the study of antiquities. Ligorio excavated Hadrian's villa near Tivoli while designing the Villa D'Este and published many works on ancient Rome, but he was notorious even among his contemporaries for his forgeries and falsification of inscriptions. For a comprehensive assessment of his work, see R. W. Gaston, ed., *Pirro Ligorio: Artist and Antiquarian* (Florence, 1988); a briefer account may be found in *EHCA*, 680–82 (R. W. Gaston). Fabretti never seems to tire of attacking Ligorio; cf. Cassio's description of Ligorio as "flagellato ad ogni passo dal Fabretti" [whipped at every step by Fabretti] (1:203).

Coin representation. Fabretti gives no reference for this coin, which seems to be identical to *RIC*², 1:128, Claudius no. 98 [= British Museum, Department of Coins and Medals, *Coins of the Roman Empire in the British Museum* (London, 1965), 1:181, no. 121]. Cf. *RIC*², 1:130, Claudius 114 [= *BM Coins, Rom. Emp.*, 1:191, no. 188, pl. 36.2]; Kleiner, *Arch of Nero*, pl. VII.1. The "very great similarity" of design cited by Fabretti is, in fact, negligible, mainly the single-span construction and the height of the attic. Gronovius (*Responsio ad cavillationes Raphaelis Fabretti* [Leiden, 1684], 13–14), replying to Fabretti's criticism—in the third dissertation (III.6)—of his textual emendation of Livy, vigorously attacked Fabretti on this very point.

Linguistic arguments. Later topographers have generally interpreted the word *Octavianus* as a reference to the emperor Augustus; see Lanciani, 265; Van Deman, 66; Ashby, 86–87 (107). However, Augustus appears never to have used his adoptive name for official purposes. It is therefore tempting to associate the branch with some other Octavius, but no censors with that name are attested in the late republic or early empire. Gronovius (*Responsio*, 53–54) criticized Fabretti for his identification of the word *Octavianus* with Augustus. Fabretti's arguments from the words *clivos* and *planitia* (from *CIL* VI, 1270) do not add much weight to his case for association with Augustus.

Guillaume Philandrier (Philander, 1505–65). Philandrier was a French philologist whose edition of Vitruvius, first published in 1544, was reprinted many times. For a recent assessment of his work and career, see F. Lemerle, *Les "Annotations,"* 11–47. Lemerle also reproduces a facsimile of an early edition of Philandrier's *Annotationes* (Lyons, 1552), from which all citations of the work in the present book have been taken. Fabretti's comparison of the arch with the instructions and proportions recommended in *De architectura* are simply another example that Vitruvius's treatise, highly theoretical in nature, frequently conflicts with what we know of actual practice in the Roman world.

Porta S. Lorenzo arch. Fabretti gives only a partial depiction of this arch, with its Augustan inscription, to strengthen as much as possible the parallel with the arch at the Porta S. Sebastiano. He does cite its later modifications and inscriptions by Vespasian and Caracalla. On the arch at Porta S. Lorenzo, see Richmond, 170–81.

Enea Vico (1523–67). An engraver and numismatist from Parma, Vico published *Augustarum imagines aereis formis expressae* (Venice, 1588), which, like de Strada's publication, depicts the arch on the Claudian coin as triple.

Octavius de Strada à Rosberg (sixteenth century). De Strada authored *Vitae imperatorum caesarumque Rom. orient. et occid. uxorum et liberorum eorundem* (Frankfurt, 1629).

Francesco Gottifredi (d. 1660), **Francesco Cameli, Francesco Mezzabarba Birago** (d. 1697). These antiquarians were connected with the circle of Queen Christina of Sweden. Cameli succeeded Gottifredi as curator of the queen's medallions and library and served as secretary to the Accademia Reale, which later became the Accademia degli Arcadi. He was forced to relinquish his appointments with Queen Christina in the late 1670s because of failing eyesight (to which Fabretti alludes in this passage) and was succeeded by Giovanni Pietro Bellori. See *DBI*, 17:163–64 (N. Parise).

Coin of Drusus with augural staff within the tympanum. Coins of this type appear in *RIC*[2] (1:129, Claudius no. 62, pl. VIII.126) and *BM Coins, Rom. Emp.* (1:181, nos. 121–23). See also Kleiner, *Arch of Nero*, pl. VII.2.

Albinovanus Pedo. An Augustan poet, Pedo was roughly contemporary with Ovid. Twenty-four lines of his epic poem on Germanicus's campaigns, describing the disaster of the expedition to the North Sea, can be found in E. Courtney, ed., *The Fragmentary Latin Poets* (Oxford, 1993),

315–19. See also C. Buechner, ed., *Fragmenta Poetarum Latinorum* (Leipzig, 1982), 147–48; H. Bardon, *La letterature latine inconnue* (Paris, 1956), 2:69–73. The elegiac couplets cited here are actually lines from the pseudo-Ovidian *Consolatio ad Liviam* (419–22), discovered in the early fifteenth century and variously attributed to Ovid and other Augustan authors. Fabretti appears to follow the attribution to Albinovanus Pedo put forward by Joseph Justus Scaliger in 1572. See H. Schoonhoven, ed., *The Pseudo-Ovidian "Ad Liviam de morte Drusi"* (Groningen, 1992), 45–52.

Andrea Fulvio (ca. 1470–1527). A poet, grammarian, antiquarian, and advisor to Raphael, Fulvio presented a verse inscription of Roman antiquities to Pope Leo X in 1513 and proposed a scheme for a plan of Rome divided into the ancient regions. See Sandys, 2:121; *EHCA*, 471–73 (R. W. Gaston). Fabretti's citation of the coins of Trajan is not meant to reject Fulvio's arguments; Mattingly and Sydenham list a coin of Trajan depicting a triple arch (*RIC*², 2:274, Trajan no. 420), but if the *Notitia* is correct in its notice that the Arcus Traiani was erected after the emperor's death in A.D. 117 (Jordan, 2:542; VZ, 1:91, 165), the coin representation, dated to A.D. 100, is more plausibly to be connected to a triumph of that year. See *LTUR*, 1:112 (D. Palombi).

Baths of Alexander. Fabretti now returns to his principal subject, identification of the aqueduct that supplied the Baths of Alexander Severus. Fabretti's mistaken interpretation of the "Arco di Druso" in the preceding discussion does not invalidate the main thesis of this dissertation.

g. The Aqua Traiana

Inscriptions on the Janiculum and Via Aurelia. Fabretti cites here two well-known papal inscriptions of the Acqua Paola, the first on the terminal fountain, or *fontanone*, built by Flaminio Ponzio on the Janiculum in 1611–12, the second on the so-called Arco dei Tiradiavoli or Arco Paolina spanning the Via Aurelia Antica outside the Porta S. Pancrazio alongside the Villa Doria Pamphilj. Both inscriptions erroneously identify Paul V's aqueduct with the Aqua Alsietina introduced, according to Frontinus (*Aq.* 11), by Augustus. A third papal inscription of the Acqua Paola near Cesano, cited by Ashby (304 [361]), repeats the error.

Fabretti's severe criticism is surprising here, since he points out in the second dissertation (II.4) that such errors are common in church documents, citing a papal bull of Paul V that makes a similarly incorrect identification. Fabretti's vigorous rejection of the papal inscriptions seems deliberate, to strengthen his argument that the aqueduct on the Janiculum is the Traiana and can in no way be connected with the aqueduct of Alexander Severus. On the Acqua Paola, see C. Cancellieri, "L'acquedotto Paolo."

Inscription cited by Ligorio. Apart from the listings of the *Notitia* (see Jordan, 2:548; VZ, 1:105) and Publius Victor (see VZ, 1:213) that Fabretti cites here, the inscription cited by Ligorio (rejected by *CIL* as a forgery) was for Fabretti the only independent evidence for a *nymphaeum* of Alexander Severus on the Esquiline. For this reason, he appears here to accept its authenticity, though not without censure of Ligorio's argument for the existence of a settling tank. Cassio (*Corso*, 1:220) describes Fabretti's acceptance of Ligorio's inscription as a "miracolo."

Where Ligorio cited a settling tank in addition to the Severan *nymphaeum* remains a mystery. The commentator of the Barbiellini edition writes, "Quae hic ex Ligorio proferuntur forte in illius manuscriptis adhuc latent; in iis enim, quae typis mandata vidimus, quorum iam meminimus, etsi aliquid argumenti ea de re citato loco exhibeant: tamen id ipsum non est, quod Fabrettus ex illius sententia hoc in loco docet" [These remarks attributed to Ligorio perhaps still lie hidden in his manuscripts; in his published works that we remember, although there is some argument along the lines presented here, there is no specific mention of the point Fabretti makes] (41 n. a).

Evidence for the Aqua Traiana on the Janiculum. Fabretti's citations of the Acta and the Instrument of Donation of Charlemagne in this section come from contemporary secondary sources, Aringhi and Nardini, both of whom he criticizes elsewhere. Only Fabretti's citation of Anastasius seems to be from an independent and separate source, and that evidence, incorrect in its details, adds little to his argument. Fabretti's debts to his contemporaries in arguing for this course of the Traiana should therefore be acknowledged.

Earlier identification of the Aqua Traiana was confused and mistaken. In 1639, Donati (298 [= Graevius, 3.836]) had identified the Aqua Traiana with Trajan's reworking of the Marcia and its extension to the Aventine, cited by Frontinus (*Aq.* 93); Angeloni (114) had made the same argument in 1641, based on the Trajanic coin representations of the aqueduct, an interpretation supported by Oisel in 1677 (see the discussion in commentary to I.3). Nardini (508–9) had associated the Traiana with an aqueduct tapped between the site of ancient Veii and Formello and introduced into "Campo Vaticano" and the Borgo. Fabretti's argument that the Traiana was to be identified with the Janiculan aqueduct reworked by Paul V, while a corollary to this dissertation's principal subject (the Aqua Alexandrina), must be recognized as a significant scholarly contribution that had far-reaching consequences.

Ligorio's Porta Aurelia. Ligorio (*Paradosse*, 37–38) did indeed argue what Fabretti attributes to him in this passage: "La porta detta hoggidì san Pancrazio non è la Porta Aurelia, come quasi tutti gli antiquarii moderni tengono: perchè Procopio la mette in Borgo, non quasi lontano dalla mole d'Adriano cio è Castel Sant'Angelo. Onde possiamo per verisimil coniettura pensare, che ella sia quella di San Spirito, o pur che ella fosse in quel contorno guasta per avventura da qualche Papa. Egli è ben vero che la Porta di San Pancrazio la sua via andava verso la Aurelia, e con essa si congiungeva, ma non era però ne la porta ne la via Aurelia, essendo nel luogo ove s'è detto" [The gate called today that of S. Pancrazio is not the Porta Aurelia, as almost all modern topographers hold, because Procopius connects the gate with the tomb of Hadrian, that is, Castel Sant'Angelo. We can therefore reasonably identify it with the Porta di S. Spirito or with another in that vicinity destroyed by the activity of some popes. It is indeed true that the street from the Porta S. Pancrazio went toward the Aurelia and joined it, but they were neither the real Porta or the Via Aurelia, since they were in the place indicated].

Modern topographers posit two gates of ancient Rome with the name *Porta Aurelia:* the first is that identified with the modern Porta S. Pancrazio at the top of the Janiculum; the second, also known as the Porta Aurelia Sancti Petri, is now destroyed but was originally situated near the present Castel Sant'Angelo. There were also two ancient roads with the name *Aurelia:* the Via Aurelia Vetus, beginning from the Pons Aemilius at the Tiber and running almost in a straight line to the Porta S. Pancrazio; and the Via Aurelia Nova, following a much more gentle ascent from its starting point at the Porta Aurelia Sancti Petri and meeting the Aurelia Vetus outside the walls of the city. See G. Gatti, "Il viadotto della via Aurelia nel Trastevere," *BC* 58 (1940): 129–41.

Fabretti does not categorically deny the existence of a second Via Aurelia, but he strenuously objects to Ligorio's location of the Porta Aurelia near Castel Sant'Angelo, to the exclusion of Porta S. Pancrazio. Fabretti introduces here epigraphical evidence (*CIL* XIV, 3610) attesting to both a Via Aurelia Vetus and a Via Aurelia Nova being under the jurisdiction of one *curator viarum*. The inscription he cites can be paralleled by others: cf. *CIL* VI, 1512, which cites a *curator* of the Via Aurelia and Cornelia Triumphalis.

Course of the Via Aurelia Vetus. Fabretti cites the evidence of tombs along the course of the present Via Aurelia Antica to strengthen his arguments about its course from the Porta S. Pancrazio. These cemeteries have been amply documented: see A. Nestori, "Osservazioni sulla topografia cimiteriale dell'Aurelia Antica," *RACrist* 40 (1964): 112–22; "La Catacomba di Calepodio all III miglio dell'Aurelia vetus e i sepulchri dei Papi Callisto I e Giulio I (I parte)," *RACrist* 57 (1971): 169–278.

Fabretti's citation of the famous passage from Livy on the Gallic sack of Rome adds little to his argument concerning the course of the Via Aurelia Vetus. Certainly, in the early fourth century B.C., the only route across the Tiber to Caere would have been via the Pons Sublicius, not a bridge located much farther to the north outside the city of the early republic.

Course of the Via Aurelia Nova. Aringhi's route for the ancient Via Aurelia Nova would follow closely that of the modern Via delle Fornaci, which meets the Via Aurelia Antica very close to the modern Via Vitelliana running to the Basilica of S. Pancrazio. Fabretti argues for a different route for the Via Aurelia Nova, one running more directly southwest, which he ascribes to Ligorio, but without reference to a specific work. He also illustrates such a course in his third dissertation, on his topographical map (III.2, fig. 31) and on his more detailed map of Rome's walls and gates (III.3, fig. 32).

The route outlined here by Fabretti corresponds quite closely to that given by Tomassetti (2:563–88). Fabretti's Taberna del Pidocchio appears to be the modern Contrada Madonna del Riposo (or del Pidocchio) described by Tomassetti (2:582–83). For an explanation of the name, see Tomassetti, 2:588. Fabretti's Vigna Carpegna is to be identified with the Villa Carpegna, property of Fabretti's patron Gaspare Carpegna, the cardinal vicar of Rome; see Tomassetti, 2:584–86. The office held by Carpegna gave its name to the modern Vicolo del Vicario running near the property (see Tomassetti, 2:551). On the Porcareccino property, see Tomassetti, 2:591. Its name and that of the property of the Bottachia are also attested in a systematic listing of estates in the Roman Campagna of the early seventeenth century. Cf. J. Coste, "I casali della Campagna di Roma all'inizio del '600," *Archivio della Società romana di Storia patria* 92 (1969): 41–115. The meeting point of the Via Aurelia Vetus and Via Aurelia Nova is identified by topographers at the present Torretta Troili in an area known as Val Cannuta, near kilometer 7.6 of the modern Via Aurelia. See Tomassetti, 2:586–87; E. Carnabuci, *Via Aurelia* (Rome, 1992), 27.

Fabretti's *On the Suburban Territory*. Fabretti appears to have planned a comprehensive work on the entire Roman Campagna. He cites this study in progress again in his third dissertation (III.6) while discussing the topography of the *ager Tusculanus*. The commentator of the Barbiellini edition writes, "De pertexendo opere, quo Agri Suburbani res expenderet Fabrettus noster, ut clare ex hoc loco intellegitur, cogitaverat; de eo tamen non modo aliquid typis non commisit, sed ne hilum quidem inter illius M.Scripta ab iis repertum est, qui ea diligentius pervolutarunt, quo maxime de huiusmodi opere aliquid cognoscerent" [Fabretti had had thoughts about bringing to completion the work in which he discussed the subject of the *ager suburbanus*, but not only did he not publish something of it, but not even a small part of it was found by those who investigated these matters to learn as much as possible about the work] (45 n. a). Ashby (3 n. 3 [21 n. 19]) notes that this uncompleted work and the illustrated map prepared for it are also mentioned by Fabretti in his *Ad Grunnovium apologema* in 1686 (99). The map (fig. 37) was also published in the Barbiellini edition in 1788 (opposite p. 90).

Conclusion. To conclude his discussion of the Aqua Traiana, Fabretti summarizes points already made in I.3. He adds only one new argument, that the coin of Trajan depicts an aqueduct drawn from a new source, not the reworking of an older line. He also includes a detail omitted in his earlier discussion, that Jacques Oisel also supported Angeloni's reading of the Trajanic coin cited before.

Fabretti might have made more of his argument from the coin representation, since Trajan's coin personifies the aqueduct as a reclining river god holding a reed in his right hand and leaning his left arm on an urn from which water flows. Such a representation would indicate water drawn from an independent source, not an addition to or reworking of another line. This omission is surprising, given Fabretti's later interpretation of a similar relief of a reclining nymph as a representation of the Aqua Claudia in his second dissertation (II.3d). At the end of the present section, he concentrates instead on attacking both Angeloni and the gullibility of Oisel for having followed Angeloni.

5. LEVELS OF THE AQUA TRAIANA AND AQUA ALEXANDRINA

Fabretti here returns to earlier arguments about the relative levels of the aqueducts entering the city at Spes Vetus, to reinforce his identification of the Alexandrina as a line separate from those listed by Frontinus and also distinct from the Traiana, the topic just addressed. His observations about the relative levels of the papal Acqua Felice and ancient Marcia and Alexandrina repeat points made earlier (I.4a) but are cogent in establishing that the Alexandrina was independent of the other lines. On the difference in height between the two aqueducts, Ward-Perkins ("Via Gabina," 123) writes: "It was this crucial factor of the height at which the Aqua Alexandriana entered the city which distinguished it from the next aqueduct to use the Pantano springs, namely the Acqua Felice. In terms of performance the former, despite the considerable proportion of its course that lay above ground, must be classified as one of the group of low-level aqueducts which could only serve the lower quarters of the city. The latter was specifically designed to bring water at a level high enough to serve almost the whole area within the city walls."

By reintroducing Frontinus, Fabretti also reiterates earlier arguments that the Traiana and Alexandrina must be considered separate from the lines listed in the *De aquaeductu*. Fabretti's citation of the number of aqueducts listed by the regionary catalogs and Procopius indicates that he was already preparing a more detailed discussion of this subject, addressed directly in the third dissertation.

6. EVIDENCE OF BRICK STAMPS AND THE TOMB OF ALEXANDER SEVERUS

Fabretti introduces his discussion of Alexander's baths (about which he says very little) and the tomb at Monte del Grano to argue for Severan construction from the evidence of brick stamps. Although Fabretti attributes the brick stamp of the Alexandrina arcade west of Centocelle to the time of Alexander Severus, the stamp is instead assigned to the first century A.D. at *CIL* XV, 2313, which also notes its inscription *litteris antiquis* [in ancient lettering]. Ashby (314 n. 6 [373 n. 31]) assigns the brick stamp to the district of Palestrina, presumably from the parallel of *CIL* XIV,

4091.24 (cited at *CIL* XV, 2313). See also T. Ashby and G. Lugli, "La villa dei Flavi cristiani 'ad duas Lauros,'" *MemPontAcc* 2 (1928): 180–81.

Monte del Grano. This funerary monument, made famous by Piranesi's study and illustrations in the eighteenth century, is no longer associated with Alexander Severus and his mother. For a discussion of the monument and evidence, see G. Pisani Sartorio, "Tomba detta di Alessandro Severo a Monte del Grano," in *Piranesi nei luoghi di Piranesi* (Rome, 1979), 5:65–71; Ashby and Lugli, "La villa dei Flavi cristiani," 179–82. For the sarcophagus described by Fabretti, presently in the Capitoline Museum, see Helbig[4], 2:73–75 (no. 1226); it is now interpreted as representing the discovery of Achilles among the daughters of Lycomedes. Pisani Sartorio ("Tomba detta di Alessandro," 65, 68 n. 8) notes that shaft E, which Fabretti describes as modern in his key to his cross section of the tomb (fig. 16), is actually an original part of the ancient monument.

Flaminio Vacca (ca. 1538–1605). Vacca was a Roman sculptor and antiquarian. His account of Roman antiquities discovered in his day was addressed to Anastasio Simonetti of Perugia. See Sandys, 2:155; *EHCA*, 1142.

Contemporary writer from France. As indicated in n. 38, the reference is to Bernard de Montfaucon, O.S.B. (1655–1741), a Benedictine scholar famous for laying the foundations of Greek paleography. Montfaucon spent three years in Italy exploring manuscripts and ancient topography. In his account of his travels, published in 1702, Montfaucon pays special tribute to the earlier descriptions of Vacca, whose *schedae* or notes he quotes (in Latin translation) throughout his text: "Eas sane, ut nova et intacta multa complectuntur, antiquitatis studiosis non displicituras confido" [I am confident that these notes will not be displeasing to students of antiquity, as they contain many things new and unexplored] (*Diarium italicum*, 105). See Sandys, 2:385–89; Pökel, 179; *EHCA*, 764–67. Fabretti's citation of Vacca here was no doubt prompted by the quotation of Vacca's account by Montfaucon himself.

7. SOURCES OF THE AQUA ALEXANDRINA AND ACQUA FELICE

Only at this point does Fabretti address the question of the sources of the Alexandrina, although he alluded to the topic in I.2, in his criticism of

Kircher's topography and discussion of the location of ancient Labicum, and cited the origin of the aqueduct in the *ager Labicanus* in I.5. Fabretti's identification of the source for the Alexandrina as one separate from those of the later Acqua Felice was rejected by Lanciani (380–81), who cites the incrustation in pipes fed by the papal aqueduct in Rome. Lanciani (381) also suggests that other springs less pure might have supplied the Alexandrina, since the water was intended for Severus's bath complex in the Campus Martius, not for drinking purposes. Modern topographers today identify the sources of both aqueducts from springs in the basin of the Pantano Borghese, recognized as the site of the ancient Lake Regillus. See FI 1.10, 879–81; Barbetta, *Via Labicana*, 47–48 and table I (opposite p. 16); G. Caruso, "L'Aqua Alexandriana," in *Trionfo*, 120–23. For an up-to-date map of the area, see L. Quilici, "Il sistema di captazione delle sorgenti," in *Trionfo* II, 47–58, especially fig. 5.

Lake Regillus. Fabretti here admits doubts about whether his own interpretation of the sources is correct, but he is fiercely critical of an unnamed topographer whose map, published in 1674, indicates the source of the aqueduct as Lake Regillus itself. The object of his censure is certainly to be identified as Innocenzo Mattei, whose *Nuova et esatta tavola topografica del territorio o distretto di Roma* was published in 1674. For discussion and reproduction, see R. Almagià, *Monumenta cartographica vaticana* (Vatican City, 1948), 2:56–57 and table XIV; A. P. Frutaz, *Le carte del Lazio* (Rome, 1972), 1:64–67 and table 154–56. Almagià (*L'opera geografica*, 134–36) demonstrates that though Mattei published the map in 1674, the cartography presented on it reflects the geographical work of Lucas Holste, who had died thirteen years earlier.

Fabretti certainly knew Mattei's publication and indeed criticizes it again in his second dissertation (II.4b). However, as Almagià points out (*Monumenta*, 2:57), Mattei's map of 1674 was quickly superseded by the more comprehensive and sophisticated cartographic works of Giacomo Filippo Ameti in 1693 and Giovanni Battista Cingolani in 1704 and became quite obscure—indeed, practically forgotten. It was certainly unknown a century later to the commentator of Barbiellini's edition, who presents a long note (51 n. a) describing a fruitless search for this reference: "Quae hoc anno charta topographica esset obsignata, quamquam a nobis diligenter conquisita, ut oculorum aciem in eam intenderemus; id tamen nullo modo licuit; nec quisquam harum rerum peritus, ex quibus non paucos, ut facem praeferrent, rogabamus, quidquid certi indigitare novit" [Although we searched carefully for a topographical map published in this year to examine it, it was not possible, nor was anyone knowledgeable about such things (not a few of whom we asked) able to indicate anything certain].

There can be no doubt about this identification: Mattei's map identifies Lake Regillus as the source of the Acqua Felice and presents another error criticized by Fabretti in his second dissertation (II.4b), situating Carseoli east of the Thoranus River, rather than west. Most interesting about Fabretti's censure here is his refusal to name Mattei himself. Fabretti seems to have been well aware that Mattei was only the publisher (the full title of the map reads "data in luce da D. Innocenzo Mattei") and that Holste himself was the scholar responsible for the topographical information presented on it. It is not surprising, therefore, that he reserves his most bitter criticism in this section for Holste himself.

The actual location of Lake Regillus remains disputed. See FI 1.10, 879–81; *Barrington Atlas*, map 43 and Directory, 1:629. Without discussion, Barbetta (*Via Labicana*, 47–48) places it "quasi certamente" at the site of Pantano Borgese. T. Ashby ("Sul vero sito del Lago Regillo," *RendLinc* 7 [1898]: 103–26, summarized in *CR* 12 [1898]: 470–72) argues for a location further southeast, at Pantano Secco, about two miles north of Frascati. L. Pareti ("Sulla battaglia del Lago Regillo," *StRom* 7 [1959]: 18–30) argues from literary evidence on the battle itself for a location near the basin of Prata Porci, about two and a half miles north of Frascati. Pareti's reading cannot be correct, because of the presence of an imperial villa on the site.

Source of the Alexandrina/Acqua Felice. Fabretti here makes reference to the inscription on the famous Moses Fountain of the Acqua Felice in Rome: "Sixtus V. Pont. Max. Picenus aquam ex Agro Columnae Via Praenestina sinistrorsum multar. collectione venarum ductu sinuoso a receptaculo mil. XX. a capite XXII. adduxit et Felicemque de nomine ante pont. dixit. Coepit anno pont. I absolvit III" [Pope Sixtus V from Piceno brought this water from the territory of Colonna to the left of the Via Praenestina from a combination of springs on a winding course of twenty miles from its reservoir and twenty-two miles from its source and named it Felice from his name before his pontificate. Begun in the first year and finished in the third of his pontificate]. (For the Moses fountain, see N. Cardano, "La mostra dell'acqua Felice," in *Trionfo*, 243–50.) Like his earlier censure of Pope Paul V's inscriptions of the Acqua Paola (I.4g), Fabretti's criticism is surprising here, since he points out in his second dissertation that such errors are common in papal documents (II.4a).

Course of the Via Labicana. Fabretti now attacks Holste in a digression concerning the route of the ancient Via Labicana east of Torrenova. On Torrenova itself, see FI 1.10, 693–95; Barbetta, *Via Labicana*, 39–41; Tomassetti, 3:485–92.

Fabretti makes only passing allusion here to what was a much debated issue. Clüver (946) had identified Lake Regillus as the Laghetto della Colonna: "Ad Praenestinam (viam) vero e regione dicti opidi, quod vulgo La Colonna vocatur, est is lacus qui tumulorum supercilio veluti amphitheatro inclusus vulgo cognominatur sanctae Praxedis: nec alius circa Tusculum totaque hac regione est lacus. Unde recte viri docti hunc etiam ante me interpretati sunt Regillum lacum" [Along the Via Praenestina, outside the area of the town already cited, which is commonly called Colonna, is a lake enclosed by a ridge of hills in the manner of an amphitheater, commonly called the Lake of S. Prassede, nor is there another lake in this area around Tusculum. As a result, scholars before my time have rightly identified it as Lake Regillus]. Holste (191–92) corrected this reading, pointing out that Clüver had confused the Via Praenestina and Labicana and therefore misidentified the Lake of S. Prassede (i.e., the Lake of Castiglione near Gabii) as Lake Regillus: "Labicanam viam cum Praenestina confundit, nam quae sub Colonna transit, non Praenestina, sed Lavicana est via. Error inde natus quod Cluverius ipse, ut nunc vulgo omnes faciunt, Praeneste Romam petens ex Praenestina in Lavicanam divertit. Lacus autem ille Regillus ad XIV. ferme lapidem ipsi viae Lavicana de proximo imminet ad sinistram Roma euntibus Lavicum hoc est ad Columnam. . . . Sed in hoc quoque fallitur, quod exiguum illum lacum, qui sub Columna tumulis amphitheatri in speciem includitur, cum lacu S. Praxedis confundit. Cum iste Lavicanam viam proxime attingat; alter autem S. Praxedis, sive, ut nunc vocant Il lago di Castiglione, non in Tusculano, sed in Gabino agro haud dubie erat" [Clüver confuses the Via Labicana with the Praenestina, for the road that passes below Colonna is not the Praenestina but the Labicana. The mistake arose because Clüver, like everyone now, when going to Rome from Palestrina, crosses from the Via Praenestina to the Labicana. However, Lake Regillus dominates the Via Labicana itself on the left for those coming from Rome at almost the fourteenth milestone. . . . In this, too, he is mistaken, because he confuses that small lake that beneath Colonna is enclosed by hills like an amphitheater with the Lake of S. Prassede, since Lake Regillus is very close to the Via Labicana, but the Lake of S. Prassede or, as they now call it, the Lake of Castiglione, is without a doubt not in Tusculan but in Gabine territory].

Fabretti does not identify here the source of the error on Sixtus V's inscription. He uses Holste's own argument to make the point of error, then singles out Holste himself for special criticism for misidentification of the Via del Finocchio (which follows the course of the modern Via Casilina) as the ancient Via Labicana. This censure seems gratuitous, since the course of the Via Labicana is really not pertinent to the main subject at hand. Moreover, Fabretti appears to base his criticism on the topographical study reflected on Mattei's map of 1674, which indeed shows the ancient Via Labicana following the course of the modern Casilina. On the course of the ancient Labicana at this point, see Barbetta, *Via Labicana*, 39–41; Tomassetti, 3:483–84.

Holste, however, had correctly recognized the Osteria del Finocchio on the modern Via Casilina as an important junction of ancient roads. Quilici (FI 1.10, 617–19), Barbetta (*Via Labicana*, 46–47), and Ashby ("Classical Topography-I," 236) make similar observations, and Fabretti himself seems to imply the same thing when he adds his own findings on the "viatrium vetus" or ancient roadway indicated on his topographic map (fig. 1), giving details about its course from the Taberna dell'Osa to the north (for its location, see Barbetta, *Via Labicana*, 47; Tomassetti, 3:576; Ashby, "Classical Topography-I," 176–77), south past the Casale of S. Antonio (see Barbetta, *Via Labicana*, 47), to the Tor Forame on the ancient Labicana (see Barbetta, *Via Labicana*, 47). Fabretti's "viatrium" seems to have been a continuation of the Via Cavona constructed to help communication between the Via Appia to the southwest and other roads to the northeast; see Ashby, "Classical Topography-I," 176–77. Fabretti's observation about the side road is therefore correct, but his attack on Holste is harsh.

CLOSING

Lucio's age. Lucio would have been seventy-three years old in 1677, when this dissertation was written. Fabretti was sixteen years younger than his addressee.

CHAPTER FOUR

Dissertation II on the Waters and Aqueducts of Ancient Rome

To the most eminent and reverend Gaspare Cardinal Carpegna, vicar of the city, his most gentle master, Raffaello Fabretti, son of Gaspare, of Urbino, sends greetings.

I. INTRODUCTION

Whoever has said that it is not at all necessary to give an account of one's leisure must have thought, when he said these things, of a man living for himself alone and focused on his own affairs. Far different, indeed, is the case of those whose course of life established by right counsel cannot be under their own control, especially those who must live or must direct their times for both working and being idle at another's bidding. This has most certainly been my experience, most eminent prince. After I declared myself among your servants and those of your office, next to the highest in the Roman church, and dedicated my effort and, at the same time, my industry to this task, I ceased to be my own man altogether. I became subject not only to your authority and will but also to your judgment. For this reason, I think it the special duty of my life to be approved by the one alone by whom I can be praised.

I therefore understand that I must make the greatest effort that you not think that the leisure you most generously grant me to interrupt my tasks in your service has been undertaken rashly and foolishly, its moments uselessly wasted. This, then, will be the result if I show you that I have advantageously spent my periods of time withdrawn from more serious concerns in the pursuit of more humane letters and careful thought, having followed you as my weightiest and greatest authority in this matter. I cite you, I say, most eminent prince, since I see that constant concerns about the special business of the Christian state have been no obstacle for indulging these same pursuits from time to time.

To you, therefore, I will give an account of those repeated excursions made in spare hours—or even sometimes during entire holidays—that make up the sum total of my leisure: the things I observed when I was setting forth toward Marsian territory on your orders—hurriedly, as you know, and on other business—and the thoughts that came into my mind from recently repeated inspection of the topography and that seemed worth knowing about the sources of the Aqua Marcia and the Claudia and about the more accurate course of the Via Valeria. I ask that you receive them kindly, as is your custom, and that you favor my enterprise, such as it is, and permit it to shine in the brightest light of your name. The result will be that the very neglect, squalor, and shadows that the injury of time has poured over the constructions and monuments providing the subject matter in my treatise will remain cast off and scattered.

2. MAP OF THE UPPER ANIO VALLEY

We have thought we should first present a topographic map [fig. 17], completed not with much skill, to be sure—modesty forbids us to praise our endeavors to the extent that others are accustomed to praise their own—but certainly with much effort. We found all the maps published thus far to be so incorrect and, if I may say so, carelessly done that they were no help at all in this presentation we are undertaking. Therefore, in comparison with very bad maps, this one perhaps may be able to seem not inattentively executed.

I bear witness that all these things have indeed been "discovered by me through direct inspection," and I am not sorry not to have believed at all the very learned Lucas Holste, who boasted the same thing in his *Annotations* on Clüver.[1] Instead, I have called back for a new assessment the

1. Holste, 165.

knowledge of all things that Holste decided himself with great and excessive (as I shall make clear) authority. Indeed, in these things, which concern scholarly knowledge of antiquity, I am confident that I will find no one better than myself. As a result, I have happily and freely put back on this map the channel of the Digentia and the shrine of Vacuna (just as elsewhere [fig. 1]), Lake Regillus and the town Labicum, and many other things, in agreement with Holste himself—indeed, some of them as a result of his arguments. However, in describing the topography itself and establishing the distances, I wish to be able to differ with so great a man and, since truth furnishes courage and strength, both to attack his interpretations frequently and to refute them.

3. DETERMINING ANCIENT MEASUREMENTS

a. The Mile

For us to have corrected the map has not seemed enough unless (after we have entered this competition) we bring the mile measurement, through which distances of places are set off, back to a fixed and unchanging calculation. Up to this point, it is nodding and wavering, as you will see.

Under the administration of Luca Peto in the preceding century, all the measurements that Rome uses were inscribed and set forth on a marble plaque on the Capitoline. Since Peto had noticed that the *palmo* of modern engineers (what they call the architectonic) does not correspond exactly to the *spithama*, or three-fourths of the ancient geometric foot, he warned his reader concerning this four times in his work *On Weights and Measures* and indicated the discrepancy in detail, first in these words at the end of his book 1: "Likewise, I also want you to know the following, that the *palmo*, which architects at Rome today use, if examined and compared with the *dodrans* (three-fourths of the legal ancient foot, i.e., nine inches), is found to be longer by one and one-third *scripula* (as much as two-ninths of an inch of the aforementioned foot), that is, one-sixth of an inch plus one and one-third *scripula*."[2] He speaks along the same lines at the very beginning of his book 2 and again within it.[3] Then, since Peto wanted the matter to be not at all hidden (as if knowing beforehand the common error of this century), he concludes that book in fixing the measure of the

2. Peto, 6 (= Graevius, 11:1619D–E).

3. Peto, 15, 18 (= Graevius, 11:1621D, 1623F).

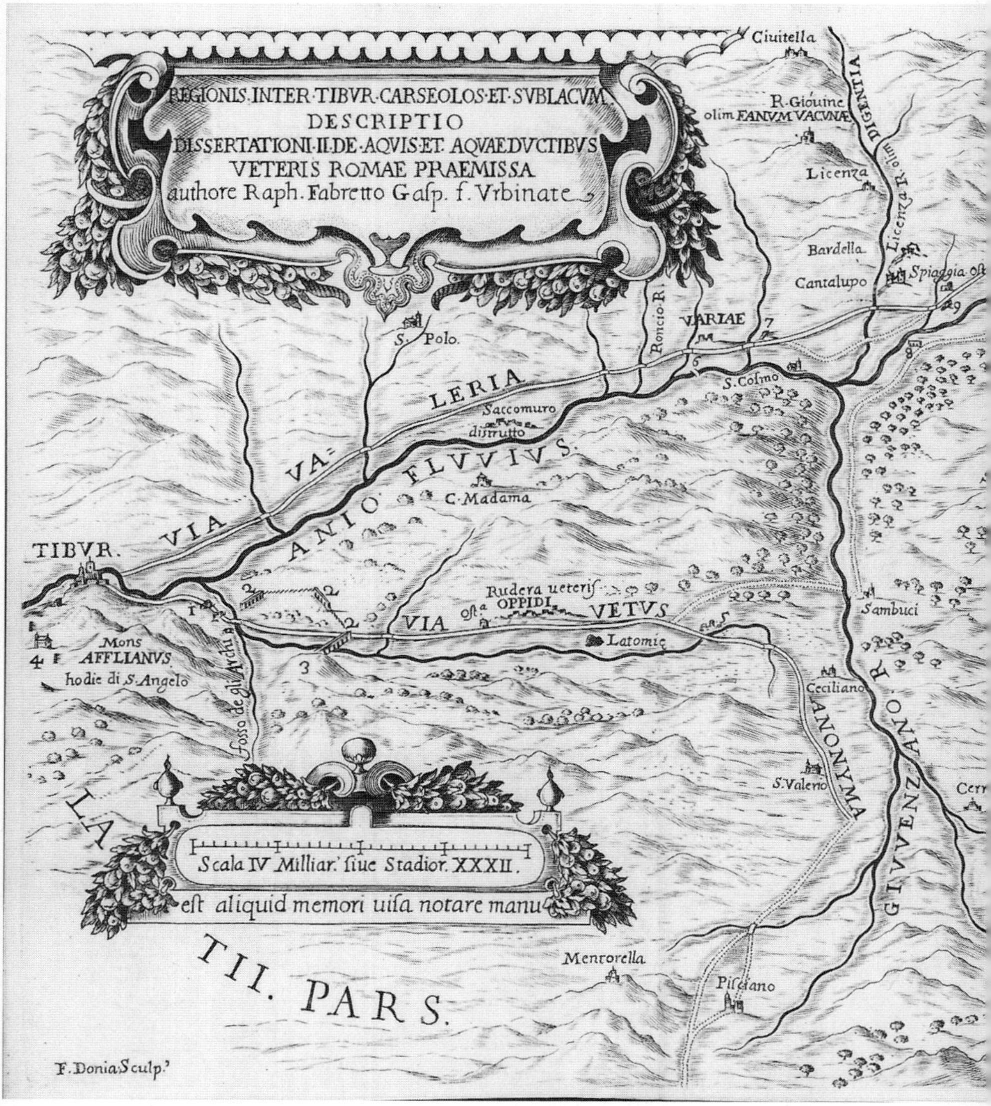

Fig. 17. Topographical map of the upper Anio Valley between Tivoli and Carsoli

1. Arch of the Aqua Claudia, appearing first. There follows the arch of the Anio Novus, toward the east. There follows the Arch of the Marcia, in the streambed itself.
2. Three sections of arcade of the same Anio Novus
3. Place where the same aqueduct emerges from the side of the mountain
4. Channel of these same three aqueducts, showing themselves at the west side of Monte S. Angelo, near the Madonna di Carciano
5. Ruins of a huge structure, perhaps above the channel of the Anio Novus, which emerges below at no. 3
6. Bridge below Vicovaro, above the remains of an ancient aqueduct
7. Traces of another ancient aqueduct, drawn off from the nearby hills from the north side
8. Aqueduct on the north side of the mountain under Saracinesco
9. Another aqueduct, near the Osteria della Spiaggia, perhaps the same as that seen under the Osteria della Ferrata in the wine cellar
10. Another, carrying the Anio Novus across from right to left under Roviano and taking itself straight into the mountain
11. The site of Somnula, where the thirty-eighth milestone of the side road from the Via Valeria into Subiaco was found

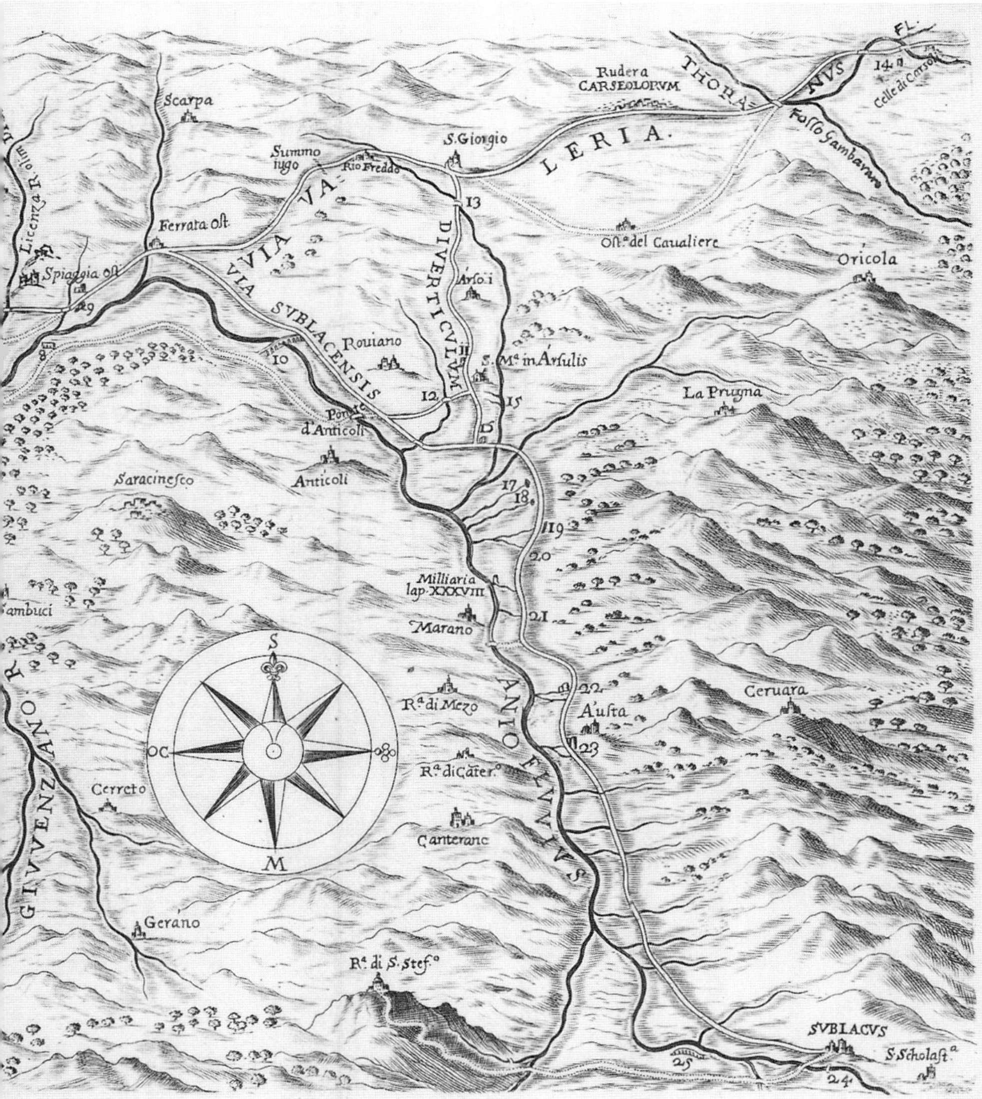

12. The Pons Scutonicus of ancient construction, called "Stratonicus" by Holste
13. Another bridge, equally of ancient construction, below S. Giorgio
14. The column of the forty-first milestone near Cellae
15. Very abundant springs under the Church of S. Maria in Arsoli—in our opinion, the sources of the Aqua Marcia
16. A dedication of Augustus, at the junction of the branch from the Via Valeria toward Subiaco cited earlier [no. 11]
17. Ruins of a structure, at the source of the aqueduct emerging underneath them
18. Ruins at the source of the aqueduct, called Lake of S. Lucia
19. Substructure, perhaps of the Anio Novus before its channeling from the lake, set against the side of the mountain; under substructure, the source of the aqueduct called Serena
20. Another channel of an aqueduct, also called Serena
21. Aqueduct, called Casa di Lemme
22. Forma della Mola, consisting of most abundant sources
23. Channel under the settlement of Agosta, the most abundant of all
24. Remains of the lakes, from the second of which the Anio Novus was drawn by Nerva, the name of Trajan having been inscribed
25. Aqueduct of the same Anio Novus, twenty feet higher than the bed of the river itself

ancient mile, understood in modern *palmi* of architects, or reeds [*cannae*] of ten palms in length: "The mile, which I said in the earlier book consists of five thousand feet, through that calculation by which we showed the *palmo* to be three-fourths of a foot plus one-sixth of an inch plus one and one-third *scripula*, will today be the measure of 649 reeds, nine palms, and three digits."[4]

But while Peto believes that he has applied medicine, he has administered poison. First, there is an insignificant error in calculating the total of his reeds, which ought to be stated not as 649 reeds, nine palms, and three digits, as he thinks, but as 650 reeds, six palms, one *sextula*, and one and one-third *scripula*, as can be seen in the following summary.

A *palmo* of nine inches, one *sextula*, and one and one-third *scripula*, reduced to so many thirds to avoid fractions, makes 664 thirds, as follows:

nine inches =	648
one *sextula* =	12
one *scripulum* =	3
one-third of a *scripulum* =	1
	664 thirds.

An ancient foot consists of twelve inches, or seventy-two *sextulae*, or 288 *scripula*, or 864 thirds of a *scripulum*. When multiplied by five thousand feet (which are one thousand geometric paces), the result is 432,000 thirds. Divide a mile, that is, 432,000 thirds, by its own palm, that is, 664 thirds:

432000/664 = 6506.

The result will be 6,506 palms plus sixteen thirds more (i.e., five and one-third *scripula*), which is the same as we have said already, one-sixth inch plus one and one-third *scripula*.

In addition to this error, Peto has committed a far more serious mistake in indicating the length itself of the modern *palmo* (i.e., the length of the already mentioned nine and one-sixth inches plus one and one-third *scripula*). From his own Capitoline marble, comparing not the inscribed foot with a *palmo* but a ten-foot length with an entire reed, we find that a reed

4. Peto, 20 (= Graevius, 11:1625B–1626A).

contains seven feet, six inches, five *sextulae*, plus one *scripulum*, with three seventeenths of a second *scripulum*, or (that which demands a shorter mathematical operation) seven feet six inches and fifteen-seventeenths of a second inch. Thus, the discrepancy of the modern *palmo* with respect to three-fourths of the ancient foot is 15/170 inch. From this, we have established that the distance of a mile is that of 660 reeds, one palm, eleven inches, and one and one-half minutes, as in the following summary.

A *palmo* of 9 and 15/170 inches, reduced to so many units of 170, produces 1545/170 in this way:

Inches nine	1530
Fifteen remaining	15
	1545.

An ancient foot (i.e., 2040/170) multiplied by five thousand feet produces 10200000/170. Divide a mile (i.e., 10200000/170) by its equivalent in palms (i.e., by 1545/170):

10200000/1545 = 6601.9300.

The result will be 6,601 palms and 1455/1545. Multiply the figure by twelve inches, and when that figure is divided by the denominator, the result will be eleven and 465/1546. Again, when the figure is multiplied by five minutes—for one inch is divided into five minutes—and the result is divided by the denominator, the result will be one and 780/1545 minutes. When each number is divided by fifteen, the number by which the fraction is reduced to smaller numbers, the result is 52/103, which we accept as very close to a half-minute.

Holste neither paid attention to Peto nor sought the truth, here not lying hidden in a well but open on the map; he instead embraced that well-worn and treacherous practice of surveyors. As a result, just as he had posited this figure in many places in his other writings, Holste announced in his dissertation *On the Golden Milestone* that the ancient mile contained somewhat less than 667 reeds.[5] This assumes that the modern architectural *palmo* would indeed be the same as three-fourths of a foot, or an ancient *spithama*, as the following summary demonstrates.

5. L. Holste, *De milliario aureo, error popularis, quem omnes fere antiquarii errant, explosus* (= Graevius, 4:1805–6).

A mile contains five thousand feet, or sixty thousand inches. Divide by nine inches, that is, by a *dodrans* or three-fourths of a foot:

60000/9 = 6666.

The result will be 6,666 2/3 *dodrantes* or three-fourths feet, or (according to Holste's hypothesis) 6,666 and two-thirds palms, that is, 666 reeds, six palms, and eight inches.

Another measurement of surveyors is through *catenae*. A *catena* contains fifty-seven and one-half palms, that is, ten *staioli*, each of which therefore comprises five and three-fourths palms. Multiply, therefore, by 116 (for surveyors wish a mile to be made up of this many *catenae*); there are 6,670 palms, in this way:

116	580
57 1/2	58
812	6670.

That is, there are 667 reeds of ten-palm length, or senatorial paces.

So that the fruit of the labors of the distinguished Holste not be ruined by an unnoticed error, it has been my pleasure to note these things for enlightenment, not criticism.

b. The Foot

Moreover, there have been different sorts of confusion about the legal foot. Some people prefer that of Statilius, others that of Cossutius or the porphyry foot to the others. However, I now am satisfied with the Capitoline foot, after I had rejected it for a long time or, rather, had believed that I had rejected it. Indeed, a strange thing happened to me in this regard; once on the Via Ostiensis near the tenth milestone, in certain ruins of an elegant villa, I had discovered the remains of a floor paved from pieces of different marble, very well fitted. As I am held by this habit, either an interest or a disease, of searching out antiquities most carefully, I extracted a piece more complete than the others and took it with me. Its two sides matched each other in a most exact way: the longer side exceeded the foot that Luca Peto represented as legal[6] by one and one-half *sextula*, and the

6. Peto, 88 (= Graevius, 11:1674).

shorter one was reduced in size by exactly half that amount. I realized that this fragment was so precisely similar to many other pieces of pavement that I reconstructed my foot from it, after it had been transferred onto a bronze sheet. It matched all the thicknesses of the walls, the widths of the flat columns, and all the larger measurements, without any fraction of *sextulae* or minutes.

As a result, although other measurements were ordered to be valid, I preferred this one. Moreover, in fixing a comparison of the Capitoline foot with the reed, I found that this foot of mine was of indistinguishable similarity with that on the Capitol. Either Peto did not faithfully reproduce it in his book, or—what I rather think—the model, incised in bronze and stamped on wet paper, contracted when the paper dried out. As a result, it is now my task to accept the Capitoline foot as my own and defend it.

Certainly (so that I not be charged with vanity for attributing anything to my observations), there is threefold evidence by which Peto confirmed his measurement, namely, the number of three different bronze feet taken together, of the same length, so shaped for use and found in different places.[7] This evidence takes precedence by far over those memorials that were carved on the two funerary monuments of Cn. Cossutius and T. Statilius. They serve to indicate the profession held by the man whose ashes are conserved within rather than to show the exact measure of a foot. This can be observed in the following foot of Aebutius [fig. 18].

> Marcus Aebutius, the freedman of Marcus, Macedo, father, Marcus Aebutius, the freedman of Marcus, Callistratus, son, Marcus Aebutius, the freedman of Marcus, Eros, Julia, the freedwomen of Lucius, Berenice, Julia, the freedwoman of Lucius, Hesuchium, Pomponia, the freedwoman of Lucius, Selene. Clodia, the freedwoman of Gaius, Antiocis, vowed it. [*CIL* VI, 10588]

On this inscription on a stone that lies in the Orti Mattei, among other construction tools, the tools themselves are quite distorted, since the carpenter's square is out of proportion, and the obtuse angle and the legs of the right angle are in error. On it, I say, not only is precise measure lacking in the foot—it exceeds the Capitoline foot by three *scripula*, the same length by which another, published by Gruter, clearly falls short[8]—but

7. Peto, 5 (= Graevius, 11:1619A–B).
8. Gruter, dcxliv.3.

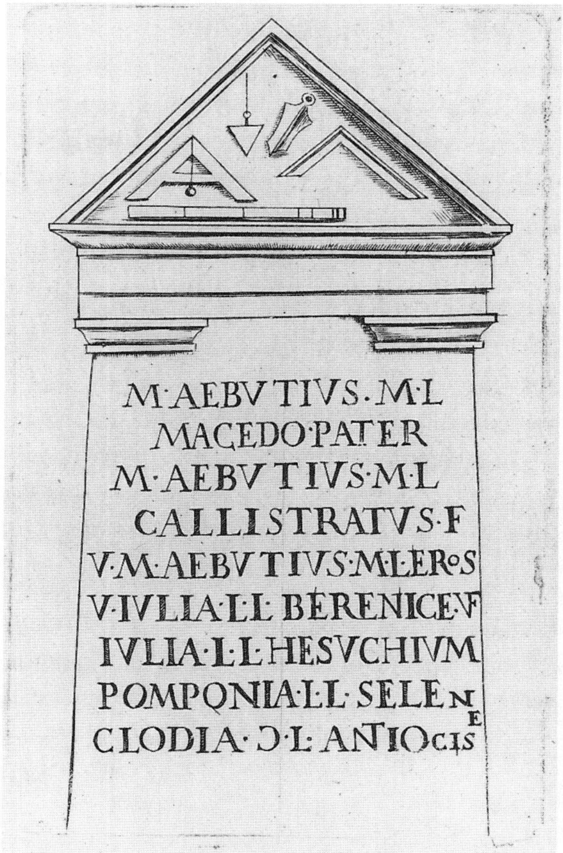

Fig. 18. Cippus of M. Albutius [*CIL* VI, 10588]

also a careful division of the parts is lacking. This is especially true because neither do the other two earlier feet, those of Cossutius and Statilius, agree at all, as Philandrier argued,[9] but there is a discrepancy (albeit modest) between each, as Peto noticed.[10]

The most distinguished and most careful Fr. Juan Bautiste Villalpando has reconstructed the measurement of the Roman foot from the Farnesina *congius*—not, indeed, from its height, as Oisel thought,[11] but from two

9. Philandrier, *Annotationes*, 117–18; cf. Lemerle, *Les "Annotations*," 165–66.
10. Peto, 5 (= Graevius, 11:1617A–D).
11. Oisel, 561–66.

lines drawn on the vessel at an angle toward each other.[12] For his commentary, Villalpando has deservedly won the greatest praise from all, in his table where he sets forth "an instrument for increasing or diminishing bodies in a given scheme."[13] But [in his measurement of this *congius*] he has won little credibility or none at all from us. Indeed, we cannot be persuaded that the bronze Farnesina vessel, established for measurement of liquids, through so many fine points would also have been applicable for measuring distances. Its dimensions do not rise up in a straight line, and—what is especially important—no circle and no projection of the vessel indicate the dimension of a half-foot. To corroborate the first, we will examine "another *congius* quite similar to this Farnesina vessel, held by a man unknown then and up to this time,"[14] as if brought on by divine intervention, better than the vessel Villalpando has shown.

The single most important factor that weakens the credibility for us of this measurement is its notable excess, that of one and one-fourth *sextulae* or 5/288 over our Capitoline foot, attested by surer evidence and confirmed a thousand times from daily measurements of ancient buildings. These discrepancies resulting from this measure and from fractions of it—already fragmentary and unmeasurable, if I might say so—would escape notice if Villalpando's foot were adopted. Moreover, there is a very easy explanation of this inscription on the Farnesina *congius:*

> Vespasian Caesar, consul for the sixth time, and Titus Caesar, son of Augustus, consul for the fourth time, measures fixed [*MENSURAE EXACTAE*] in the Capitol, ten pounds. [*ILS* 8623]

If the phrase "MENSURAE EXACTAE" should be interpreted as nominative plural, it would then have to be inferred that "in one measure only, reference is being made to fixed measures, so that a scheme of all measures should be sought from this measurement alone."[15] Clearly and in plain

12. J. Prado, S.J., and J. B. de Villalpando, S.J., *In Exechielem explanationes* (Rome, 1604), 3:499–502.

13. Prado and Villalpando, *In Exechielem*, 3:316.

14. Cf. Prado and Villalpando, *In Exechielem*, 3:499C–D: "cum casu illac pertransuit vir quidem mihi tunc ignotus, qui simile quoddam vas inter pretiosa maximi cuiusdam principis . . . asservaret" [when there passed by chance a man at that time unknown to me, who was keeping a watch on a certain similar vessel among the treasures of a certain prince . . .].

15. Cf. Prado and Villalpando, *In Exechielem*, 3:500: "Iam, ni fallor, aperte videre potuit lector, quomodo hoc uno vase mensurae omnes contineantur, ac pondera" [The reader has been able to see now, unless I am mistaken, how in this one vessel all the weights and measures are contained].

Latin, we interpret the phrase as genitive singular, as if the name of the *congius* vessel itself were understood and the inscription had denoted "a *congius* of fixed measure." In this way, the phrase is clearly to be interpreted in an inscription cited by Gruter, "of fixed *stater* on the Capitol."[16] The word "pound," "ounce," or something similar, as if superfluous, is supplied by the context. So, likewise, in another inscription, "of fixed *sextarius* . . . of our lords Arcadius and Honorius,"[17] we see the word "measures" omitted, like the word *congius* here, for the sake of brevity and elegance in the Latin; and in the inscription that precedes it,

> C. Umbrius Edrastus Fortunatus, quattuorvir, legal measures of the city at his own expense . . . [*CIL* IX, 980]

the word "renewed," "set up," or something similar, which the grammarians call "principal," is lacking.

Now, as an appendage, let us indicate some accurate representation of the Capitoline foot to foreigners and outsiders; it is found nowhere else more accurately and, moreover, divided into its *sextulae* than in the work of the same Fr. Villalpando, precisely in 72 of those 125 spaces in which he divided diameter AB of his semicircular instrument.[18] To avoid error, they should not trust in this regard the reconstruction of Fr. Giambattista Riccioli in his *New Almagest* [Comprehensive Treatise] more than is right.[19] They should observe, I ask, that Riccioli's half-foot exceeds our Capitoline half-foot by three-eighths of an inch. They should not think a deviation of one-sixteenth in a whole foot, or an entire digit in a foot, something very small or negligible. Again, they should consider that the same Riccioli sets forth in his *Revised Geography and Hydrography* another variant from his own (one not at all different from the Farnesina measure, which he believes he is following), two *sextulae* longer than the Capitoline foot.[20] But, finally, let us return to our subject.

---

16. Gruter, ccxxii.8.

17. Gruter, ccxxiii.2.

18. Prado and Villalpando, *In Exechielem*, 3:316.

19. G. B. Riccioli, S.J., *Almagestum novum astronomiam veterem novamque complectens* (Bologna, 1651), 58.

20. G. B. Riccioli, S.J., *Geographiae et hydrographiae reformatae nuper recognitae et auctae libri duodecim* (Venice, 1672), 32.

4. TOPOGRAPHY OF THE UPPER ANIO

In a different way, in that outstanding work on the aqueduct system—we owe the one copy of it surviving barbaric devastation to the diligence of the Florentine Poggio [Bracciolini], who witnesses in the *Description of the Ruins of the City of Rome* that he "found this book shortly before hidden away and concealed in the Casino monastery"[21]—in a different way, I say, does Frontinus speak about the source of these aqueducts. In designating the sources of the Aqua Marcia, Frontinus cites the Via Valeria and the side road from it onto the Via Sublacensis. However, in discussing the sources of the Aqua Claudia, he cites only the Via Sublacensis: he says, "The Marcia begins on the Via Valeria at the thirty-third milestone on a side road for those coming from Rome three miles to the right on the Via Sublacensis, first paved during Nero's principate, at the thirty-eighth milestone on the left, within a distance of two hundred paces, bounded by substructures, almost standing still, with a very dark green color" [*Aq.* 7.6–7], and "The Claudia takes its beginning on the Via Sublacensis, at the thirty-eighth milestone, on a side road to the left within three hundred paces, from two very abundant and beautiful springs, the Caerulean, which is named from its appearance, and the Curtian" [*Aq.* 14.1].

a. Holste's Theories

Holste paid no attention at all to that careful statement of Frontinus, nor in any way to the Marcia's distance from the city, shorter than that of the Claudia. In his *Annotations* on Clüver, he indiscriminately locates the sources of each aqueduct on the Via Sublacensis,[22] indeed, for the Claudia, assigning the Aqua Serena (no. 20 on our map) as the Curtian spring and the Lake of S. Lucia (no. 18) as the Caerulean and allotting confidently as source to the Marcia the Forma della Mola (no. 22), which issues a full mile above the Serena. Holste in fact says that the name of the Aqua Augusta still remains in a certain very rich source (no. 23) that issues in a double spring under a settlement called L'Austa.[23]

---

21. P. Bracciolini, *Historiae de varietate fortunae libri quatuor* (Forni, 1969), 16.
22. Holste, 130–31.
23. Holste, 129–30.

Frontinus, however, refutes him in this, expressing the distance of the source of the Augusta in these terms [*Aq.* 12.1–2]: "Likewise, whenever dry spells required assistance for resupply of the Marcia, Augustus brought another source of water of the same high quality by underground conduit to the Marcia's channel, which is called the Augusta from its builder. The Augusta arises beyond the source of the Marcia, and its conduit, until it reaches that of the Marcia, runs eight hundred paces." Indeed, in very fact, that source of the L'Austa settlement, even when each of its springs is taken at its origin—although Frontinus fixed his measurement from the source of the Augusta not to that of the Marcia but to its conduit—is not five hundred whole paces distant from the other spring of the Mola (which, according to Holste, is the Marcia). From another argument to be made later, both the Marcia and the Augusta itself must be established as coming toward Rome and the Via Valeria below the sources of the Claudia.

I might concede that the locality L'Austa could contribute something of importance in another context; however, we should not think the name of the Aqua Augusta is surely detected in it, with other better conjectures scorned. Holste himself notes how little faith, indeed, one should have in those common names of places in the case of the town Arsoli, which Clüver assigns to the Colonia Carseolana, an error for which Holste scolds him.[24] We have by obvious demonstrations corrected Holste elsewhere—indeed, in a letter on his *Annotationes* just cited to the distinguished Lorenzo Panciatichi, that portent of erudition and memory while the fates allowed. Holste moves the site of Laurentum to the Torre S. Lorenzo because of the similarity of name,[25] and indeed, shortly afterward, he pushes in S. Anastasio too, having contradicted Clüver and himself at the same time.[26] Apart from this, this two-syllable word *L'Austa* has some similarity in spelling with the word *Augusta* but none in sound and derivation. The inhabitants render it in such a way that we would pronounce it "hausta," that is, "a channeled aqueduct." From this perhaps is more closely derived the name for the place and the spring.

So that I not seem to be suppressing by cleverness the things that can be claimed in support of Holste, let me say also that in a certain constitution of Clement III in the year 1189 (included in the new *Bullarium casinense*), mention is made of an Aqua Augusta and of a settlement of

24. Holste, 165.
25. Holste, 172.
26. Holste, 202.

Augusta, in the following words: "Concerning the lake or the river flowing from it, let no one, apart from the permission of the abbot and the brothers, be allowed to fish or to build a millwork up to the arch that is called 'de Ferrata' in the territory of Roviano. In addition, concerning the ancient conduit that brings water from the river to the congregation of S. Lorenzo, from this aqueduct, which is called the Augusta, let no one, apart from your permission, be permitted to draw water, except as much as is sufficient for irrigating lands and replenishing the baptismal font in the same church";[27] and later, among different possessions of the Subiaco monastery is listed "the settlement of Augusta, with its farms and dwellings."[28] All this is repeated word for word in another bull of Honorius III of the year 1217. No, indeed, these provisions are said to be derived in part from a certain higher source, from a privilege granted by the Most Holy Pontiff Gregory the Great, in which the same Aqua Augusta is named. I have seen it on page 2 of the copy taken from the registry of the holy monastery of Subiaco, kindly shown to me by the Most Reverend Abbot Cornelio Margarini, the compiler and editor of the *Bullarium* cited earlier.

But not for this reason will I withdraw from attacking Holste's opinion. Indeed, these apostolic documents or similar indulgences of proven note—for concerning the authenticity of this one, which is attributed to Gregory I, you will summon someone else than me as a supporter—include whatever territory lies from the monastery itself up to the arch of Roviano. (This, I believe, is the same aqueduct that is seen to cross the Anio below Roviano, no. 10 on the map.) As a result, in so wide a territory, they therefore leave uncertain the location of the source or the conduit of the Aqua Augusta itself, to the extent that they made reference to the Augusta mentioned by Frontinus, according to both our reasoning and other arguments.

It is indeed clear that the bulls cited above move each aqueduct far from these places around the thirty-eighth milestone and bring it close to the monastery itself: the conduit, which is said to bring "water from the river to the congregation of S. Lorenzo" (just like another Augusta) is said "to bring water from the lake" in an older bull of John VI of July 21, 704. I will cite its words more willingly for the reason that they lead to an understanding of the origin of the Augusta settlement built at a distance from here: ". . . the settlement that is called Augusta, with its entire mountain,

27. C. Margarini, O.S.B., ed., *Bullarium casinense, seu constitutiones summorum pontificum* (Todi, 1670), 218.2.2.

28. Margarini, *Bullarium casinense*, 219.1.

for construction of a fortress. Likewise, the entire Aqua Augusta and the ancient aqueduct, commonly called the conduit, through which water is drawn from the lake and flows into the river at the Church of S. Lorenzo, which is called 'of the people,' so that no one at all should have the opportunity of drawing water from it, except for the baptismal font, or the irrigation of gardens, or useful purpose . . . ." Here, indeed, there is by chance a certain mention of this settlement of Augusta, far distant from that Aqua Augusta about which the bulls speak; the proximity of each aqueduct line and the congregation of S. Lorenzo to the lakes or Simbruine pools is also assumed.

But on this point, we might take confirmation from Holste himself, who makes reference as follows in his *Annotations* about the second and higher conduit of the Anio Novus—after its water was drawn "from the lake, which is above the Neronian villa at Subiaco," according to Frontinus [*Aq*. 93]—and about the lakes themselves: "The traces of this aqueduct are seen now below Subiaco itself, and it is commonly called 'il buco della Cartiera'; it runs from there below the Osteria of S. Antonio and from this point continuously along the left bank of the river. Its height at this time is twenty feet higher than the bed of the river itself; from this, one may conjecture about the height of the lake. However, those three lakes were not at all natural but artificial, the first, indeed, under the monastery of S. Scholastica, the very narrow mouth of which was closed around by a very strong wall of eighteen or twenty feet. From here, water flowed down into a second lake, which afterward received the first, the mouth of which seems to have been somewhat more open. The third lake was under the town of Subiaco itself, where even now a large part of the Anio is dammed by a wall, for driving various mills. Here, the Church of S. Lorenzo, built by Narzio under Pope Damasus, rises above it; the church used to be called 'at the high waters,' as ancient documents of donations of the Subiaco monastery bear witness. However, the lakes were destroyed in the huge flood of the river on February 20, 1305."[29]

Moreover, if the bulls must be thought to refer to an Aqua Augusta below a settlement with the same name, we shall understand that they are concerned not with the aqueduct about which we are speaking, which Octavian named from himself, but with another, which Antoninus Augustus added to the Marcia.

---

29. Holste, 128–29.

[We make] one last point: because all those names have been suggested by the monks themselves, who perhaps at leisure made up modern names as a substitute for ancient erudition, there is no reason why a papal document should be presented as an argument against us. The Holy See ought to be concerned not with the authenticity of some names set forth for itself but about showing in the best way its generosity most broadly and the Christian state in a most deserving order. Indeed, not here alone have papal writers, having followed common phraseology, been deceived in these minute and secondary matters. The famous Pietro Bembo, the glory of letters and restorer of Latin purity, in his eagerness to gratify the Flasconians, cites them in the name of Pope Leo X among the Faliscans many times.[30] How incorrect he is in doing this Massa will prove in his *Book on the Origin and History of the Faliscans*,[31] with confirmation from the younger Nardini in his very learned *Apologetic Discourse*.[32] Moreover, to avoid looking for an example at greater distance, you will find errors in a bull of Pope Paul V concerning the Acqua Traiana, a most well known work of the same emperor, in which it [the Aqua Traiana] is attributed to Caesar Augustus and in which the same aqueduct is confused with the Aqua Alsietina.[33] Indeed, there are wrong statements in a bull of Pope Sixtus V concerning the Aqua Appia and the restoration of the Marcia restored to Rome, aqueducts from different regions combined into one and the same channel of the Acqua Felice, although their conduits are separated.[34] You will wonder, indeed, and you will confess that the words here do not agree with the accepted facts of antiquity. Yet you will not be offended by these very slight blemishes when more very important things shine brightly, namely, the greatness of public benefit and foresight, combined with scarcely believable magnanimity. In view of this, it comes to the minds of readers to bless eternally the memory of the lawmakers, rather than, like critics, to make a judgment concerning words of a law adopted

30. P. Bembo, *Opere* (Venice, 1728), 4:103 (*Epistolarum Leonis Decimi pont. max. nomine scriptarum libri* 12.9).

31. A. Massa, S.J., *De origine et rebus Faliscorum liber* (Rome, 1546), 4.

32. N. Nardini, *La catedra vescovile di S. Timoleo in Nepi: La pentapoli Nepesina e il vero sito degli antichi Veienti, Falisci, e Capenati* (Rome, 1677), 120–30.

33. L. Cherubini, ed., *Magnum bullarium romanum* (Lyons, 1692), 3:307 (inter Constitutiones Pauli V ad annum VIII, 1612, quae incipit, "In Sede B. Petri.")

34. Cherubini, *Magnum bullarium romanum*, 2:681–82 (inter Constitutiones Sixti V ad annum V, 1590, quae incipit, "Suprema cura Regiminis.")

inadvertently, not over the substance of the matter, but over a certain ornamentation.

Even though we have said that these very small things must be overlooked in the writers of bulls, it is nonetheless not appropriate to have the same indulgence toward those for whom the sum of their work consists of learned discourse. As a result, we will not permit to escape censure Dumolinet, the author of the *History of the Supreme Pontifs Taken from Coins*, who in more than one place mixes very well known facts with blatant errors. With respect to the Aqua Alsietina, climbing to the top of the Janiculum, he

> lay down on the skins spread out and sought sleep,
>
> [Verg. *Aen.* 7.88]

on a coin of Pope Paul V,[35] where in fact reference was made to the Acqua Felice. By his own correction, Dumolinet reduced to an impossibility the declaration of a bull of Sixtus concerning the restoration of the Aqua Marcia to the city (not, indeed, impossible, since the Marcia, at the beginning of the conduit of the Acqua Felice, surpassed this aqueduct in altitude and crossed over to that point). In place of "Marcia" he proposed the emendation "Virgo," citing an aqueduct far distant from here and lying hidden in a very low region, as we observed in our first dissertation [I.3], contrary to the evidence of Fr. Kircher's topographical map.

No, indeed, Dumolinet does not fear to take away from a most generous pope the glory of having introduced an aqueduct—how great has been his pride—and having constructed a new bridge on the Tiber near the ancient Via Flaminia below Ocriculum, when he [Dumolinet] says concerning a coin, "The Aqua Virgo is indicated, which Sixtus V brought into the city in greater capacity and called the Felice from his own first name."[36] Afterward, on other coins, he tells us that the reconstruction of a bridge below the Janiculum is represented.[37] So much for historic credibility and a most scrupulous author of the greatest affairs of Rome!

---

35. C. Dumolinet, *Historia summorum pontificum a Martino V ad Innocentium XI per eorum numismata* (Paris, 1679), 147–48.

36. Dumolinet, *Historia*, 118.

37. Dumolinet, *Historia*, 120.

b. Course of the Via Valeria

But finally, to return to Holste, the particular cause of error for him was his ignorance of the ancient course of the Via Valeria, a subject on which he differs and is self-contradictory. At one place, he recognizes the crossroad of the Via Sublacensis and a side road from it at the Osteria della Ferrata, at which he by chance made sacrifice for his safe journey.[38] Elsewhere, he moved the Via Valeria back to the Osteria "la Spiaggia," which is almost eleven stades before the side road.[39] Afterward, on the same page and in the same context,[40] he extends the Via Valeria almost four milestones after the crossroads established by himself, within the area of the Via Sublacensis and, indeed, the area of the side road from it onto the Valeria, when he says that traces of it appear below Roviano, in the bridge that he himself calls the "Stratonicus" [fig. 17, no. 12], and he identifies as the Via Valeria itself the entire connecting road and the side road of the Valeria onto the Sublacensis. We will prove Frontinus was thinking about this road when he indicated the intake of the Aqua Marcia [*Aq.* 7.6].

Indeed, a man of admirable learning and most experienced in antiquities ought never to have believed that the ancient Romans would have sought a somewhat wider and curving course where they could lay out a road in a straight line and would have extended over a short course the length of a road by more or less three miles. All the roads throughout the Roman world can be an example of the straightness with which their engineers were accustomed to plan them, as Plutarch witnesses that Caius Gracchus once established. "In rebuilding of roads," he says, "he applied special care, having a plan both of usefulness and beauty and charm. The roads used to be led out straight through regions, paved partly with rock that had been cut and hewed for this purpose and partly with heaped up piles of sand. Empty spaces were filled; where they might be interrupted by valleys or torrents, they were connected by bridges, and having reached the same height on each side, they were especially beautiful to see. In addition, he set off individual miles, measured carefully, with stone columns" [*C. Gracch.* 7.1].

Persuaded by this argument, I and a young man cultivated in Latin and

38. Holste, 165–66.
39. Holste, 35.
40. Holste, 165.

Greek learning, Giuseppe de Giuli, with hard work climbed the ridge that you see noted on the map—to be sure, at an unsuitable time of the year, since the heat of the oncoming summer scarcely permitted a journey on foot. Indeed, the horse, over a steep and continuous climb of two and one-half miles, was hardly able to draw himself and the carriage. With the greatest admiration and joy, we found sure traces of the road over the ridge and through the settlement "di Riofreddo," in the terracing of the slopes, ambitious and truly most worthy of Roman spirit, and in the excellent bordering of the road in its lower part, and elsewhere also in the paving of the basalt. An enthusiasm quite similar for each of us deceived our exertions: de Giuli himself was searching for a better attested branch of the Via Valeria, that very road, indeed, that at a short distance opens itself from the consular roads into the territory of the Aequicolae, the Marsi, the Peligni, and the Marrucini. He is writing their history and that of the surrounding peoples, so that he might revive memories of his native soil, greatly confused in the works of other authors.

Therefore, over this true and straight Via Valeria, from Tibur to the shrine of S. Giorgio (beneath which the side road to Subiaco descends), there are sixteen miles and three stades besides; you may attribute them to the somewhat different orientation of the modern road from the ancient, especially beneath Cantalupo. Tibur is said to be twenty miles distant from Rome, both by Antoninus [*It. Ant.* 309.1] and by the Peutinger Table correctly amended (as is done by Clüver in his *Italy*),[41] as well as by Martial in these verses to Faustinus [4.57.3–4]:

> You inhabit the kingdom of the Argive settler, Faustinus,
> Where the twentieth milestone leads from the city . . .

That side road from the Via Valeria to the Via Sublacensis must therefore be put back at the thirty-sixth milestone, and the number "XXXIII" of Frontinus [*Aq.* 7.6] must be emended to "XXXVI," by joining the first two units of the number "III" at the bottom. On this point, many arguments are persuasive.

Persuasive, I say, is the distance already cited from Tibur of sixteen

41. Clüver, 955. *Tab. Peut.* IV.5 gives the distance as sixteen miles. Cf. L. Bosio, *La Tabula Peutingeriana* (Rimini, 1983), 84; O. A. W. Dilke, *Greek and Roman Maps* (Ithaca, N.Y., 1985), 113–18.

miles, the measure of which we correctly fixed by a device attached to the wheel of the wagon. Although this measurement is not at all consistent with mathematical precision, I learned that it is very close to and not significantly different from the truth.

Two further proofs openly establish the matter. The first is the milestone erected at no. 14 [fig. 17] near the town called "Le Celle di Carsoli," in front of the doors of the Church of the Blessed Virgin de Carmelo, with an inscription indicating the forty-first milestone. A transcription of it can be read there, for it is damaged on top and eaten away:

> . . . of tribunician power, consul for the third time, undertook the building of the Via Valeria, milestone forty-one. [*CIL* IX, 5966]

Indeed, by the modern road that turns off by the Osteria del Cavaliere, from the intersection to Cellae, there are more than five and one-half miles; by straight path, through the ruins of the Colonia Carseolana, there will be five miles exactly, which, added to the well-known "XXXVI" of Frontinus [*Aq*. 7.6], as corrected by us, correspond exactly to the number "XXXXI" of this milestone.

Another proof is supplied by the thirty-eighth milestone on the side road itself from the Via Valeria to the Via Sublacensis, once standing at the Fons Somnulae, as Gruter reports, but now erected in the piazza of the nearby settlement of Arsoli, with this inscription:

> Milestone thirty-eight. The emperor Nerva Caesar Augustus, pontifex maximus, with tribunician power, consul for the third time, father of his country, undertook its construction. [*CIL* IX, 5963]

Indeed, the site of Somnulae, even now retaining the same name, is exactly two miles distant from the intersection below S. Giorgio. As a result, the crossroad is fixed correctly at the thirty-sixth milestone.

Gruter also wrote that that milestone stood in front "near the spring of Somnula along the Via Valeria,"[42] nevertheless by the authority of Ercole Ciofano[43] (I wish this to be stated for the sake of excusing the very distin-

---

42. Gruter, clv.4.

43. E. Ciofano, *In omnia P. Ovidii Nasonis opera observationes* (Antwerp, 1581–83), 127, reprinted in *Publii Ovidii Nasonis opera omnia*, ed. P. Burmann (Amsterdam, 1727), 3:643 (at *Tr*. 4.10.4).

guished Gruter). Ciofano, with Clüver himself,[44] believed that the town Arsoli, very close by there, came from the ancient Carsioli. There is no support at all for this identification, according to Holste, who quite rightly observed the ruins of the ancient colony at approximately one mile beyond the Osteria del Cavaliere on a slightly elevated hill, to the left of the Via Valeria for those going from Rome, as he sets forth at great length in his treatise.[45] However, not with equal accuracy but with the same error through which he extended the Via Valeria by leading it over the Pons Scutonicus and through Arsoli, Holste, from the itineraries, fixed the distance of Carsioli from the city at forty-two or forty-three miles.

Our correction of Frontinus, as a result of which the side road to Subiaco is fixed at the thirty-sixth milestone, two miles this side of Carsioli, also makes necessary the following correction of the Peutinger Table:[46]

From the city on the Via Tiburtina to

Aquae Albulae	16
Tibur	4
Varia	8
ad Lamnas	5
Carsioli	5
	38

Despite Holste, however, the author of the *New and* (if we believe him) *Exact Map of Roman Territory Published in the Year 1674* keeps the wild nourishment of the ancient acorn, even though soft food has already been discovered. While avoiding the mistakes of earlier writers, he runs in a different direction. Tricked by the sound of its name, he has moved Carsioli to the other side of the River Turano and in that very place where today we declare that the Celle di Carsoli and the forty-first columnar milestone stand. Yet we know that Holste's work was not unknown to this man (although to an equal extent badly understood): he steals the name *Lake Regillus*, which Holste correctly attributed to that lake that is seen under Monte Falcone and the town Colonna, without acknowledging his debt, so that he may rashly apply it to another, the Lake of S. Prassede, which Holste names Buranus and Gabinus.[47]

44. Clüver, 475.
45. Holste, 164.
46. *Tab. Peut.* IV.5 (cf. Bosio, *La Tabula Peutingeriana*, 84).
47. Holste, 192.

With the same neglect, the recent author of *Latium*, in his "Most Accurate Delineation of Sabina Ancient and Modern" (as he himself names it), puts Arsoli (formerly Carsioli) not far from the Anio and above Agosta, and he so identifies it everywhere.[48] To be sure, I am quite disturbed that this author, who treats similar subject matter, either did not read or ignored such a man as Holste, the praise of his Germany, who has contributed many excellent comments on the ancient *Latium* of Clüver. In other respects, this author did not begrudge in the least to admit and even insert, up to the point of excess, written accounts of the destruction of Ampliglione,[49] old tales about Evander's Palatium, and confusions of Lanuvium with Lavinium, from the accounts of the archbishop of Albano and an epistolary report from a doctor of Castel Gandolfo, a most wise investigator of antiquities who used to catch their breezes between the vineyards.[50]

If, indeed, following Holste,[51] scholars have said that the thirty-eighth milestone near Somnula must be counted by a turn to the left from the bridge of Anticoli and over the "Pons Stratonicus" [fig. 17, no. 12], this is refuted for many reasons. First, Holste posits that the Pons Scutonicus there—which he calls the "Stratonicus," honoring it by a Greek name—is on the Via Valeria, an argument we have proved to be false from the course of the Valeria shown earlier. Moreover, Holste computes an incorrect measurement of its route, since from Vicovaro to the Osteria della Ferrata, which he himself declares as the fifth mile—even though, as we have shown earlier, Holste's mile measurements are looser than is accurate—there are no more than four miles and since there are just as many, more or less, at the Pons Scutonicus. Beyond this bridge, there once stood, not a whole mile distant, the thirty-eighth columnar milestone along the road, at the spring of Somnula. With this sum subtracted, at that point, the thirty-seventh milestone was not yet reached. Add the fact that the Via Valeria, which Strabo (who lived under Augustus) mentions [5.3.11], is older than the Via Sublacensis itself, "first paved under the emperor Nero," as Frontinus describes it [*Aq.* 6.1]. Moreover, the Via Valeria was important among the roads in Italy, namely, that which stretched to the mouth of the Aterno and the Adriatic. From the Via Valeria, rather than from the Sublacensis, which could be called its appendage, it is right that

48. Kircher, 188, 208.
49. Kircher, 176–82.
50. Kircher, 52–54.
51. Holste, 165–66.

the milestones of side roads be counted, from one to the other. Finally, since the figure from the Via Valeria, taken from the crossroad below S. Giorgio to the right, corresponds splendidly, whereas the figure from the Via Sublacensis appears to be twelve stades short, it seems that nothing must be sought further or left to be in doubt.

Last of all, there is added a clear demonstration of the matter, which could not have occurred to Holste, since it came to light just recently, namely, that of a second thirty-eighth columnar milestone of the Via Sublacensis, three miles (less a stade) distant from the intersection of the side road already cited, that which joins the Sublacensis with the Valeria beyond the Anticoli bridge [*CIL* IX, 5971]. Through the location of this newly discovered column, it is clear that this intersection was at the thirty-fifth milestone, and therefore the number thirty-eight does not, as Holste himself wrongly believed, agree with the column at the spring of Sonnoletta, a mile and one-half distant from here over the side road and the Pons Stratonicus (to the extent that we, with Holste, would call that road the name *Via Valeria*).

This new column of the thirty-eighth milestone, about which we are speaking, still stands in the bed itself of the river, fixed on top of its base, receiving the force of the current—the course of the Via Sublacensis has been changed because of flooding—with the following inscription on the upstream side.

> The emperor Caesar Trajan Augustus, son of Nerva, Germanicus Dacicus, pontifex maximus, with tribunician power, imperator for the fourth time, consul for the fifth time, undertook the rebuilding of the road, milestone thirty-eight. [*CIL* IX, 5971]

Evidence of this I freely confess that I owe to Giovanni Battista and Pietro de Massimi, the sons of Fabrizio, a gentleman of the town of Arsoli and one to be counted among the patricians of traditional integrity. They, with that generosity with which they excel, offered themselves as guides of my journey and comrades of labor in freeing the column from the mud and cleaning it, not without danger of the floods of the Anio below us; we were scarcely a palm's length from the river itself, on slippery and wet ground, holding on to the branches of nearby trees.

Therefore, with the Via Valeria and Via Sublacensis having been separated, with the discovery of the side road from the one road to the other taken by Frontinus as designating the Aqua Marcia, and with the column of the thirty-eighth milestone remaining even now, the result is a plain and easy understanding of Frontinus and an appropriate determination both of these aqueducts and of the sources of the Augusta.

c. Sources of the Aqua Claudia

Indeed, if you should look for the sources of the Claudia at the thirty-eighth milestone (now certain), "within three hundred paces to the left," with Frontinus, they cannot be other than those "very extensive and attractive springs" (so one may call them with the same Frontinus) marked nos. 19 and 20 on our map [fig. 17]. I do not know why Holste omitted the second of them (for both are called by the common name *Acqua Serena*).[52] Each is most abundant, and they are 120 paces distant from each other, emerging at the side road to the left, more or less within three hundred paces. It is hardly strange that no further trace of this side road remains, if we consider the huge buildup of that level ground, which extends almost two miles in length and width. As a result, the column, which certainly ought to have stood along the road, now does not emerge from the water except in the great dry season of the year.

Which of the two springs was the Caeruleus and which the Curtius, however, I will leave undecided. The genius of the spring, or the naiads inhabiting it, were not so favorable to me—even though I am most devoted to water and for a long time abstemious, in the same way as Holste, a man from most penitent Germany—that they removed this ambiguity by advising me openly, or from their appearance. Holste indeed identifies the Curtian and Caerulean springs by name,[53] aided by some mark of distinguishing them, which did not at all offer itself to us.

Even so, from the text of Frontinus, it is well enough established that the source of the Marcia was in those most abundant springs under S. Maria in Arsoli [fig. 17, no. 16], at a distance of three miles from the Via Valeria and within a space of two hundred paces to the left from the side road by which a crossing is made from the Via Valeria to the Via Sublacensis. But from careful consideration of another passage of Frontinus, it is understood that the Marcia's sources could scarcely have been fixed on the banks of the Anio, as Holste argues, above those of the Claudia.[54] In his

52. Holste, 130.
53. Holste, 130.
54. Holste, 129.

review of the heights of the conduits, Frontinus states, "The Anio Novus is the highest of all, next the Claudia; the Julia holds third place, the Tepula fourth, then the Marcia, which at its source even matches the elevation of the Claudia" [*Aq.* 18.4]. But if we consider with Holste that the source of the Marcia is at the Forma della Mola [fig. 17, no. 22],[55] what wonder will there be if the Marcia, which Holste makes seven stades higher at its source, in the downward course of a very swift and rapidly falling river, should equal the elevation of the Claudia? The Claudia is set below it and issues in a low place between the modern road—which goes back a bit to the left, as the position of the column teaches us—and the river itself.

Yet the location of our source of the Marcia, although two miles and more below the sources of the Claudia, actually equals its elevation, as we learned from the plumb line, since the valley in which it issues clearly has a steep downward slope. It could therefore have been drawn from the same elevation as the Claudia, "had not the older engineers," as Frontinus adds at once, "laid out their aqueducts at a lower elevation, either because the fine points of leveling had not yet been ascertained or because they deliberately made it their practice to bury aqueducts underground to prevent them from being cut easily by enemies, since a good many wars were still being fought against the Italians" [*Aq.* 18.4]. How accurately, however, Frontinus noted this old ignorance of leveling can be easily seen in the conduit of the Marcia itself, near the city. In the place where its structure on arcade begins, near the Via Latina around the fifth milestone from Rome, when the Marcia is compared with the conduit of the Claudia, it is found to be twenty feet below it. And afterward, at the Porta Maggiore, it is pushed down below it by a good twenty-five and a half feet, as is clear from our first dissertation to Lucio, from the cross section of the conduits [fig. 7], unless the downward slope after its settling tank contributed something to so great a fall in elevation.

d. Representations of the Aqua Claudia

What if the following image [fig. 19] were to show to us dramatically the low level of the Claudia, which we have said arises in the lowest plain, at the foot of the Simbruine mountains? Among the ancients, you would find nothing empty and without some significance in monuments of this kind.

55. Holste, 130.

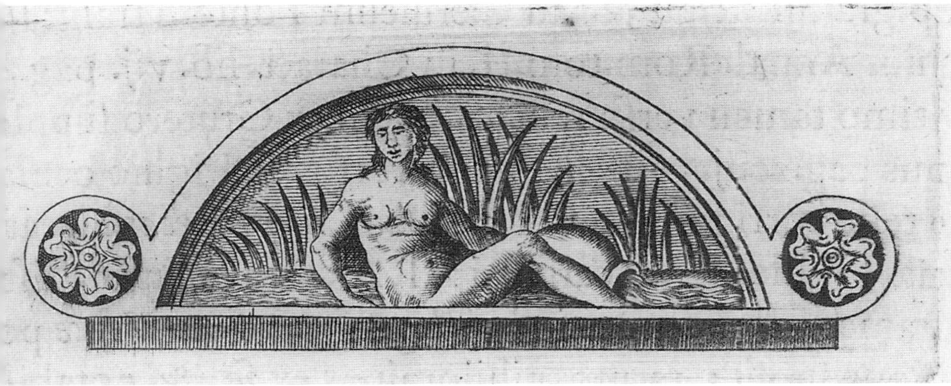

Fig. 19. Possible representation of the Aqua Claudia

Not below it "do laughing waters leap down" [Hor. *Carm.* 3.13.15–16], nor does it cast down its stream, as other divinities of nymphs and rivers do. Instead, the urn, as if dipped in water sluggish and standing, seems to drink in the liquid to convey it somewhere else, rather than to pour it forth. The relief also seems to make clear reference to the following description on a very ancient altar dedicated to these Caerulean nymphs—for the Caeruleus (which means *cyaneus*, or "dark blue," in Greek and poetically), as we saw, was one of the chief sources of the Claudia—that of the same water rising and falling into itself, in these verses:

> To the Cyanean waves, Contuccius has made a pleasing gift, an altar to be revered in return for soothing waters. From here, the flowing water arises and continues, then will fall into itself, giving thanks to the nymphs who renew the splendor of the green bank with their power born from the springs. It makes safe its comrades and refreshes them with its water. [*CIL* VI, 555]

Pighius cites the epigram and connects it with the Caerulean spring, yet with the last verse omitted,[56] which we supply from Gruter;[57] however, we correct Gruter from Pighius's text that we have consulted, when Gruter calls a "marble tablet" that which Pighius and the poem itself call an altar.

56. S. V. Pighius, *Annales Romanorum* (Antwerp, 1599–1615), 2:4.
57. Gruter, xciii.11.

Nevertheless, even if there were nothing unusual in this relief and the position of its urn, this very image is still rare and noteworthy in its depiction of the Aqua Claudia itself, through the inscription presented below it and reported here.

> To the spirits of the dead, to Clemens, imperial slave, *castellarius* [reservoir man] of the Aqua Claudia, Claudia Sabbathis made it for herself and her own. [*CIL* VI, 8494]

The following stone tablet from your collection will provide another likeness of the same Aqua Claudia, indeed carved artlessly but not to be overlooked, now set forth in its entirety for the awareness of the learned public.

> To the spirits of the dead, Sabbio, imperial bailiff of the Aqua Claudia, made this for himself and Fabia Verecunda, his most upright wife, with whom he lived twenty-four years, and his freedmen and freedwomen and his *vikarii* and the descendants of them all, in half part his own.

> To the spirits of the dead, Sporus, imperial bailiff of the Aqua Claudia, made this for himself and Claudia Hermione, his most upright wife, and for his freedmen and freedwomen and his *vikarii* and the descendants of them all, in half part his own. [*CIL* VI, 8495]

Here, you see evidence of something new, the duty of *vilicus*, or bailiff—not known to Gruter and different from that of *castellarius* cited earlier—which, according to Frontinus [*Aq.* 105.3], belonged to the gang assigned to guard the conduits. Moreover, there is included a designation of *vicarii vilicorum*, or deputy bailiffs, not mentioned by Frontinus himself. Each term indeed differs from the correct spelling: in *vikarius*, there is a clear misuse of the letter *k* for *c*, which Gruter, in his grammatical index, has noted in many other words;[58] [there is a misspelling] in *vilicus*, even though Varro [*Rust.* 1.2.14] tells us that this noun is to be written with a double *l*, when he derives its etymology from *villa*. Nevertheless, I do not know why the word is found more frequently in inscriptions with a single *l*, as it is here, both in Gruter[59] and in these inscriptions that follow. The

58. Gruter, "Totius indicis harum inscriptionum conspectus," chapter xix, letter K.
59. Gruter, lxxix.4 (= *CIL* VI, 203), cvii.9 (= *CIL* X, 1561), cccxxxix.5 (= *CIL* VI, 10046).

first of these is among my unpublished inscriptions; the second, Dausque reports as *villicus*.[60]

> Sacred to Silvanus, the slave Speratus, bailiff of the emperor Caesar Hadrian Augustus. [*CIL* VI, 619]

> Sacred to Silvanus, Onesimus the bailiff l.v.p. [*CIL* VI, 31010]

Here, Dausque makes an effort to prove that this word from the beginning demands one *l*, perhaps more cleverly than accurately.

Still rarer is the representation of a solitary image of a reclining nymph that we have introduced, because we see that ancient reliefs and paintings always show them in groups of three. I have collected into one place examples of this phenomenon, observed by me and new, as I think, beyond those that are obvious everywhere.

The first will be this one, which I have transferred here from your *Antiquitatum recollectiones* [fig. 20]:

> To the divine power of the nymphs, the *aquarius* [waterman] Augustalis, freedman of the Augusti. [*CIL* VI, 30791 = 547]

Its original, said to have been at Naples, in the home of Adriano Guglielmo, is conspicuous in every detail: first, from the waterman sacrificing at a crowned altar, with his head indeed uncovered, as a result of which the sacrifice is being made to divinities in the middle rank and common, as I have proved was the ritual in my study of Trajan's Column;[61] next, from the relief of the snake, that very well known symbol of health in imperial coinage, through which well-being is suggested, as is read in this inscription:

> Sacred to the health-bringing nymphs, M. Lucilius Lucilianus, of the College of the Augustales, for his own health and that of Lucius Antistius Onesimus, of the College of the Augustales, paid his vow freely and deservedly. [*CIL* VI, 31010]

Similar to this, another spring is celebrated for "curing diseases" [*CIL* VI, 61*]. Finally, there is the ornamentation of the broad-leaved plant on the

60. C. Dausque, S.J., *Ortografia Latini sermonis vetus et nova* (Paris, 1677), 330.
61. Fabretti, *De columna Traiani syntagma*, 169.

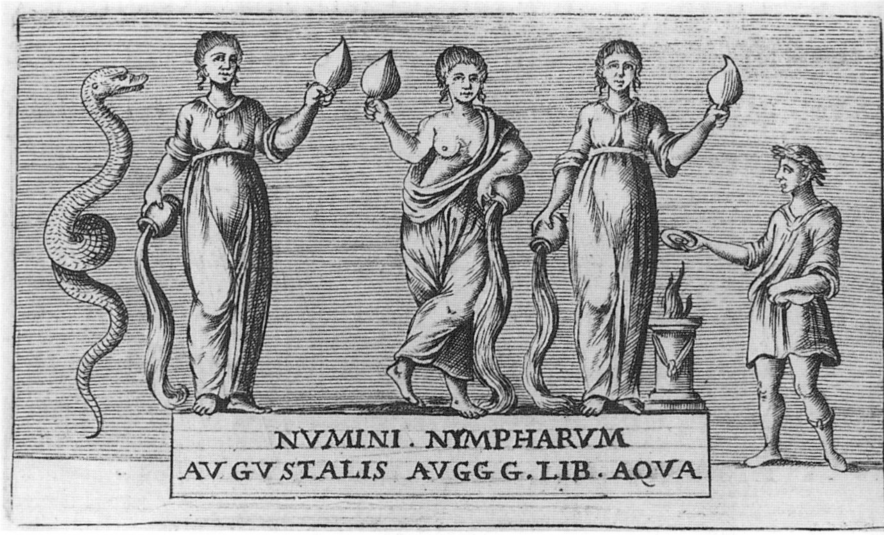

Fig. 20. Votive relief to the nymphs [*CIL* VI, 30791 = 547]

relief, for the reason attested by Pliny in his description of it: "According to tradition, Nymphaea was born of a nymph who died of jealousy toward Hercules, wherefore some call it [the plant] Heracleon; it grows in watery spots, with large leaves on the top of the water and others coming forth at the root" [*HN* 25.75–76].

I have observed another example, which follows, and I took it from a *cippus* of two-palm length, inserted in the wall of a certain vineyard on a path that leads from the Porta S. Paolo to the Via Appia. The marble is in very poor condition, as you see [fig. 21]:

> . . . [ ] ebius Eutyches, freedman, made . . . [*CIL* VI, 554]

However, we prefer to render it faithfully, rather than, as others do with coins and reliefs, to deceive those absent by misrepresentation. When scholars attempt to make corrections, they frequently corrupt the ancient evidence. Indeed, most accurately does Ianus Gruter charge in his *Corrigenda* that license, even to the smallest degree, goes badly in these matters.[62]

62. Gruter, "Corrigenda et animadvertenda toto opere inscriptionum."

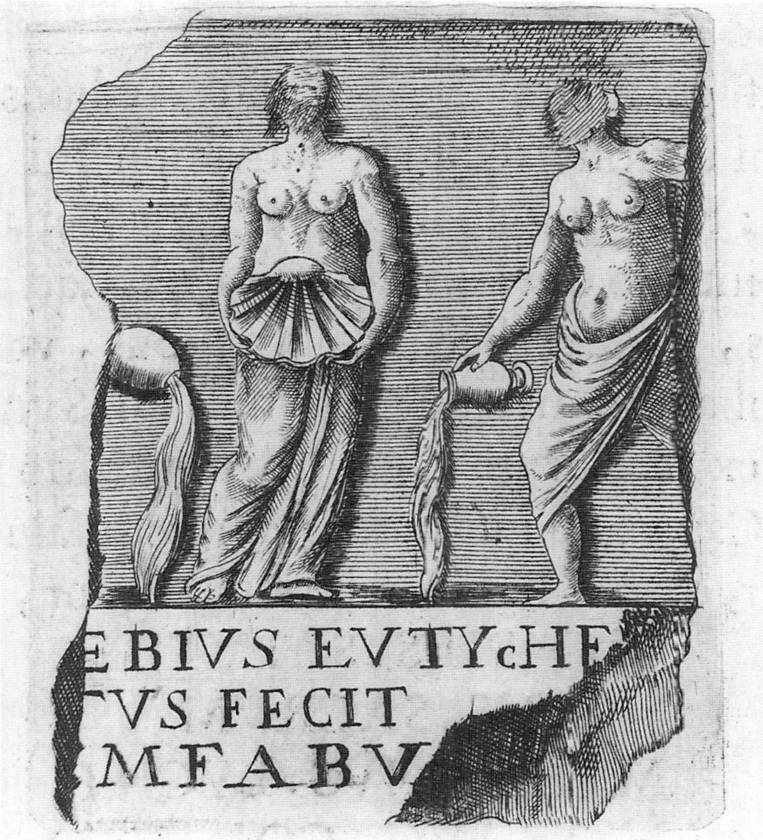

Fig. 21. Fragmentary votive relief to the nymphs [*CIL* VI, 554]

This votive tablet [fig. 22], which is seen in the Orti Mattei, furnishes a third example:

> Titus Claudius Asclepiades and Caecilius Asclepiades, from a vow, gave and dedicated to the nymphs. [*CIL* VI, 549]

On this, in addition to those details I have set forth to be noted especially concerning the threefold number of nymphs, that common cult of Silvanus and Hercules recurs, cited in one and the same temple of Region XIII, according to Victor (= Graevius, 3:109) and five times illustrated in

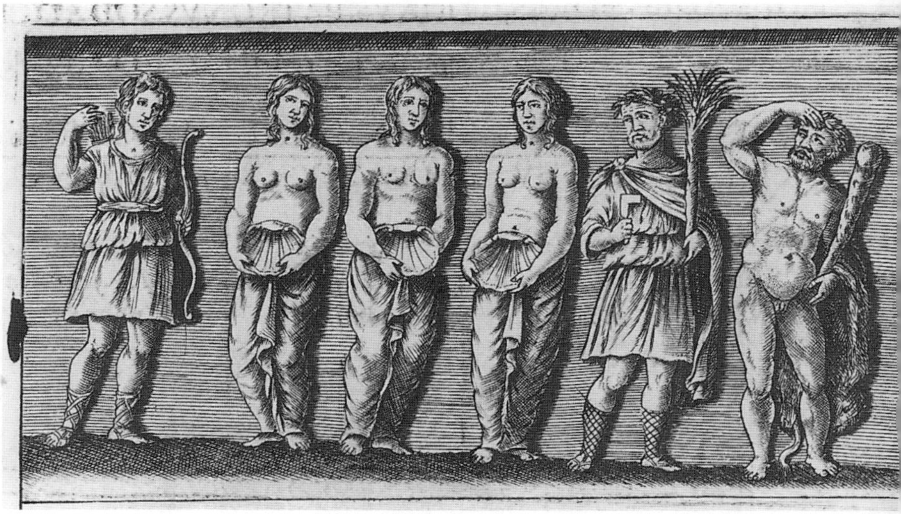

Fig. 22. Votive relief to the nymphs, Silvanus, and Hercules [*CIL* VI, 549]

Gruter;[63] from a common epithet, we will connect this very cult with a new and tighter bond.

Just as we have seen, up to this point, the title *Pollentis* given only to Hercules [*CIL* VI, 328], so the following fragment of a recently discovered inscription in my collection shows that both the epithet *Pollentis* and (that which is synonymous) *Valens* or *Valentius* was attributed also to Silvanus. I have divided this inscription into two parts, so that I might accommodate myself to the space.

> C. Julius Helpidephorus Cyrinus, patron of the association of the powerful god Silvanus, since he himself brought for construction two thousand sesterces with those who had planned and dedicated the temple from the ground up, from the Clymbim . . . (There follow ninety-nine names, from three different decuries.) [*CIL* VI, 647]

There is also another inscription on the Via Labicana, in the vicinity of Torrenova.

63. Gruter, xlii.4 (= *CIL* VI, 367*), xlii.5 (= *CIL* VI, 297), xlii.6 (= *CIL* VI, 296).

> To Silvanus Valentius, A. Flutius Athenaeus, in free payment of his vow, gave and dedicated. [*CIL* VI, 698]

Another example is unearthed from a stone.

> Batinia Priscilla dedicated this, sacred to the nymphs. [*CIL* VI, 548(1)]

On it, below this inscription, were erected statues, not, indeed, of the nymphs, but of the three Graces. From this, it will have been sufficient for our present purpose to infer our threefold number, indicated for how many other divinities. In summarizing the very many things that might be said about the association of the Graces and the nymphs, scholars have cited Horace [*Carm.* 1.4.6, 1.30.6, 4.7.5].

There is, in addition, the same triple number contained in an ancient painting that was recently moved from the tomb of the Nasonii, at the fifth milestone of the Via Flaminia, to the Orti Alterii. In it, nymphs, perhaps the Hippocrenides, are caressing Pegasus, the producer of the stream itself or of the inhabitants of the Tritonian swamp (in which Pegasus is said to have been born), by pouring water on him or patting him with their hands [fig. 23].

Having made a transition from marble reliefs to paintings, we shall present for examination this other portrait, gleaming not with fleeting and short-lived colors but with the tint of gems themselves, again with three nymphs, from Theocritus [*Id.* 13.36–45]:

> And fair Hylas departed, to bring water for dinner, holding a golden vessel. But at once, he noticed the spring in a low place; there were many leaves around it, as well as dark blue celandine, green maidenhair, sprouting parsley, and curly dog's tooth grass. In the midst of the water, however, the nymphs were holding a chorus, watchful nymphs, divinities to be feared by rustics—Eunica, Malis, and Nycheia, the friend of springtime.

Our observation is not refuted by a passage of Pausanias concerning Elis [6.22.7], where he says that near the River Cytherus was erected a temple of four nymphs, who were called the Ionides, namely, Calliphaea, Synallaxis, Pigaea, and Iasis. Although we would confess that those Ionides were water nymphs, for that reason, accordingly and in order, the name *nymphs* would be suitable for goddesses who are protectors of waters, as Vossius, a

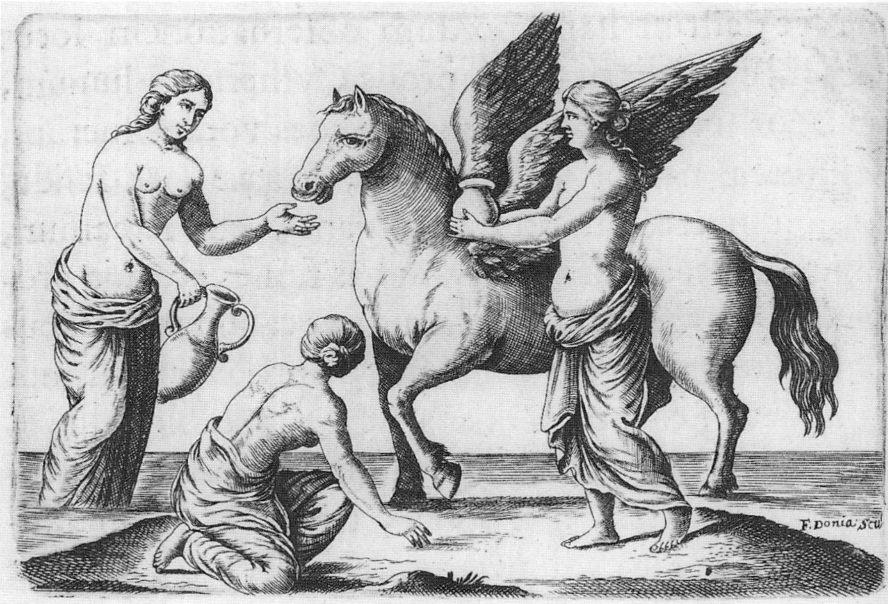

Fig. 23. Tomb of the Nasonii: painting of Pegasus and the nymphs

most illustrious author of this generation, proves.[64] Nevertheless, to support the agreement of so many monuments, it can be argued that some heroine was joined to the three nymphs and, as it were, added on to them, in the same manner of Greek flattery on a tablet, a copy of which we give from the original [fig. 24]:

To the four sisters. [*CIL* VI, 10036]

Here, with the three Graces embracing themselves according to custom, a fourth is inserted in their chorus, just as if she were worthy to increase the number of the Graces and to be worshiped in common with them. I think that she is a new bride, because of the twin birds playing in the pediment, not noticed by Smet,[65] and because of the veil with which she modestly

64. G. J. Vossius, *De origine et progressu idololatriae* (Amsterdam, 1668), 1:255.

65. M. Smet, *Inscriptionum antiquarum quae passim in Europam liber* (Leiden, 1588), fol. xxxii.8.

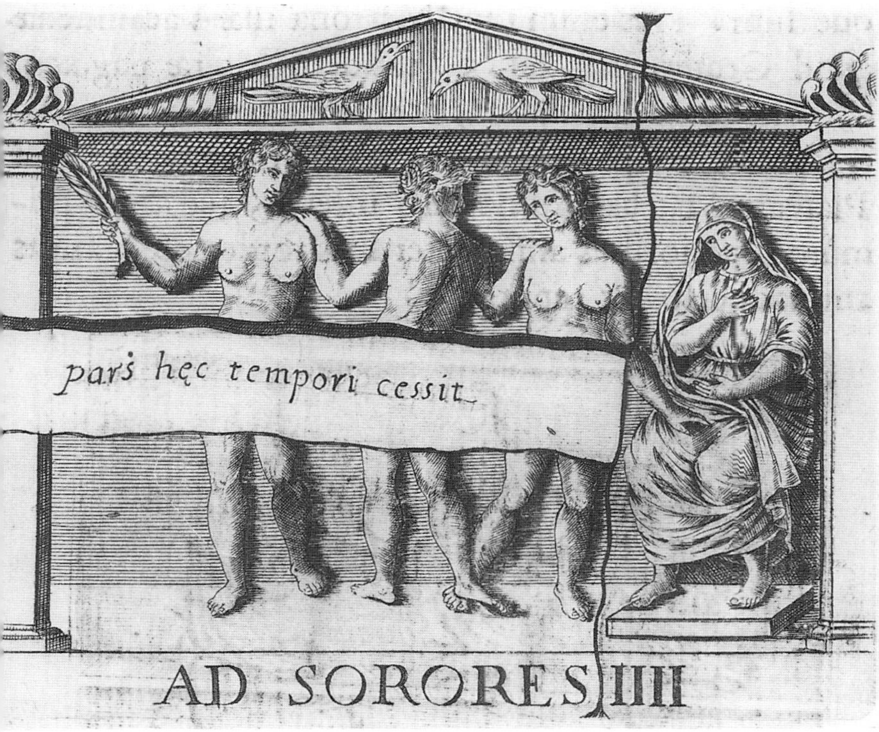

Fig. 24. Funerary relief depicting the three Graces [*CIL* VI, 10036]

covers and veils (*obnuit*, from which the word *nuptials* is derived) her head. Similarly, the inscription of this same relief is to be explained by these verses of Ausonius (*Epigrams* 23.13 Peiper):

> There had been three Graces, but as long as my Lesbia lived, four; when she perished, they are numbered three, just as before.

Not only among water nymphs, but in other rustic divinities of the same sex, I have noted that the same practice of a threefold number flourishes and have wondered at it. For there are those three Vacallinehae Matronae [*CIL* XIII, 7952], the same number of Mairae [*CIL* XIII, 4303], and the same number of Suleviae and Campestres on the following relief, for some time now inserted in a wall in the courtyard of a private house in

Piazza Mattei, but noticed by me and cleaned off for the first time amid its squalor and cobwebs [fig. 25]:

> Sacred to the Suleviae and Campestres. L. Aurelius Quintus, centurion of the seventh legion Gemina, happily and freely [*LAETVS LIBENS*] paid his vow on August 24th, in the consulship of Bradua and Barus. [*CIL* VI, 768 = *ILS* 4776]

I have brought to light this relief but am not confident that I can shed some light on the obscurity of that word *Suleviae*—indeed, not at all of corrupted Latinity, since the consulship of Appius Annius Bradua and Titus Vibius Barus is fixed in the 912th year of the city in the *Fasti* (A.D. 160). Accordingly, I will concern myself with other noteworthy things in this relief. Among them, there appears a boar girded with a lustral garland, just as Trajan's Column, on its first Suovetaurilia, shows this victim with this same adornment; afterward, it attributed to it a fillet, or you might call it a sash, as a substitute for a garland of this sort. Writing about Trajan's Column, I have proved that among the ancients, the use of now a crown, now a garland, both in victims as in sacrificers, was indeed common;[66] here, too, the centurion, indicated by his own staff of vine wood, no less than by the title of the inscription, is distinguished by the fillet or sash hanging down from his left shoulder. But what makes this inscription especially noteworthy is the formula "LAETVS LIBENS." From it, students of abbreviations have finally understood *L.L.* It was interpreted as *libens libens* or *libentissime* previously and—what surprises me more—even after the inscription was seen in Gruter,[67] where the phrase "LAET.LIB.," slightly less extended than in this inscription of ours, seemed to lead directly to the true understanding of the abbreviation.

Why go on at great length? Might we be able to say in one word that almost all examples of the feminine sex that have a number greater than one, whether divinities or monsters, unfavorable or favorable, are included in this threefold number? Ancient mythology has indeed introduced such creations as the Gorgons, the Graiai, the Parcae and the Fates (for Ausonius [*Griphus ternarii numeri* 19] distinguishes each of the two), the Sirens, the Harpies, the Hesperides, the Stymphalides, the Eumenides, the

66. Fabretti, *De columna Traiani syntagma*, 162.
67. Gruter, mlxxiv.4 (= *CIL* V, 5502).

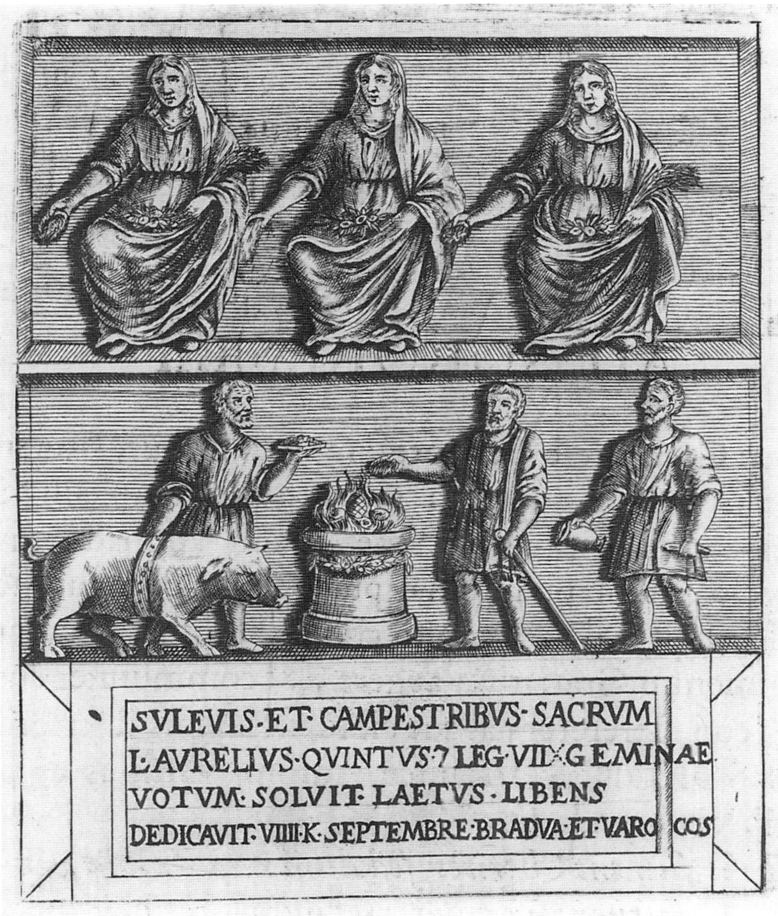

Fig. 25. Votive relief to the Suleviae [*CIL* VI, 768]

Graces, the Sibyls (if we again believe Ausonius [*Griphus ternarii numeri* 85]), and the Muses themselves. The Muses, I said, are three in number, according to the opinion of more ancient sources, concerning which there is ample witness from Plutarch [*Quaest. conv.* 9.14.744C–D]; Pausanias [9.29.2], on Boeotia; Varro, cited by Servius [at *Eclog.* 7.21] ("Nymphs of Libethra, our love"); Ephorus, cited by Arnobius [*Adv. nat.* 3.37]; and, with him, Augustine [*De doctrina Christiana* 2.68]. Indeed, the "statue of Apollo dedicated on Delos, with that arrangement, that he holds the bow with his right hand, the Graces with his left, although each one is grasping some musical instrument (for one holds the lyre, another the flute, and the one in the middle holds a pipe brought up to her mouth," concerning which Plutarch speaks [[*De mus.*] 1136A], we shall argue has confirmed the number of three Muses, from the instruments appropriate to each but unsuitable for the Graces. We will declare that Ausonius, already cited, made reference to this same statue when he sang, "The three alone, whom once the right hand of Phoebus held . . ." [*Griphus ternarii numeri*, 31].

But at long last, while we, "as fashioner of a crown, owed to the nymphs" [Hor. *Carm.* 3.27.30], around the road have deliberately gathered flowers for the nymphs, the Muses, and the others at their Caballine spring, I seem to have strayed from the road itself. I shall seek again our flowing waters around the Anio, which we were surveying.

e. Source of the Aqua Marcia

Because of its proximity to the location of the Marcia's source, it is probable that the fragment of a stone with the inscription

> With our lords safe [SALVIS DD. NOSTRIS] . . . Perpet [ ] . . . the arcade . . . [*CIL* IX, 4051.6]

in the altar of the nearby Church of S. Maria in Arsoli makes reference to this conduit of the Marcia and its reconstruction. The formula "SALVIS DD. NOSTRIS" is seen to be peculiar to Honorius and Theodosius.[68] Moreover, Honorius, who ruled in the west, in the same way he issued a law concerning the Aqua Claudia along with Arcadius [*Cod. Theod.* 15.2.2], perhaps in like manner undertook with Theodosius II some work on the Aqua Marcia here.

This Aqua Marcia then, "the clearest," as Pliny says, "of all the aqueducts in the whole world, with the first prize for chilliness and healthfulness, by proclamation of the city" [*HN* 31.41], and likewise preferred to all others for mixing with wine, as Tibullus [3.7.58] reports ("May the water of the Marcia temper your ancient wine"), Gerolamo Mercuriale unfairly censures, with all the other waters of that region. Relying on the authority of Galen, Mercuriale denies that the Marcia was once in use for sustenance

68. See Gruter, clxx.5 (= *CIL* VI, 31911 = 1718[2]), cclxxvi.7 (= *CIL* VI, 1659).

and drink: "Although by no means easily do I believe that those waters, which were brought to Rome from the Tiburtine mountains, were much in use as sustenance or drinking, because Galen, the most careful observer of all conditions at Rome, asserts that all of them were raw and were unsuitable for cooking food, adding that that city, like other excellent gifts, so had very many and very beautiful springs, from which the sick as well as the healthy might drink."[69] These things Galen does indeed say. But Mercuriale does not face up to the brilliant testimony about the good quality of the Marcia from Pliny and Frontinus, who states that its water was entirely assigned for drinking [*Aq.* 92], and likewise that of Arrian, who seems to prefer it to the most celebrated water of Dirce near Thebes: "Is the Marcia's water worse than that of Dirce?" [*Epict. diss.* 2.16.31].

Mercuriale, however, has as his opponent and adversary Galen himself. In the middle of common praises for all the city aqueducts, from which neither his intention nor his words permit the Marcia to be excluded, Galen mixes in some trifling matter pertinent to these aqueducts from Tiburtine territory that seems to indicate not a fault but high quality. However, let Galen himself be heard in the passage cited by Mercuriale: "At Rome, just as there are many other extraordinary things in that city, so there is a remarkable elegance and multitude of springs, with none of them giving forth water foul or polluted and muddy or harsh and raw, just as neither at Pergamum in our country. In many other cities, however, by no means few corrupted waters are found. Indeed, those waters, indeed, that are brought down from the Tiburtine mountains through stone pipes in the Roman city without other faults are nevertheless somewhat raw, with the result that they do not become heated rapidly, like the springs of the city, nor are they chilled, nor in them, as in the waters of the springs, will whatever beans, vegetables, and meats you have thrown in be cooked quickly. Some of these waters, however, are much more raw in some places, others in other places, and they are called harsh and heavy by the inhabitants themselves, since they block the stomach and since many people experience a certain heaviness from drinking them" [*Commentary on Epidemics* 4.4 (XVIIB, p. 159 ed. Kühn)].

Indeed, these things that Galen has said in the words cited above against the waters of that region certainly do not pertain to the Marcia and

---

69. G. Mercuriale, *Variarum lectionum in medicinae scriptoribus et aliis, libri quinque* (Paris, 1585), 98.

the other aqueducts brought to Rome, as perhaps Mercuriale understood. Otherwise, Galen, the most eloquent of physicians, would almost be laboring in a contradiction in practically the same phrase; he allows to issue forth in those first words "no water at Rome that is polluted and muddy, or harsh and raw," words to which Mercuriale's paraphrase is diametrically opposed.

But now it is the time that we bring forth a conjecture, new and unusual, by rarity of argument, if we are not mistaken, through which we may fix our sources of the Marcia in the place already mentioned and prove that the reading we have given to Frontinus agrees with the epigraphical evidence. Where, therefore, the side road, proceeding from the Via Valeria toward the south, runs into the Via Sublacensis on a straight course—for another road, on which stands the Pons Scutonicus, bends off to the right from this side road toward the southwest, perhaps opened for the sake of visiting the Aqua Marcia and its sources, which it looks on directly—vaulted ruins of an ancient structure are seen [fig. 17, no. 16]. Among these ruins, when a certain farmer, who had run on it plowing, showed us, we discovered both a togate statue and a marble tablet inscribed as follows:

> The emperor Caesar Augustus, son of a god, by decree of the Senate, 1242, 240 feet. [*CIL* VI, 1251b]

It is already well known that this formula of inscriptions, defining 240 feet in its numbering, pertains to aqueducts, since in other inscriptions, which we will now cite, the name of the aqueduct, to which each one was connected, had been expressed—to the first two, namely, that of the Virgo; to the third, those of the Julia, Tepula, and Marcia.

> The Aqua Virgo. Tiberius Caesar Augustus, pontifex maximus, in the thirty-eighth year of tribunician power, consul for the fifth time, imperator for the eighth time. I. 240 feet. [*CIL* VI, 1253b]

> The Aqua Virgo. Tiberius Claudius Caesar, son of Drusus, Augustus Germanicus, pontifex maximus, in the fourth year of tribunician power, consul for the third time, imperator for the eighth time, father of his country. I. 240 feet. [*CIL* VI, 1254]

(Chifflet has shown these two inscriptions in corrupted form in his treatise on the Aqua Virgo, for he has misrepresented the final number 240 as 211.[70] I wanted this to serve as a warning, so that all the force not be removed from inscriptions; for them, like the Scorpion, as you will see, the power is in the tail.)

> The Aqua Julia, Tepula, Marcia. The emperor Caesar Augustus son of a god, by decree of the Senate, 25, 240 feet. [*CIL* VI, 31561c = 1249b]

Ligorio, who perhaps had seen one of the preceding inscriptions but had not understood it, passed over it in silence. But as he was an imitator of scholars, he made up the following inscription out of his head, so that he might in turn set a puzzle for learned men and show at the same time that he was in no way bereft of a literary secret.

> For the Aqua Julia [and] Tepula, the emperor Caesar Augustus, son of a god, pontifex maximus, consul for the twelfth time, in the nineteenth year of tribunician power, imperator for the thirteenth time, with the curule aedile Marcus Vipsanius Agrippa as administrator, 950 feet, tenth milestone. [*CIL* VI, 800*]

Ligorio said this inscription stood on a *cippus* of travertine found on the Via Latina, on the third milestone from the city. But the more monuments he attempts to stand on, the more he reveals his fraud and counterfeiting. The third milestone of the Via Latina, around which Ligorio claims that the *cippus* had been found, could not have been marked with the abbreviation of the tenth milestone if it designates a spot distant not from the sources (as we shall see afterward) but from its distribution in the city. Moreover, inside the seventh milestone from the city, the Julia and Tepula were joined with the Marcia, resting, as it were, on substructures of the Marcia above ground and its arcade; therefore, at the third or fourth milestone, these first two lines could not be named separately from the other.

There is, in addition, the spurious citation of the aedileship, so that Ligorio might seem to agree with Pliny [*HN* 36.24], who speaks about

---

70. J.-J. Chifflet, *Aqua Virgo, fons Romae celeberrimus, et prisca religione sacer: Opus aedilitatis M. Agrippae in vetere annulari gemma* (Antwerp, 1657) (= Graevius, 4:1791F–92A).

works of this sort completed in Agrippa's aedileship. But in this, too, "he [Ligorio] did not cook up credibility, and he revealed himself abandoned, both by the suitable foresight of old-fashioned learning and skill of judgment," as Exechial Spanheim elegantly puts forth, among other remarks about him.[71] Agrippa's aedileship, according to Frontinus [*Aq.* 9.1], occurred "after his first consulship, when the emperor Caesar Augustus and Marcus Laelius Volcatius were consuls," which was the year of Augustus's second consulship, as all the *Fasti* agree. From this, it is obvious how ignorantly that aedileship is linked with the twelfth consulship of Augustus, undertaken twenty-eight years later, at a time, moreover, when Agrippa had died: from Dio Cassius [54.28], Panvinio relates his death to the twelfth year of Augustus's tribunician power, that is, the seventh year before this public office.[72]

After these errors so thick and weighty, Ligorio's fault will be minor in listing the acclamation of imperator as the thirteenth, since he links this acclamation with Augustus's twelfth consulship; coins everywhere have it as the fourteenth, and it is also linked with the eleventh consulship and the sixteenth year of tribunician power by Occo.[73] Likewise, there is another error in expressing the family name of Marcus Vipsanius, never taken in public monuments or in many coins of Agrippa himself, as if it made reference to a family of which he was ashamed. Concerning this fact, Seneca [*Controv.* 2.4.13] gives credibility in these words: "Agrippa had been a Vipsanius, and he had dropped the name, as if proof of his father's humble station, and was called Marcus Agrippa. When he was defending a client, there was a prosecutor who said, 'Marcus Agrippa and that which is in between,' intending that *Vipsanius* be understood."

After we have therefore removed this small stumbling block of a fictitious and uninformed inscription—perhaps at greater length than was necessary—you can recall, Your Eminence, that we observed another in the vineyard of Bartolomeo Virgilio, at two miles, less a stade, from the Porta Maggiore, lying between the ruins of the arcades of the Marcia and Claudia. This inscription, with the owner's permission, has been added to your well-equipped collection. I have reviewed the place from which it

71. Spanheim, *Dissertationes de praestantia*, 141.

72. O. Panvinio, *Fastorum libri V a Romolo rege usque ad imp. Caesarem Carolum V* (Venice, 1557), 296 "*In Fastos Consulares appendix, 24; Commentarius in Librum II. Fastorum.*"

73. A. Occo, *Imperatorum romanorum numismata ab Adolfo Occone olim congesta, illustrata a Francisco Mediobarba Birago et expurgata curante Philippo Argelato* (Milan, 1730), 38.

was excavated in detail and not in vain (as you will soon see). So it presents itself:

> The Aqua Julia, Tepula, Marcia. The emperor Augustus Caesar, son of a god, by decree of the Senate, 63, 240 feet. [*CIL* VI, 31561g = 1249f]

Indeed, that ritual formula of 240 feet in all these inscriptions—you have feet expressly stated in the third one [*CIL* VI, 31561c = 1249b], whether of two land measures doubled or its equivalent, the measure of one *iugerum* [Roman acre] lengthwise—now leads me to believe that the ancient curators of aqueducts were accustomed to measure the lengths of the conduits of each line by that standard. We find this most consistent in this first of the two earlier inscriptions, discovered in the very first *iugerum* of the Aqua Virgo's distribution, namely, below the Church of S. Trinità dei Monti, where Luca Peto recognizes the terminus of that aqueduct (he also presents the same inscriptions in his book *On the Restoration of the Aqua Virgo*).[74]

There is also additional confirmation in the *cippus* just now presented, at a distance of sixty-three *iugera*, that is, 3,024 paces. At this distance, the place of the Marcia's distribution ought to be removed from the spot where the *cippus* was found, a place that, as we know from Frontinus [*Aq.* 19.4], was at the Porta Viminalis—namely, in the middle of the *agger* of Tarquinius, where ancient writers place it—where we also see the conduit of the Marcia itself was directed. After the arch at the Porta di S. Lorenzo, the Marcia bends off to the left and leaves the modern walls of the city to the right. Since there are twenty-four hundred paces from the spot where the *cippus* lay, along the line of conduits, up to the Porta S. Lorenzo, the remaining 624 paces for completion of the sixty-three *iugera*—to the extent that we could measure them from a distance, not, indeed, exactly, on account of the obstacles of vineyards and the bending of the conduits themselves—will indicate the site of the ancient Porta Viminalis in the Villa Peretti, through which part of the *agger* used to run. This will be the result if a line has been extended from the straight course of the Via Collatina. This road could not go out anywhere else than from this Porta Viminalis, and as we have said elsewhere, it ran along the sides of the arch of

74. Peto, *De restitutione*, 119.

the Porta S. Lorenzo slanting off to the right, as can be seen in the plan presented by us elsewhere [fig. 14].

As a result, we would not hesitate to affirm that those 1,242 *iugera* of our inscription indicate the very spot on which Augustus erected his monument and agree very well with the measurement of Frontinus, since 1,242 *iugera* multiplied by 240 feet (or forty-eight geometric paces) produce 59,616 paces. Frontinus [*Aq.* 7.6], in like manner, also assigns to the entire conduit of the Marcia 60,710 and a half paces. When there is added the distance of this boundary stone from the sources themselves, from where Frontinus takes the limit of his measurement, and when the curvature of the conduit through that valley has been taken into account, this difference of approximately one thousand paces must be supplemented and made equal. As a result, we can rightly and deservedly say that the mile measurement of Frontinus agrees exactly with our measurement of the *iugera*.

Why, however, no mention of this measurement is made in Frontinus, who wrote so accurately about the aqueducts, results, indeed, I think, from this: in his time, that type of measurement remained uncertain and misleading. Indeed, as Frontinus himself observes [*Aq.* 18.5], "Yet now in certain places where the conduit has collapsed from old age, with an underground circuitous route avoided, the stretch of a valley is crossed to shorten their length with substructures and arcades." As a result, with the circuitous course of the conduits shortened, there is no wonder that the old measurement by *iugera* became obsolete.

Moreover, when the two other aqueducts of the Anio Novus and the Claudia were added after Augustus, this measurement by *iugera,* as well as the erection of so many boundary markers and *cippi,* could have seemed too troublesome to the superintendents of the water system under Claudius and his successors. Indeed, Frontinus, in his performance of office, employed another easier method to indicate and delineate all the conduits, as he himself has written [*Aq.* 17.1–4]: "Nor has it seemed unfitting to me to include as well a description of the lengths of the courses of each aqueduct, according to the classifications of construction. Since the greatest part of this duty lies in maintenance of the lines, the person in charge must know what things demand greater outlays. My sense of responsibility has not been satisfied with personal examination of particular items, but we have also taken care to prepare maps of the lines"—understand a trace or a cross section, with language used in agriculture—

"from which it is clear where there are valleys and how great they are, where rivers are crossed, where the channels attached to the sides of mountains demand greater and constant concern for guarding and strengthening. As a result, we gain the advantage of being able to deal with a matter at once, as if under our own eyes, and make a decision as though we were on the spot." So [writes] Frontinus.

f. Source of the Aqua Augusta

Now that the channel of the Marcia has become known to us below the ruins mentioned earlier, the source of the Augusta can no longer lie hidden. It will be without a doubt at a distance of eight hundred paces "from the channel of the Marcia"—not from its source, for so Frontinus [*Aq.* 12.2] has indicated—at the Lake of S. Lucia, where remains of a substructure appear. It is, however, in the vicinity of 225 paces of the other spring [fig. 17, no. 17], around which there still stand traces of ancient construction not to be scorned. Because the spring and that conduit would have otherwise cut across and blocked the conduit of the aqueduct of S. Lucia, I believe that the water of that spring was also tapped in the Augusta. Moving the source of the Aqua Augusta to this position is quite consistent with another passage of Frontinus [*Aq.* 14.3], which reports that the Augusta was made available to each aqueduct, the Claudia and the Marcia; this could best be done if the Augusta stood in the middle between the same lines.

g. Origin of the Aqua Marcia

Holste correctly reckons among impossibilities the source that Pliny attributes to the Marcia. Pliny reports that "the same aqueduct from the Fons Piconia in the farthest mountains of the Paelignians crosses the territory of the Marsi and the Fucine Lake" [*HN* 31.41], from which [report] Statius [*Silv.* 1.5.16] has borrowed the following as his own about the Baths of Etruscus: ". . . the Marcia, drawing the Marsian snows and chillness . . ." I second his words: "I think," Holste says, "that this story is to be held in no other place than the secret passage of the lover Alpheus."[75]

75. Holste, 146.

Georg Fabricius has certainly tried to lessen the improbability of this, but with a farfetched, unfortunate, and embarrassing argument, when he converts the Fucine Lake into a river and makes up a crossing in lead pipes.[76] The same author, nonetheless, proudly makes such promises about himself in his preface: "Indeed, even though this is a bit arrogant, yet, since it is true, I will say it; I can refute the various errors of others who have recently written about the city through the sure testimony of ancient writers."[77] Nevertheless, whether the opinion of Pliny is true or improbable, it is all the same for us. We do not conduct our water inspections with the eyes of Lynceus in the bowels of the earth, but we have explained the place where, again with the same Pliny, "the Marcia reveals itself in Tiburtine territory" and from where it was brought to Rome, in agreement with the thinking of Frontinus.

5. SETTLING TANKS

a. The "Villa delle Vignacce"

It will not be inappropriate here to report something about a settling tank of this Aqua Marcia, which I indicated with hesitation on the topographical map of my first dissertation [fig. 1, no. 27], even if with the same qualification. On the Via Latina itself, which advances to that place shortly after the fourth milestone, between the arches of the Marcia and the Claudia, on the property "dello Spedaletto" and near the ruins of several structures—which today, because of their great number, are commonly called "le Cento Celle;" from their location, we believe them to be the remains of the Pagus Lemonius—there is a cistern by good luck somewhat more intact than others, with a wellhead remaining between it and the ancient conduit of the Marcia. With these combination of two parts, we think that a complete settling tank has resulted, as the following plan [fig. 26] and the three keys to its parts make clear.

In one of the bricks mentioned is seen the fragment of the following brick stamp: "RAED.DOM.LV . . ." on the outer circle, "PAETIN . . ." on the inner circle, which I interpret as "from the estate of Domitia Lucilla," the mother of the emperor Marcus, "during the consulship of Paetinus and

76. Fabricius, 200 (= Graevius, 3:526D–E).

77. Fabricius, 9 (= Graevius, 3:459).

Apronianus" [*CIL* XV, 1913]. Indeed, I observed more bricks with this stamp in the ruins nearby. From this, I conjecture a reconstruction by Hadrian, into the sixth year of whose principate this pair of consuls falls, and perhaps the work of the emperor himself in the vicinity of his villa near the city. But [I will write] more about these things elsewhere.

The position of a similar wellhead and, next to it, of a triangular cistern [fig. 1, no. 28] supports the idea of a settling tank by no means different in structure, with my reconstructions confirming each other. Indeed, this twin construction [fig. 27], inside the distance of seven miles from the city that Frontinus fixed for the settling tanks of all the aqueducts [*Aq.* 19.1], lends the greatest probability to our argument.

As to why, indeed, I have not been concerned to demonstrate certainty and am using inferences in a matter that some will say ought to be shown by fact, rather than through words, my excuse will be the failure to explore even once the connection of the wellheads with their cisterns. My witness for this, also the authority for the structures cited, has been the most noble Marchese Camillo de Astalli, the owner of this property. Our effort was always in vain, since these places, very close to the conduit of the Marcia, have been built up to a very great extent from ruins and since that larger cistern [fig. 17, no. 27], with dirt heaped up within, is found filled with mud for the manufacture of potash. This, at least, we accomplished by our diligence: when the sides of the wall *DE* were uncovered in the lowest chamber, we came on no channel through which the water of the wellhead might flow in but evidence of a still lower and underground series of chambers. There are also traces of drains for the discharge of the muddy water settling in this underground tank (in a triple opening of the ground between nos. 27 and 28), first in the direction of the other settling tank for receiving also the discharge from it, then in the direction of the mausoleum of Alexander and those lower places, indicating a certain continuous channel under the earth.

The larger cistern of the two, I believed, served the Marcia, since its larger wellhead seemed to be owed to its greater supply of water. There is a still stronger reason, since I saw that the wellhead of the second cistern exceeded its opening almost by double; as a result, it served some structure connected to one of its sides or even to each of them (since, from here to there, remains of walls are seen). From these [cisterns], the common flow of the Julia and Tepula (from a shared wellhead before the settling tank—

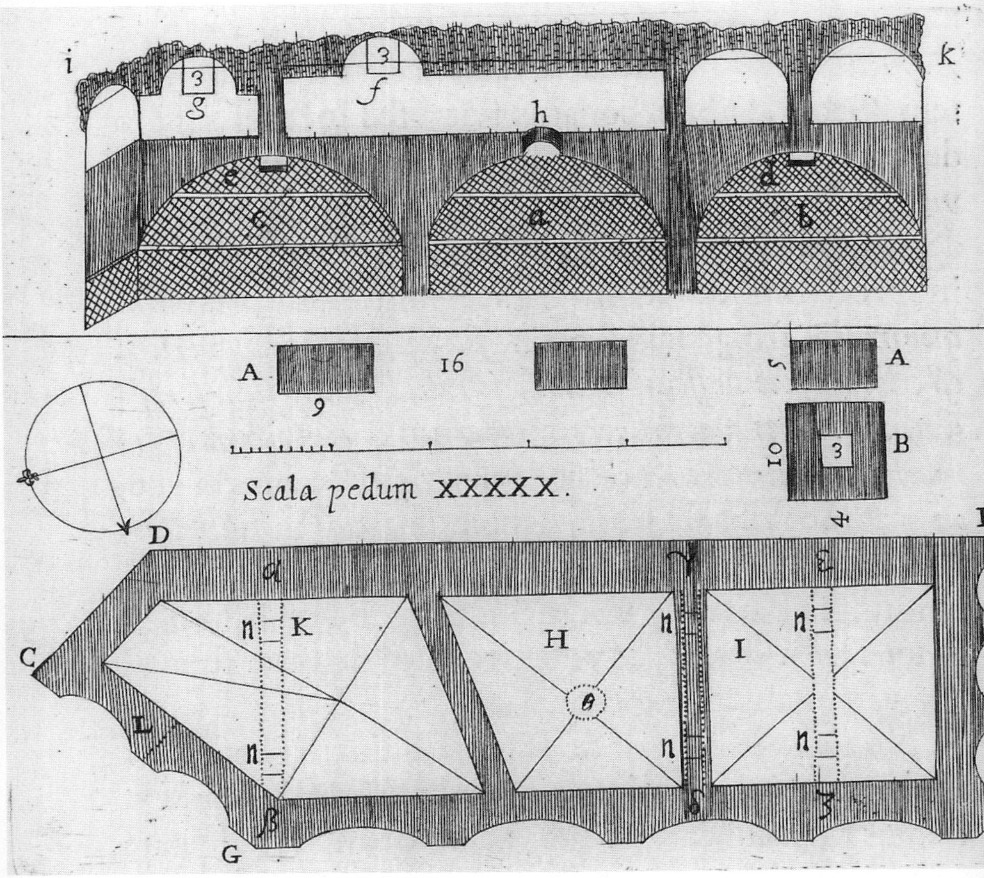

Fig. 26. Settling tank at the "Villa delle Vignacce": cross section and plan

Ground plan of the lower cistern

A. Conduit of the Aqua Marcia, carrying also the Julia and Tepula

B. Wellhead, receiving water from the conduit and transporting it into the settling tank by an underground channel, still unexplored by us

CDEFG. Cistern, or settling tank, divided in its lowest part by three basins

H. Middle basin, closed off from all sides, receiving water from the wellhead and transporting it through the opening into the upper cistern

I. Side chamber, built for support of the structure above it, although it was not part of the settling tank, since at its top it had a window on the south side (*DE*)

K. Another chamber, serving for the intake of that portion of the water that was carried into the nearby settlement through a round terra-cotta pipe eight inches in diameter in *L*, at the level of the window or intake at the top of the south side wall

Ground plan of the upper cistern, the divisions of which are marked by points and Greek letters

αβ.γδ.εζ. Three walls in this upper cistern, which divide its area into four chambers

η. The same number of windows in the walls already mentioned, at the same height from the floor, through which water, rising from the opening θ, was distributed throughout those upper chambers

θ. Round opening, three and a half feet in diameter, corresponding to the size of the wellhead, four square feet in width *(caption continued on facing page)* concerning this, see Frontinus [*Aq*. 8–9]—and then from a double conduit here), the appropriate channel of each one, and each one's appropriate name (about which, see Frontinus [*Aq*. 19.1–3]) could be drawn.

Nevertheless, there are (I confess) two disturbing factors: both the appearance of the place and, more exactly, the level of each cistern examined by us on this occasion are causes for doubt. That larger one at no. 27 [fig. 1] is recognized as being approximately four feet higher than the other, and there is also the account of Frontinus himself: according to it, the Marcia's settling tank is said to have been closer to the city than that of the Julia and Tepula. Talking about the Tepula, Frontinus says [*Aq*. 68.4], "From here"—that is, from the common settling tank of the Julia—"it first receives 190 *quinariae*, then immediately from the Marcia ninety-two *quinariae*," and before this, he includes these ninety-two *quinariae* in the total that was distributed before the Marcia's settling tank [*Aq*. 67.3]. If these things are to be understood in this way and there is no error in Frontinus and if by some scheme a supplement could be added from the Aqua Marcia (at a lower level) to the Tepula (which ran at a higher level), then our order of settling tanks would be reversed. Nevertheless, because I am not for the moment much concerned about this inquiry, it will have been sufficient to have shown at least the form and the function of the settling tanks, if not a sure identification of them individually and assignment of them to each aqueduct.

I will add finally, for confirmation of the preceding, that the structural appearance of our settling tanks agrees with other ancient works of that sort that we know were established for filtering aqueduct water. Many wit-

---

(caption continued from facing page)

Cross section of each chamber, upper and lower

abc. **Three chambers, of level footing, in *opus reticulatum*, in which brick is mixed for reinforcement**

d. **Window at the top of the vault, facing the well shaft**

e. **Another window, from the area of which, in wall *CG*, a portion of the water was channeled by pipe into *L* for the use of the settlement nearby**

f. **A window, raised six feet from the ground, on the side facing the aqueduct, through which water now filtered again sought its own channel**

g. **Another similar window, six inches lower and at the same horizontal height as the six windows**

h. **Opening, built up with bricks of approximately two feet, through which the water from basin *a* rose to the upper cistern**

ik. **Line of the height of the water in these upper chambers still conspicuous up to this point, from a very thin incrustation by which the walls are covered to that height**

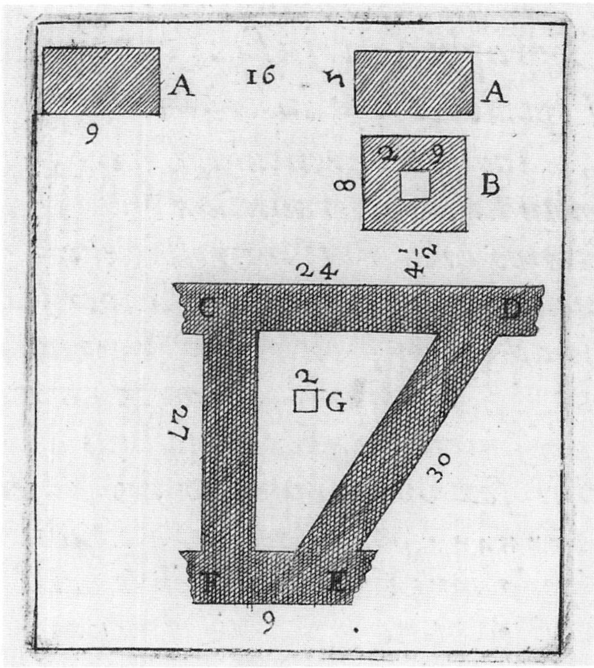

Fig. 27. Settling tank on the Via Latina

A. Conduit of the Aqua Marcia, atop which the Tepula and Julia were carried
B. Wellhead, providing the same function in the filtration of the water as the other shown in fig. 26
CDEF. Cistern, into which the well shaft delivered the water
G. Opening, through which the water rose to the floor of the upper cistern, or settling tank, and was restored to its own level

Projections and markings of the walls in *CDEF* were connecting other cisterns to it, where the remaining water of the wellhead could be admitted.

nesses who have looked firsthand have reported that something similar is seen in the settling tank of the aqueduct at Pisa. We have learned from experts that here, too, at Rome, below the Pincian Hill, the Aqua Virgo once put off its mud and dirt in a manner not much different, and as permitted, we have given a rough sketch, not according to an accurate scale of measurements [fig. 28]. The place, for a long time now inaccessible, being filled with refuse, remains useless for its function, because our watermen and reservoir men make offerings very prettily, not to the nymphs, but to Mercury.

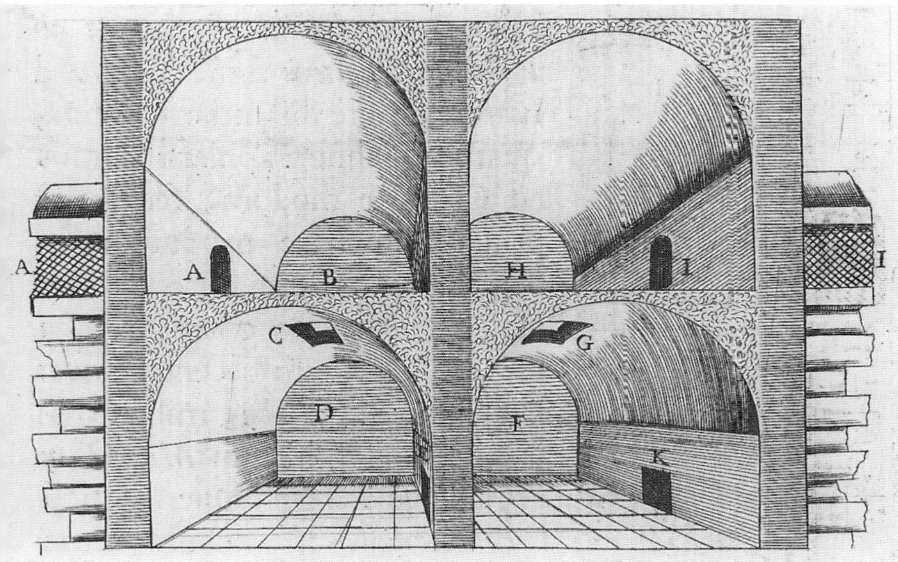

Fig. 28. Aqua Virgo: cross section of settling tank in the Campus Martius

A. Conduit of the Aqua Virgo, emptying itself into tank *B*
C. Opening through which the water goes down into underground chamber *D*
E. Gate, from which the water is carried down into another underground chamber, *F*
G. Opening through which the water again rises to tank *H* and restores itself to conduit *I*
K. Sluice gate, from which mud and waste were discharged into the sewer

b. The Aqua Claudia/Anio Novus

In searching out the settling tank of the Claudia (to move forward the explanation already begun about settling tanks of this sort and amplify it as much as possible), Frontinus is our authority and guide. Frontinus [*Aq.* 72.3] says that it was "at the seventh milestone from the city." At exactly this distance, on the left-hand side of the modern road leading to Marino, at a mile and a half beyond the Osteria Mezza Via di Marino, I found ruins, a plan of which we give [fig. 29], in the shape of a cistern, but far exceeding the scale of a private cistern, constructed of very hard *selce* [basalt] and covered with *opus reticulatum*.

Although too little remains of this structure—it ought to have been underground, since the arches of the Claudia end shortly beyond the fifth milestone from the city—there is certainly enough to provide a place for a

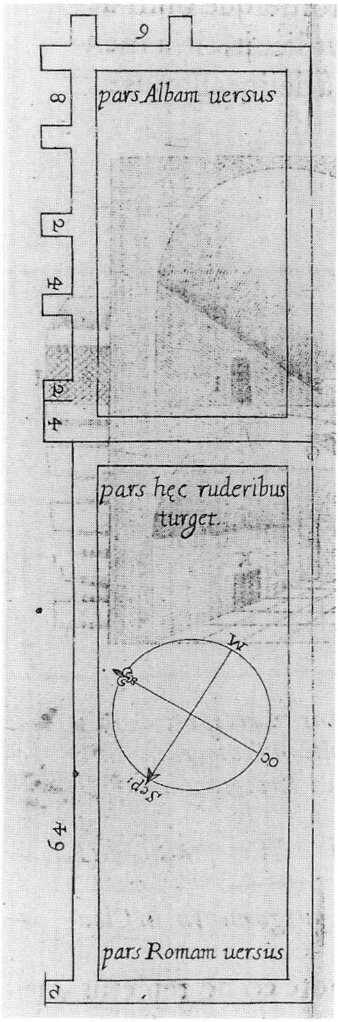

Fig. 29. Settling tank: plan

not improbable reconstruction. Fragments of a very large amount of squared stone around this structure and the huge quantity of incrustation that once had built up in its channel lie scattered, not quite everywhere, but mixed in with the aggregate of modern walls nearby. Both the double nature of construction and another passage of Frontinus [*Aq.* 19.1], where he includes all six aqueducts that had their own settling tanks within this same area of the seventh milestone, argue that the settling tanks of the Claudia and Anio Novus were joined. We are prepared to believe those for whom these arguments of ours will seem shapeless and weak if they will bring forward something better. In the meantime, let them at least judge our attempts as good and just, and in such obscurity of facts and antiquity, let them bestow on our labor the following word of Horace, "It is no small thing to advance so far, even if it is not allowed to go further" [*Epist.* 1.1.32].

c. Measurement of Water

Frontinus tells us that settling tanks had another function, besides the well-known one of filtering aqueducts, namely, in fixing the surest possible measurement of each and every aqueduct at them, "where the measurements are not to be doubted," as he himself says [*Aq.* 72.3]. The explanation for this accuracy could be this: "when the course of the conduits was taking its breath there" [*Aq.* 19.1] and when the water, deprived of all its force, was seeking again its own conduit, it was least subject to distortions resulting from a swifter or a slower flow (a fact very well known to Frontinus, as is clear from various passages of his treatise).

As a result, we declare the necessity of rejecting Fr. Castelli's criticism of this outstanding author in his book *On the Measurement of Running Water*,[78] as if Frontinus were ignorant of that great theory of his "that from the swiftness of water its measurement varies" and did not, in accordance with this more accurate reason, report the discrepancy of volume of the aqueducts in intake or in distribution. Indeed, whoever will have read Frontinus and not with bleary eyes (like Castelli) will recognize that the theory is no longer new but restruck from an ancient one. "Let us remember," says Frontinus [*Aq.* 35], "that whenever water comes from a higher elevation and falls into a distribution tank within a short distance, it not only corresponds to its original level of measurement but also exceeds it; in comparison, whenever water is brought from a lower elevation (i.e., under less pressure) for a longer distance, it also decreases its delivery because of the slowness of its passage. For this reason then, according to this reckoning, its delivery figures must be supplemented or decreased."

---

78. B. Castelli, O.S.B., *Della misura dell'acque correnti*, 3d ed. (Bologna, 1660), 29–30 (corollary 16, app. 1).

Here is a second example [*Aq*. 70]: "For the Aqua Virgo there was a capacity of 652 *quinariae* listed in the record books. It was not possible to measure its capacity at its intake, since it is drawn from several tributaries and enters its conduit too slowly. Close to the city, however, at the seventh milestone, on property now belonging to Ceionius Commodus, where it has a swifter current, I took a measurement amounting to 2,504 *quinariae*, 1,852 more than in the record books. Our confirmation is very much at hand, since the Virgo delivers all the volume we determined by measurement, that is, 2,504 *quinariae*." Finally, [Frontinus speaks to his own defense] with these words [*Aq*. 73.5–6]: "Moreover, I have discovered that there are illegally tapped not only 528 *quinariae* (of the Anio Novus), the difference between our measurements and the scheduled delivery, but a much larger amount. As a result, it is evident that the actual volume is even in excess of our measurements. This can be explained because the rather swift force of the water, taken from a bountiful and fast-flowing river, INCREASES ITS VOLUME BY ITS VERY SPEED."

From the preceding passages, it is clear, therefore, that Frontinus was not ignorant of that change experienced by volume because of velocity of water. Even if these quotations have stated, according to the same authority, that volume is increased from swiftness but is not decreased, as Castelli has taught, there is nevertheless a common explanation according to which what is increased in intake is in turn, through an alternative principle, decreased in delivery.

But ignorance of so great an axiom could not have been harmful to Frontinus, nor, for him, could anything be lost in the new measurement he undertook of the aqueducts at the city on account of increased velocity and subsequently diminished capacity, as Fr. Castelli thinks that he deduces from Frontinus's treatise. Not only the measurement taken at the settling tanks but also that taken at Rome in distribution went back of necessity to the measurement "at the source of the aqueducts." Indeed, water inside and outside the city had to be tapped not from public conduits but from distribution tanks and reservoirs, after it had recovered its earlier state of rest, by the decree of the Senate cited by Frontinus [*Aq*. 106.1], as he had also earlier said was done [*Aq*. 27.3].

However, with only that single geometric proposition that makes up a single page of his book, Castelli indeed claims the right to attack outstanding writers with a scolding. Readers should know that the very principles from which he says that his observation—not new, as we have seen—was brought forth are read in Frontinus, with the same meaning: [Castelli writes,] "Let the water issue through two channels of equal width, one placed in the lower part of the receptacle; the other in the upper part; it is clear that in the time in which there will issue from the upper part a certain measure of water, from the lower part will issue four, five, and even more of the same measures, in proportion to how great the difference will be of the height of the channels, and the distance of the upper channel from the surface, and level of the water in the receptacle," and so on.[79] These words, expressed by Castelli in Asiatic fashion, Frontinus shortens in Laconic style [*Aq*. 113.1–2]: "Concerning the positioning of connecting pieces, it is necessary to observe that they be set in a straight line and that the connecting piece of the first not be placed too low and that of the second too high. The lower one draws more, the higher one less, since the flow of water is taken from the lower."

Finally, Frontinus is more precise and more careful in this matter. In the case of pipes of equal size and arranged on the same surface, Frontinus [*Aq*. 105–6] reports that by decree of the Senate it was provided that men not install a larger pipe within fifty feet of the distribution tank from which they were drawing water, "to prevent the water, not, indeed, restrained by its legal space but squeezed out through smooth"—*leves*, to be emended to *breves*, "short," I think—"narrow passages, from easily filling the larger pipe nearby," as he states in *Aq*. 112.5. With this notice, as if concerning a matter most useful and often pertinent in practice, Castelli must be corrected. Castelli introduced something of significance to this investigation with his axiom,[80] but he did not solve it in this area; instead, he thought in vain that he was refuting Frontinus about a matter that we have proved Frontinus was not ignorant of, rather than taking from someone else a proposition he himself did not know.

6. THE BUILDER OF THE AQUA MARCIA

After so many topics, it will be our final task to make a judgment between the opposing opinions of very well known writers concerning the origin of the name and the introduction of this Aqua Marcia. On this point, Fron-

79. Castelli, *Della misura*, 4–5.
80. Castelli, *Della misura*, 33.

tinus, to whom we have been especially accustomed to defer in this discussion of aqueducts, argues that Marcus Titius introduced it and gave it its name from his own praenomen [*Aq.* 7.1]. The authority of Pliny, however, takes us in an opposite direction: "Ancus Marcius, one of the kings, first began the introduction of it into the city; afterward, Q. Marcius Rex in his praetorship" [*HN* 31.41]. There is added the not insignificant testimony of Plutarch in *Coriolanus*, at the very beginning of the book (1.1), "It was also the Marcii, Publius and Quintus, who built the greatest and most beautiful aqueduct at Rome." In this doubtful matter, the example of the Aqua Appia, named from the praenomen of Appius Claudius, seems to remove uncertainty, in support of Frontinus. Nevertheless, for me, the credibility of Pliny and Plutarch is more influential, supported especially by this coin of the gens Marcia [fig. 30].

Who, indeed, would suspect in a publicly minted coin a fraudulent claim and pretense? Instead, I think that the authority of this coin had such great weight for Pliny that he concluded, I say, from this perhaps—and not from somewhere else, since, rightly, Fr. Donati considers this evidence hardly persuasive[81]—that Ancus, whom Philippus rendered on the reverse of the coin for the royal distinction of common kinship with him, was the builder of the aqueduct depicted on the obverse.

Whoever of these men it pleased to have left his name for this aqueduct, it is clear that it must not be called "Martia," as commonly all writers and a very faulty edition of Frontinus otherwise persist in keeping it, but "Marcia," as the two inscriptions of Titus and Caracalla cited in our first dissertation [*CIL* VI, 1245, 1259] show, as well as this other:

> To the spirits of the dead. To Timbraeus, freedman of Augustus, reservoir man of the Aqua Marcia, Claudia Dyname made this dedication to her excellent husband. [*CIL* VI, 3193]

Quite a few other things—about the uncertain source of the Anio Vetus and the channels of these aqueducts and the Anio Novus; about the Antoninian, Herculanean, and Albudine springs; and about the measurements of the conduits in Frontinus, expressed with obvious error in their numerical figures, and their leveling—remain to be discussed. However, so that we not mix things more certain—such as we believe those topics

81. Donati, 296 (= Graevius, 3:835E–F).

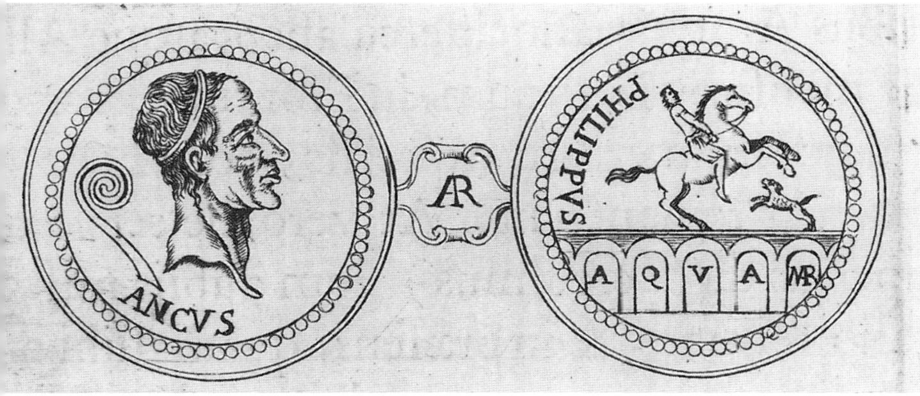

Fig. 30. Coin of L. Marcius Philippus, with arcade of Aqua Marcia

treated earlier to be—with what is doubtful, I will put off the matter for another time, and for the present, I will not go beyond the subject I have set forth for myself. Instead, I will lift my hand from the table and put an end to a letter that has grown into nearly a book, having prayed to Almighty God for all favorable things for Your Eminence. This alone administers to my weakness, that I pay back my greatest benefactor and patron, who is like my guardian divinity, with most devoted training of my mind. I write this from your library, on August 12th, the year of redemption 1679.

Commentary

Fabretti's second dissertation was published, like the first, without section headings, but it is here organized as follows:

1. Introduction
2. Map of the Upper Anio Valley
3. Determining Ancient Measurements
   a. The Mile
   b. The Foot

4. Topography of the Upper Anio
   a. Holste's Theories
   b. Course of the Via Valeria
   c. Sources of the Aqua Claudia
   d. Representations of the Aqua Claudia
   e. Source of the Aqua Marcia
   f. Source of the Aqua Augusta
   g. Origin of the Aqua Marcia
5. Settling Tanks
   a. The "Villa delle Vignacce"
   b. The Aqua Claudia/Anio Novus
   c. Measurement of Water
6. The Builder of the Aqua Marcia

This dissertation centers on the topography of the upper Anio Valley and the sources of the Aqua Marcia and Aqua Claudia, but Fabretti also includes in it a digression on Roman measurements and observations on settling tanks of the aqueducts discussed.

OPENING

Gaspare Cardinal Carpegna (1625–1714). Carpegna was cardinal vicar of Rome under five popes, from 1671 through 1714, and Fabretti's patron, who secured for him the position of custodian of sacred relics and cemeteries, an appointment that enabled Fabretti to excavate where he wished and obtain access to antiquities. Carpegna himself was highly regarded as a man of learning and a collector of books and ancient coins, as well as a member of the Accademia degli Arcadi. See *DBI*, 20:589–91 (G. Romeo).

1. INTRODUCTION

Fabretti's introduction to his patron is appropriately deferential, stressing his gratitude for the freedom to explore antiquities and Carpegna's encouragement of his efforts.

2. MAP OF THE UPPER ANIO VALLEY

Fabretti's map, entitled "Plan of the Region between Tivoli, Carseoli, and Subiaco," gives a detailed presentation of the topography of the upper Anio Valley east of Tivoli, including the major tributaries of the Anio River, the Licenza and Givenzano; the courses of the ancient Via Valeria and Via Sublacensis; settlements in the area; and remains of the Roman aqueduct lines. Given the extensive area covered, its scale is considerably larger than those of the two topographical maps of the other dissertations.

This map, one of the earliest detailed plans of the upper Anio Valley, was the first to depict remains of the Roman aqueduct lines and to fix their sources. Because earlier maps of the area were quite vague and general in their coverage, Fabretti's praise for the cartography presented here is not without merit, and his map, once published, had an enormous influence on later topographical study of the region. Poleni followed Fabretti on the sources of the Marcia and Claudia and reproduced Fabretti's map of the upper Anio Valley with the same title in his 1722 edition of Frontinus (Poleni, fig. V, opposite p. 90). For a historical account of the cartography of the upper Anio, see F. Crainz and C. F. Giuliani, "I due tracciati della Via Valeria fra *ad Lamnas* e *Carseoli*," *AMST* 58 (1985): 78-80; for a more general survey of the area, see Z. Mari, "La valle dell'Aniene nell'antichità," *AMST* 68 (1995): 25–52.

Fabretti's listing of data in his legend to this map is selective, as in his legend to his map in the first dissertation, and omits many topographical details depicted on the map itself. His numbering corresponds to identifications and descriptions of modern topographers as follows:

1. "Valle degli Arci" aqueducts on the modern Via Empolitana. For the Anio Novus, see Ashby, 273–74 (331–34); Van Deman, 327. For the Marcia, see Ashby, 109–10 (133–34). For a general description of the area, see Aicher, 139–41.

2–3. Arcade of the Anio Novus along the Via Empolitana: Ashby, 267–73 (318–31); Van Deman, 289–99; C. F. Giuliani, *Tibur* (*Pars altera*), FI 10.7 (Rome, 1966), 86–87; Aicher, 139–41.

4. Via di Carciano segments. For the Anio Vetus, see Ashby, 64–65 (81–82); for the Marcia, Ashby, 112 (137); for the Claudia, Ashby, 208–9 (243). See also Aicher, 136.

5. Remains of Roman villa west of Ciciliano, commonly named the "Ruderi di Saxula." Ashby (266 n. 1 [315–18 n. 60]) observes that Fabretti is incorrect in his reconstruction of the course of the Anio Novus in this area, since the actual villa is situated "a long way to the east" of the real line of the aqueduct.
6. Vicovaro bridge: Ashby, 3 (21); Van Deman, 196–98; Aicher, 154.
7. Ancient aqueduct near Vicovaro. Ashby (99 [121]) identifies this conduit as part of the Marcia.
8. Traces of another ancient aqueduct, perhaps independent of the Roman system, to be connected with supply of Tibur itself. For other segments of this Tiburtine conduit, see M. G. Fiore Cavaliere and Z. Mari, "Acquisizioni lungo il tracciato degli acquedotti aniensi," *QArchEtr* 24 (1995): 470–72; Panimolle, 53–55; C. Roncaioli Lamberti, "Il percorso dell'*Anio Vetus* e la Porta Trebana di Tivoli," *Giornale italiano di filologia* 40 (1988): 93 n. 32.
9. Conduit of the Aqua Marcia in the Osteria della Ferrata: Ashby, 3 (21), 99 n. 3 (121 n. 71).
10. Ponte Arconi: Ashby, 258–59 n. 3 (309–10 n. 40); Van Deman, 73.
11. Thirty-eighth milestone of the Via Valeria (*CIL* IX, 5963) near Somnula (the modern Sonnoletta): Crainz and Giuliani, "Via Valeria," 77; Ashby, 96 n. 2 (118 n. 55).
12. Ponte Scutonico: Crainz and Giuliani, "Via Valeria," 84–85; Mari, "La valle dell'Aniene," 29, table IVa.
13. Ponte S. Giorgio: Crainz and Giuliani, "Via Valeria," 85.
14. Forty-first milestone of the Via Valeria: *CIL* IX, 5966.
15. Springs below the Church of S. Maria in Arsoli: Lanciani, 276–77.
16. Aqua Marcia *cippus* of Augustus: *CIL* VI, 31562g = VI, 1251b = XIV, 4074; Ashby, 93 (114); Lanciani, 279.
17. Ancient construction near aqueduct source. Fabretti's notice is too vague to be identified securely but may be a vaulted structure cited by Ashby (92 n. 1 [113 n. 39]) and Fiore Cavaliere and Mari ("Acquisizioni," 465 n. 17), who suggest the identification of the nearby source of water with that of the "Acqua Santa."

18–20. Basin of the springs identified as Serena I and II: Ashby, 96–98, fig. 6 (116–18, fig. 14); Fiore Cavaliere and Mari, "Acquisizioni," 464–65 n. 16; Aicher, 159–62.

21. Spring identified as Casa di Lemme: Lanciani, 276.
22. Spring identified as Forma della Mola: Lanciani, 276.

23. Channel of spring below Agosta: Lanciani, 276.
24. The lakes above Subiaco: Ashby, 253–56 (304–8); Aicher, 162–64.
25. Conduit of the Anio Novus west of the Anio River. Ashby (257, 259–60 [308, 310–11]) notes Fabretti's error in describing the Ponte Arconi (no. 11 on his map) as this aqueduct's crossing point of the Anio River.

3. DETERMINING ANCIENT MEASUREMENTS

Fabretti now introduces an involved discussion of Roman linear measurements, a problem that greatly engaged the interest of the scholarly world at his time. For a summary of such investigations, focusing particularly on the earlier related research of John Greaves, see A. E. Berriman, *Historical Metrology* (London, 1953), 121–24. Greaves, the Gresham Professor of Astronomy at Oxford, published a treatise focusing on many of the same issues Fabretti raises in this section, *A Discourse on the Romane Foot and Denarius* (London, 1647), reprinted in *The Miscellaneous Works of Mr. John Greaves* (London, 1737), 1:165–233.

A preliminary review of Fabretti's terminology follows. For more detailed discussion, see Berriman, *Historical Metrology*, 128–29; O. A. W. Dilke, *Mathematics and Measurement* (Berkeley and London, 1987), 26–27.

foot (*pes*) = the Roman standard of measurement
palm (*palmus*) = span of four digits (*digitos*, or fingers), one-third of a Roman foot
inch (*uncia*) = span of three-fourths of a digit, one-twelfth of a Roman foot
dodrans = nine inches, three-fourths of a Roman foot
spithama = a Greek measurement indicating the span between the thumb and little finger = Roman *dodrans*
sextula = one-sixth of a Roman foot = two inches
reed (*canna*) = measurement ten palms in length
scripulum = one twenty-fourths of an inch
pace (*passus*) = five Roman feet
mile (*milliarium*) = one thousand *passus* = five thousand Roman feet
staiolo = five and three-fourths palms
chain (*catena*) = ten *staioli* = fifty-seven and a half palms
palmo (palm) = the palm measurement used by Italian architects in the sev-

enteenth century, somewhat longer than the ancient palm. To distinguish this measurement from the ancient palm, the *palmo* (which Fabretti cites consistently as *Palmus*, capitalized in his Latin text) appears throughout the translation in its Italian form.

a. The Mile

Fabretti opens his discussion of the mile measurement by citing with approval *De mensuris et ponderibus Romanis et Graecis cum his quae hodie sunt collatis libri V* (Venice, 1573), a treatise by the jurisconsult Luca Peto (cited in the first dissertation [I.4d]), who determined the length of the Roman foot through comparative study of several examples of ancient bronze feet and inscribed its length and that of other ancient measurements on a marble plaque on the Capitoline. For a description of Peto's measurements, see M. Folkes, "An Account of the Standard Measures preserved in the Capitol at Rome," *Philosophical Transactions of the Royal Society of London* 442 (1736): 262–66. On the publication of *De mensuris*, see del Re, "Luca Peto," 325–26.

Fabretti's first quotation from Peto focuses on the discrepancy between the length of the seventeenth-century *palmo* and that of the ancient *dodrans* or *spithama*, which Peto expresses in measurements of *scripula*. We can reconstruct the discrepancy demonstrated by Peto as follows:

one *palmo* = one *spithama* + two-ninths of an inch (.2222 inches)

or

one *palmo* = one *spithama* + one-sixth of an inch + one and one-third *scripula* (.1666 + .0416 +.0139 =.2221 inches)

Fabretti's next citation from Peto concerns the length of the Roman mile. Through his own calculations, based on thirds of a *scripulum* (one seventy-seconds of an inch), Fabretti first points out a minor error in Peto's calculation of the mile:

Peto:	one mile = 649 *cannae* + nine palms + three digits
Fabretti:	one palm = 664 thirds of a *scripulum*
	one Roman foot = 864 thirds of a *scripulum*
	five thousand feet = one mile = 432,000 thirds

one mile therefore = 6,506 palms + sixteen thirds (sixteen thirds = five and one-third *scripula* = one *sextula* + one and one-third *scripulum*)

Fabretti next cites a more serious error in the length of the *palmo* given by Peto (nine and one-sixth inches plus one and one-third *scripulum*). Fabretti compares the length of the modern *palmo* presented by Peto with the length of the *canna* represented on Peto's Capitoline plaque, demonstrating that the *canna* measures seven feet plus six and fifteen-seventeenths inches, resulting in a discrepancy of 15/170 from the *dodrans* of the Roman foot (or, in decimal equivalents, 7 feet 6.8823 inches); the modern *palmo*, Fabretti argues, must therefore be .8823 inches longer than the ancient *dodrans*. As a result, the Roman mile, by Fabretti's calculations, contained 6601.9357 palms, approximately 102 palms more than the length stipulated by Peto.

Fabretti's calculations now permit him to introduce criticism of Lucas Holste's reconstruction of the Roman mile in his treatise *De milliario aureo*. Fabretti is vague in his account of Holste's methodology, which he describes as derived from surveying practices, but he criticizes the resulting measurement that Holste presents, a length of 6,666 and two-thirds palms for the mile. Finally, Fabretti presents still another mile measurement, one based on *catenae* (ten *staioli* in length), used by surveyors (who are not specifically identified). We may summarize the four mile measurements presented here as follows:

Peto:	one mile = 6499.3333 palms
Fabretti:	one mile = 6601.3957 palms
Holste:	one mile = 6666.6666 palms
surveyors:	one mile = 6,670 palms (= 116 *catenae* or 1,160 *staioli*)

b. The Foot

Fabretti next turns to contemporary controversy over the measurement of the Roman foot, citing the different epigraphical evidence for its length before turning to his own investigation of the problem. Fabretti cites, in passing, the following evidence presented in earlier scholarly reconstructions of the Roman foot:

Pes Statilianus. A foot measurement represented on the funerary monument of the *agrimensor* (or land surveyor) Statilius Aper, now in the collection of the Capitoline Museum (*CIL* VI, 1975 = *ILS* 7737; cf. Helbig[4], 2:59–61 (no. 1214).

Pes Cossutius. A Roman foot represented on the funeral monument of Cossutius (*CIL* VI, 16534) in the Horti Colatiani, also known as the *pes Colotianus*.

Pes porphyreticus. A foot based on the evidence of a porphyry column, two examples of which are cited in Rome by the notes in the Barbiellini edition (62 n. d), one at the Basilica dei SS. Apostoli and a second on the Via Lata. Each column carried at its base an inscription in Greek indicating its length. This measurement was accepted as accurate by the topographer Bartolomeo Marliani (*Antiquae Romae topographia libri septem* [Rome, 1534] = Graevius, 3:115–263) and by Guillaume Philandrier in his *Annotationes*, 117. For a translation and commentary on this passage of Philandrier, see Lemerle, *Les "Annotations,"* 165–66.

Fabretti presents his support for the length of the **Capitoline foot** fixed by Peto, confirmed by his own investigations of the problem. Citing his measurement of a marble paving stone found in the remains of an "elegant villa" near the tenth milestone of the Via Ostiensis, Fabretti points to the exact correspondence between the dimensions of his sample and the length of the Roman foot inscribed on Peto's plaque as confirmation of its accuracy. Fabretti's description here is extremely vague, probably because the matter was not of crucial importance to the main subject of the dissertation. Fabretti's comments on the discrepancy between Peto's foot inscribed on the Capitoline and that published in his treatise echo observations made by another author cited in this section, **Giambattista Riccioli, S.J.,** who describes the distortion of published measurements of Roman feet by contraction of wet paper after printing, in his *Almagestum novum astronomiam veterem novamque complectens* (Bologna, 1651), 59.

Fabretti also mentions, in passing, two other attempts to fix the Roman foot, one based on study of the **Farnesina *congius*** by the Jesuit **Juan Bautiste Villalpando** (1552–1608) and one by Riccioli. Villalpando, with his mentor Jerome del Prado, S.J., was commissioned to write the three-volume biblical commentary *In Exechielem explanationes* (Rome, 1604), on chapters 40–42 of Ezekiel, which present an architectural description of

Solomon's Temple. After Prado's death in 1595, Villalpando completed three volumes of his study, containing information on astrology, music, and mathematical theories, as well as reconstructions of Hebrew, Greek, and Roman systems of measures and currency. See *DSB*, 14:29–31 (M. T. Ryan); R. Taylor, "Hermeticism and Mystical Architecture in the Society of Jesus," in *Baroque Art: The Jesuit Contribution*, ed. I. B. Jaffe (New York, 1972), 63–97. Fabretti cites a section of Villalpando's third volume, *De apparatu urbis ac templi Hierosolymitani, pars i et ii*, treating ancient metrology, in which Villalpando argued that the Farnesina *congius* documents all ancient Roman measurements; for more detailed discussion of the technical procedures and other later experimentation with the *congius*, see Berriman, *Historical Metrology*, 125–26.

The **Farnesina or Vespasianic *congius*** was a volume measurement based on the capacity of a bronze vessel with an inscription (*ILS* 8628) dated to 75 A.D., attesting a measurement of ten *librae*. Interpretation of its inscription, however, is problematic, as Fabrétti points out in his discussion here. The original vessel, part of the Farnesina collection at Rome, appears to have been lost, and copies in the Villa Giulia Museum at Rome and at Dresden are considered spurious; Berriman (*Historical Metrology*, 125) also reports a copy in the British Museum. For illustrations of the *congius*, see Berriman, *Historical Metrology*, 125; Graevius, 11:1674.

In this discussion, Fabretti rejects an erroneous interpretation of the *congius* by the numismatist Jacques Oisel and Villalpando's reconstruction of the Roman foot: Villalpando argued specifically that half a Roman foot was represented by a straight line drawn from the bottom of the lip of the *congius* to the top horizontal band incised in its midsection (cf. Prado and Villalpando, *In Exechielem*, 3:501–2, presenting a double image of the *congius* entitled "Forma aerei congii quo ab antiquis Romanis mensurae et pondera exigebantur ad exemplum duorum similium quos Romae habuimus expressa," shown both from the exterior and in cross section). The direct quotations in this passage are paraphrases of Villalpando's text. Villalpando also included in his treatise a detailed account of the proportionality of linear measurement based on Vitruvius 9.1.2–3 and an elaborate chart illustrating the relationship of cubic and linear measurements "augendorum minuendorum in data ratione corporum instrumentum" (Prado and Villalpando, *In Exechielem*, 3:316–17), to which Fabretti makes reference in this section.

Giambattista Riccioli, S. J. (1598–1671), the second metrologist criticized in this discussion, was an astronomer and geographer who spent most of his scholarly career seeking to disprove Galileo's theories. Fabretti much more briefly rejects measurements of the Roman foot presented in two of Riccioli's major works, the *Almagestum novum* and the *Geographiae et hydrographiae reformatae nuper recognitae et auctae libri duodecim* (Venice, 1672). In a review of Roman metrology within his *Almagestum novum*, Riccioli discussed both the Capitoline foot of Peto and Villalpando's Farnesina *congius* and reproduced the dimensions of a *semipes Romanus* [a half Roman foot] and a modern half-foot (*semipes Boniensis recentior*). For Riccioli's career, see *DSB*, 11:411–12 (L. Campedelli).

4. TOPOGRAPHY OF THE UPPER ANIO

To provide a clear break from the preceding digression, Fabretti begins his discussion of the topography of the upper Anio with well-known citations of Frontinus on the sources of the Marcia and Claudia.

Poggio Bracciolini (1380–1459). Bracciolini is the Renaissance humanist credited with the discovery of many Latin classics, among them Frontinus's *De aquaeductu*. See Sandys, 2:25–34; *EHCA*, 907–8 (P. W. G. Gordan).

Via Valeria. An important Roman road running east as a continuation of the Via Tiburtina to Alba Fucens and the Adriatic coast, it was probably first paved in the censorship of M. Valerius Messala (154 B.C.). For its route in the area under discussion, see Crainz and Giuliani, "Via Valeria," 71–88.

Via Sublacensis. A branch of the Via Valeria paved by Nero to provide access to his villa at Subiaco (see Plin. *HN* 3.109; Tac. *Ann*. 14.22).

a. Holste's Theories

Fabretti had earlier criticized the topographical work of Lucas Holste in general terms (I.7); he now focuses on particular errors concerning the sources of the Aqua Marcia and Aqua Claudia, as well as Holste's attempt to identify the Aqua Augusta, added as a supplement to the Marcia by Augustus (Frontin. *Aq*. 12.1–2), with springs below a settlement named Austa (the modern village of Agosta).

Fabretti's citation of Frontinus here is selective, omitting the additional notice in *De aquaeductu* 14.3 that the Aqua Augusta served as supplement for the Aqua Claudia but remained as a potential reserve for the Marcia. As Ashby (88 n. 8 [109 n. 8]) observes, Fabretti makes a special effort here to refute Holste's identification of the name of the Aqua Augusta with that of the village of Agosta, calling attention to other errors in Holste's published work and to the vagueness of citations of the Aqua Augusta in documents of the Subiaco monastery.

Fabretti does repeat here, however, without objection, an important notice from Holste's *Annotationes* concerning the intake and conduit of the Anio Novus from the lakes above Subiaco. Holste's description of the three lakes and his topographical notice concerning the fourth-century Church of S. Lorenzo "ad altas aquas" are repeated by later topographers: cf. Lanciani, 352; Ashby, 254–55 n. 4 (305 n. 19); Panimolle, 151–52.

Fabretti cites earlier criticism of Holste made in his unpublished letter to **Lorenzo Panciatichi** (1635–76), a priest and author of two works published posthumously, the novella *La barba fatta per carità* (1856) and *Scherzi poetici* (1729). Three years before the completion of this dissertation by Fabretti, Panciatichi had committed suicide by jumping into a well, "trasportato da furore più che poetico" [carried away by more than poetic madness] (G. Negri, S.J., *Istoria de Fiorentini scrittori* [Ferrara, 1722], 378–79); Panciatichi's recent death no doubt prompted Fabretti's praise of his learning and comments about fate. For an assessment of Panciatichi's work, see C. Guasti, ed., *Scritti vari di Lorenzo Panciatichi* (Florence, 1856), i–lxv.

The *Bullarium casinense, seu constitutiones summorum pontificum* (Todi, 1670), cited by Fabretti at length in this section, was a recently published compilation by Cornelio Margarini, O.S.B. (1605–81), whom Fabretti mentions warmly. Margarini was also the author of a *Dictionarium lombardicum* (Todi, 1670) and editor of *Inscriptiones antiquae basilicae S. Pauli ad viam Ostiensem* (Rome, 1654). It should be noted, however, that the second and third bulls cited by Fabretti here, those of Honorius III in 1217 and John VI in 704, do not appear in the *Bullarium casinense;* cf. Barbiellini, 71 n. b, 72 n. a.

Fabretti argues that the papal documents in the *Bullarium casinense* are notoriously unreliable as topographical documentation and are therefore to be discounted as evidence concerning the source and route of the Aqua Augusta. He adds as well that references in them to that aqueduct might well be references to a later supplemental source added to the Aqua Marcia by Caracalla in conjunction with the building of the Thermae Antoninianae in the early third century. This second argument is not particularly convincing; Caracalla's supplement is identified only as the "novus fons Antoninianus" (*CIL* VI, 1245) and is never attested by the name *Aqua Augusta*. However, Fabretti expands on his first argument, pointing out similar topographical errors in other papal documents, specifically a bull of Paul V concerning the Aqua Traiana (cf. Ashby, 88 n. 8, 304 [109 n. 8, 361]), and citing corrections of other misidentifications in the works of Pietro Bembo by two other scholars, **Antonio Massa, S.J.** (1500–1558) and **Niccolò Nardini.** For Massa's career, see N. del Re, *Antonio Massa da Gallese, giurista e litterato* (Naples, 1992). Nardini, the son of the topographer Famiano Nardini, had published his study of the Faliscans much more recently, in 1677, two years before the completion of this dissertation.

Fabretti dismisses topographical errors found in papal bulls as trivial, only to introduce strong censure of the numismatist **Claude Dumolinet** (1620–87), whose *Historia summorum pontificum a Martino V ad Innocentium XI per eorum numismata* appeared in 1679, the year this dissertation was written. As the commentator of the Barbiellini edition observes, Fabretti's criticism of Dumolinet's association of the papal Acqua Paola with Augustus's Aqua Alsietina is unfair: Dumolinet nowhere asserts this identification in his *Historia* but cites only the erroneous papal inscription on the Acqua Paola fountain on the Janiculum.

b. Course of the Via Valeria

Fabretti's principal topic in this section is the original course of the ancient Via Valeria in the region of the sources of the Aqua Marcia and Aqua Claudia. He cites what he describes as conflicting testimony about the Roman road in Holste's *Annotationes* and argues instead for a direct course for the Via Valeria from the Osteria della Ferrata northeast to Riofreddo, as shown on his topographical map (fig. 17).

Fabretti's discussion addresses a longstanding topographical problem not resolved until recently: what was the course of this Roman road through the territory between the Anio Valley and the Piana del Cavaliere six kilometers distant to the northeast and 260 meters in elevation above it? As Crainz and Giuliani point out ("Via Valeria," 78–80), earlier cartographic evidence for the region is of limited value in determining how the original Via Valeria ran: an anonymous map of around 1650 in the British Library indicates two separate courses for the road, a direct route northeast from the Osteria della Ferrata through Riofreddo, as argued by Fabretti here, and a longer route following the course of the Anio past the bridge to Anticoli Corrado, then turning sharply north past Arsoli (corresponding generally to the course of the modern Italian state highway [S.S. no. 5] and the route of the Rome-Pescara autostrada). Another seventeenth-century map, the Catasto Alessandrino of 1661, shows only the branch through Riofreddo as the Via Valeria. For an extensive review of earlier cartography, see Crainz and Giuliani, "Via Valeria," 73–88, especially 78–82 and tables XII–XIII.

To defend his arguments for positing a more direct route through Riofreddo, Fabretti cites a passage of Plutarch on the Roman practice of planning roads along straight courses. However, he gives much greater emphasis to his autoptic investigation of the area, describing a journey undertaken with **Giuseppe de Giuli** (presumably in the late spring/early summer of 1679). De Giuli, whom Fabretti describes as a man of scholarly interests, appears to have been the author of a *Compendaria et facilis ad linguam graecam manuductio* (Rome, 1681) and a commemorative poem, *Pacis augurium sub Clementis X pont. max. patrocinio* (Rome, 1670). See Barbiellini, 77 n. b.

Fabretti's account shows that already in his time the course of the road was difficult, given the steep rise in elevation between the Anio Valley and Riofreddo (see Crainz and Giuliani, "Via Valeria," 80 n. 38). The route through Riofreddo described here appears to have been abandoned in the early nineteenth century, although its course can still be traced on the ground (see Crainz and Giuliani, "Via Valeria," 82–84, tables XV–XVII). Fabretti's identification of this route as the ancient Via Valeria was uncontested until the early nineteenth century, when Gell and Nibby argued for the alternate route, the longer course running southwest along the Anio past Anticoli, then directly north past Arsoli (see Crainz and Giuliani, "Via Valeria," 80–81). Their arguments were also supported by Rodolfo Lanciani (278–79 n. 1) in his reexamination of the problem in connection with the sources of the Aqua Marcia and were further strengthened by the discovery along the Anio, in August 1889, of the *bivium* (intersection) of the Via Valeria and Via Sublacensis located five hundred meters southeast of the Anticoli bridge, along with three milestones of the Via Valeria dating from the fourth century A.D., all bearing the number "36" (see Ashby, 95–96 [118–19]; Crainz and Giuliani, "Via Valeria," 75–76, table V).

Only recently have Crainz and Giuliani reopened a general examination of the problem, arguing that there was indeed a "Via Valeria Vetus," the direct route running past Riofreddo posited by Fabretti, which represents the original course of the Roman military road first built in the late fourth century B.C., in conjunction with the establishment of Latin colonies at Alba Fucens and Carseoli. This earlier route through Riofreddo served as the principal course of the Via Valeria from the Anio Valley to the Piana del Cavaliere until the late second century B.C., when the introduction of the Aqua Marcia into Rome prompted a major change in the route of the road to avoid the steep mountain climb northeast of the Osteria della Ferrata and to provide easy access to the sources of the aqueduct and for maintenance of its conduit (see Crainz and Giuliani, "Via Valeria," 86–88). Frontinus took his data concerning the distances of the sources of the Aqua Marcia, Aqua Claudia, and Aqua Anio Novus in the *De aquaeductu* from the second Via Valeria—running southeast along the Anio River, then turning north after Anticoli to run past Arsoli toward S. Giorgio—which was in primary use during the first century A.D.

Fabretti's observations about the course of the Via Valeria are therefore correct in part, even if his arguments for the location of its *bivium* with the Via Sublacensis are now recognized to be in error. Had Fabretti known about the *bivium* of the Via Valeria and Via Sublacensis near the Anticoli bridge and the three milestones unearthed there two centuries later, his reconstruction and reading of the evidence would no doubt have been far different.

To support his arguments here, however, Fabretti cites further evidence. First, he cites the distance of sixteen miles on the Riofreddo route from Tivoli to S. Giorgio confirmed by a measuring device installed on his carriage, which Fabretti argues corresponds to the figure of thirty-six miles given by Frontinus at *De aquaeductu* 7.6 (cited earlier in this dissertation). This point is far from certain, as Fabretti himself seems to recognize in this passage, where he alludes to the deviation of the modern Via Valeria from the course of the ancient route below Cantalupo and to discrepancies inherent in measurements of this sort. Lanciani (278–79) describes Fabretti's figures from his wagon measurements as "un pò alla buona" [a bit exaggerated]. It is interesting to note that Poleni (33–34 n. 25) accepts

Fabretti's argument for the emendation of "XXXIII" to "XXXVI" but also points out that there is manuscript authority for the figure "XXXVI."

Second, Fabretti cites the forty-first milestone of the Via Valeria (*CIL* IX, 5966) erected at Carsoli. Although Fabretti makes much of the number "XXXXI," calculating the distance from Tivoli according to his reconstruction of the route of the Via Valeria, this *cippus* is not in situ and therefore offers no convincing evidence about distances and the course of the road.

Third, Fabretti cites the thirty-eighth milestone erected by Nerva (*CIL* IX, 5963) for an unnamed route (presumably the Via Valeria), found in the locality of Sonnoletta northwest of Arsoli and erected in the piazza of Arsoli (cf. Ashby, 96 n. 1 [118 n. 55]). Lanciani (278) fixed its location at the Vigna della Corte, twenty-one hundred paces from the *bivium* of the Via Valeria and Via Sublacensis near the Anticoli bridge. In citing the information given by Gruter on its location, "ad fontem Somnulae secundum Viam Valeriam" [at the spring of Somnula, along the Via Valeria], Fabretti takes special pains to attribute that notice to **Ercole Ciofano,** whom he identifies as Gruter's source. According to Fabretti's interpretation of the evidence, the milestone could not have stood on the Via Valeria proper. However, the criticism presented here of Ciofano's commentary on Ovid (Antwerp, 1581–83) is not accurate: Ciofano, in his notes on *Tristia* 4.10.4, did not specifically identify ancient Arsoli with the ancient Carsioli but equated Carsioli with the nearby *Piano di Perito e Celle*. See P. Burmann, ed., *Publii Ovidii Nasonis opera omnia* (Amsterdam, 1727), 3:643; Burmann incorporates Ciofano's commentary.

Fabretti's criticism of Ciofano also permits introduction of additional criticism of Holste and of a recently published map of the region based on Holste's topographical work. This map, already cited (and criticized) in the first dissertation (I.7), is attributed to Innocenzo Mattei (cited in its title), who appears to have published it thirteen years after Holste's death. For a discussion, see Almagià, *L'opera geografica,* 134–36; the map is reproduced in Almagià, *Monumenta,* 2:56–57, table XIV.

Fabretti's final censure in this passage is reserved for **Athanasius Kircher, S.J.,** whom he does not name directly here, but whose *Latium* he criticized sharply in the first dissertation (I.2). Fabretti's criticism here is accurate but again fierce, calling attention to Kircher's excessive reliance on local antiquarians for his topographical arguments. Kircher (52a–b) does indeed cite the archbishop of Albano as one of his authorities, describing him as a "virum haud vulgari doctrina indagatorem" [investigator of extraordinary learning], and mentions letters from a doctor of Castel Gandolfo, one Valentinus Steboerus, praised as "doctrina et eruditione eximius de Lanuviniis Albanisque antiquitatibus" [distinguished in learning and erudition concerning the antiquities of Lanuvium and Albanum].

Fabretti also censures Holste's misnaming of the Pons Scutonicus, a prominent landmark in the region, criticizing once more Holste's theories about the course of the Via Valeria. On this point, Holste was correct and Fabretti mistaken: the Pons Scutonicus near the present village of Roviano indeed carried the Via Valeria on its course north from the Anio toward Arsoli. See Crainz and Giuliani, "Via Valeria," 84–85; Mari, "La valle dell'Aniene," table IVa.

Fabretti reserves for fullest discussion one final piece of epigraphical evidence, the thirty-eighth milestone of the Via Sublacensis that he found in situ in the bed of the Anio River in 1679. He describes this dramatic discovery in considerable detail to provide what he considers decisive proof of his reading of the evidence for the course of the Via Valeria. In actuality, Fabretti's discovery, while extremely important in its own right in demonstrating the changes of the course of the Anio River itself since antiquity, confirms the route of the Via Valeria outlined by Holste, as well as the location of its *bivium* with the Via Sublacensis, attested by the discoveries in 1889.

c. Sources of the Aqua Claudia

In contrast to his earlier discussion of the route of the Via Valeria, Fabretti treats the sources of the Aqua Claudia very briefly, identifying them as the springs of the Acqua Serena at kilometer 60.5 of the modern Via Sublacense (fig. 17, nos. 19–20). Fabretti's basic agreement with Holste on this point is surprising, given his earlier objections to Holste's work on the route of the Via Valeria and its *bivium* with the Via Sublacensis. Fabretti's comments here are in fact limited to criticism of Holste's attempt to distinguish between the Caerulean and Curtian springs, identifications that he questions skeptically but does not dispute.

In actuality, however, both Holste and Fabretti were mistaken about the sources of the Claudia. The springs less than three hundred paces from Fabretti's milestone are actually the sources of the Aqua Marcia, an identification now secure from the thorough investigations of Lanciani (270–86), later endorsed by Ashby (95–98 [115–20]) and recently confirmed by the topographical work of Crainz and Giuliani, 80–82. Fiore Cavaliere and Mari ("Acquisizioni," 464–65) present a convenient summary of the evidence. First, in fixing the sources of the Marcia at *De aquaeductu* 7.6, Frontinus based his notice of its location "three hundred paces from the thirty-eighth milestone of the Via Sublacensis" from the course of the Neronian road running along the Anio River southeast of its *bivium* with the Via Valeria. Second, the other figure given by Frontinus in *De aquaeductu* 7.6, that of the distance of three miles from the *bivium* itself, was based on an older, pre-Neronian routing of the Via Sublacensis; this older route served as a service road to the sources of the Marcia and Claudia before Nero paved his shorter, more direct route to his villa at Subiaco. Third, the pre-Neronian route of the Via Sublacensis followed an arclike course further east, along higher ground at the base of Monte La Prugna, and approached the sources to the northeast; the third mile of its course from the *bivium* with the Via Valeria therefore falls at the site of the Serene springs, directly east of which, along the Anio River itself, Fabretti found the thirty-eighth milestone of the Trajanic Via Sublacensis in situ (see Fiore Cavaliere and Mari, "Acquisizioni," 464, fig. 1).

Fabretti's error in identifying the sources of the Claudia is a logical result of his mistaken reconstruction of the course of the Via Valeria. This led to his error in identifying the sources of the Marcia, a topic he introduces in this passage but reserves for fuller discussion in a later section. His brief treatment here focuses only on Frontinus's notice about relative levels of the conduits (*Aq.* 18.4), to argue against Holste's location of the Marcia's sources at the Forma della Mola (fig. 17, no. 22).

As Poleni observed (64 n. 12), Fabretti's printed text in this passage, which reads that the Marcia ran five and a half feet below the Claudia at Spes Vetus, must reflect a typographical error; as is shown in Fabretti's own cross section presented in the first dissertation (fig. 7), the Marcia is a good twenty-five and a half feet below the Claudia at this point. The translation presented here adopts Poleni's correction.

d. Representations of the Aqua Claudia

In this lengthy digression on the iconography of the Aqua Claudia, Fabretti argues that the aqueduct is represented in a lunette relief of a solitary nude female reclining in a pool or stream, into which she is pouring or drawing water through a partially submerged urn. To support this reading, Fabretti first cites epigraphical evidence to demonstrate that the Aqua Claudia was named and depicted in inscriptions and received votive offerings.

Interpretation of the first inscription cited here, that of a votive altar to the "Cyanean waves" by a certain Contuccius (*CIL* VI, 555), is problematic at best. The translation presented in the text follows the reconstruction of Moritz Haupt published in *CIL*. Whether the "Cyanean waves" are to be equated with the springs of the Fons Caeruleus that, as Frontinus notes (*Aq.* 14.1), were one of the Claudia's sources is problematic, as is also Fabretti's attempt to connect the "rising" and "falling" of the water with details of the relief presented here. This discussion of the Contuccius inscription permits Fabretti to introduce references to Gruter, as well as the scholarship of **Stephanus Vinandus Pighius** (Stephan Wynants Pighe, 1520–1604). Pighius, a Dutchman who spent eight years in Italy and later became a canon at Xanten on the Rhine, produced an edition of Valerius Maximus (Antwerp, 1575) and the *Annales Romanorum* (Antwerp, 1599–1615) cited here by Fabretti. See Sandys, 2:217.

The other two inscriptions cited (*CIL* VI, 8494–95) are much more straightforward but provide no substantial support for Fabretti's identification of the female relief as that of the Aqua Claudia. Introduction of these inscriptions permits Fabretti to include a further digression on the term *vilicus*, appearing in *CIL* VI, 8495, and on the variant spelling there of the word *vicarius* with a *k*. The citation of *vilicus* introduces a correction of the Latin linguist **Claude Dausque, S.J.** (1566–1644), whose *Ortographia Latini sermonis vetus et nova* appeared in 1677.

Fabretti devotes far more discussion to the iconography of the relief itself, taking special pains to point out what he considers its unusual features: the position of the urn held by the female figure and the fact that the figure, who has many of the characteristics of a nymph, is solitary, not appearing as one of a group of three. The first argument, that the urn appears to be drawing water as much as pouring it out, does not seem conclusive, since the figure of a reclining nymph resting on an urn is frequently attested in the iconography; see *LIMC*, 8:893, "Nymphe couchée" 86 (M. Halm-Tisserant, G. Siebert).

Fabretti's second observation—that the figure is solitary, not part of a group—introduces a much longer digression on the tradition of threefold representations of nymphs and other goddesses, with a survey of six reliefs and paintings. The first relief surveyed is a votive relief (*CIL* VI, 30791) depicting three nymphs pouring water from urns and holding leaves of an aquatic plant, which Fabretti appears to identify as *Nymphaea heraclea*, cited by Pliny the Elder (*HN* 25.75); the nymphs are flanked by a snake on the left and the freedman Augustalis of the inscription, sacrificing at a round altar on the right. Discussion of this relief introduces comparisons on the salutary attributes of snakes and *Nymphaea heraclea*. Fabretti cites a manuscript of Cardinal Carpegna (his addressee) as his source for this relief but provides no definite information on the whereabouts of the relief itself.

The second item surveyed by Fabretti here is a fragmentary relief with a votive inscription (*CIL* VI, 554), depicting two nymphs pouring water from urns, flanking a third who stands holding a seashell. The third item surveyed is a well-known votive tablet now in the Vatican Museum (*CIL* VI, 549), depicting, from left to right, Diana, three nymphs holding seashells, Silvanus in a tunic, and Hercules. See *LIMC*, 2:833, "Artemis/Diana" 299; 7:766, "Silvanus" 49 (G. Bauchhenss). For a possible identification of the nymphs with the Silvanae, nymphs associated with Silvanus, see P. F. Dorcey, *The Cult of Silvanus: A Study in Roman Folk Religion* (Leiden, 1992), 42–44. This relief introduces another digression on the cultic associations of Hercules and Silvanus and the attributes *valens* and *valentius* given to each god.

The fourth item surveyed here is a wall painting depicting three nymphs and Pegasus from the tomb of the Nasonii outside Rome; see *LIMC*, 8:896, "Nymphai" 72. The fifth item is a well-known funerary relief now in Berlin (*CIL* VI, 10036), depicting three Graces and a bride, perhaps recently deceased; see *LIMC*, 3:206, "Gratiae" 48 (H. Sichtermann). Fabretti's discussion introduces a correction of **Martin Smet** (d. 1578) who discussed this relief in his *Inscriptionum antiquarum* (1588), a forerunner of Gruter's epigraphical corpus. On Smet's career, see Sandys, 2:145.

The sixth item surveyed here is a two-part votive relief from the headquarters of the Equites Singulares in Rome (*CIL* VI, 768 = *ILS* 4776), depicting, in its top panel, the three Suleviae Matres holding ears of grain and baskets of fruits (or flowers?) in their laps and, below, a sacrifice carried out by the centurion Aurelius Quintus. For detailed discussion of this relief, see *LIMC*, 4:1593–1600 (M. Ihm); *RE* 4A (1931): 725–27 (F. Heichelheim). Fabretti claims credit for the discovery of the relief and provides a detailed description of the sacrificial scene and the formula

"LAETVS LIBENS" in the inscription, but he gives no details about the Suleviae themselves.

Fabretti's presentation of these six reliefs and paintings is supplemented by numerous citations from classical literature (including Theocritus, Pausanias, Ausonius, Plutarch, and Varro) illustrating the triple appearance of female divinities. His introduction of a passage of Pausanias permits a complimentary reference to the Dutch polymath and classical scholar **Gerard John Vossius** (1577–1649). See Sandys, 2:307–8.

e. Source of the Aqua Marcia

Fabretti now returns to the topic touched on briefly in II.4c, the source of the Aqua Marcia. Arguing from his reconstruction of the route of the ancient Via Valeria that the intake of the aqueduct is to be identified with springs below the Church of S. Maria in Arsoli, he first cites a fragmentary late inscription making reference to an arcade that he found in the altar of the church (*CIL* IX, 4051.6). The inscription, which does not cite the Aqua Marcia by name, gives little support to Fabretti's argument concerning the location of the Marcia's source but is important evidence for late restorations of the aqueduct by Honorius and Arcadius. The fragment, published here for the first time by Fabretti, was supplemented in the nineteenth century by other pieces of the same inscription discovered in the castle at Arsoli by C. L. Visconti. For a full discussion, see Lanciani, 284–85 (Lanciani quotes Visconti's report at length); Ashby, 92 n. 1 (113 n. 39).

Fabretti next spends considerable effort refuting the published work of **Gerolamo Mercuriale** (1530–1606), a doctor and scholar of ancient medicine. Mercuriale, who produced an edition of the Hippocratic corpus (Paris, 1585), as well as extensive other works on ancient medicine, had argued from the text of Galen that the water of the ancient Aqua Marcia was heavy and therefore considered unfit for drinking. Fabretti takes pains to refute this, through extensive quotations from Galen himself, as well as ancient praise of the Marcia from Pliny and other authors. For Mercuriale's career and contributions to scholarship on ancient medicine, see A. Simili, *Gerolamo Mercuriale lettore e medico a Bologna* (Bologna, 1966).

The most lengthy discussion within this section concerns the *cippi*, or boundary stones, marking the courses of aqueducts, which Fabretti introduces through a notice of one such marker discovered by him near the

Pons Scutonicus (*CIL* VI, 1251b), along with a togate statue and remains of a vaulted structure. Ashby (93–94 [114]) observes that the *cippus* of travertine was not found in situ and that Fabretti's notice that it was marble is incorrect.

Much more important, however, are Fabretti's observations about the listings of 240 feet found regularly on *cippi* along the aqueduct courses. Citing parallels of boundary markers of the Aqua Virgo (*CIL* VI, 1253b, 1254) and one of the Aqua Marcia/Tepula/Julia discovered by himself (*CIL* VI, 31561g = 1249f), Fabretti argues that these standard intervals of 240 feet represent the length of a Roman *iugerum*, regularly employed in measuring the lengths of the conduits from their distribution points in the city.

The number 240 is not canonical in *cippi*, since examples have been found indicating different distances (*CIL* VI, 31558g = VI, 1243e = XIV, 4080 [320 feet]; *CIL* VI, 31558a [230 feet]; *CIL* VI, 31562f [250 feet]). But Fabretti's observations, made here for the first time, were well founded and have been generally accepted by later students of the aqueduct system. Cf. Lanciani, 557–61; Ashby, 57–58 (73–74); Hodge, 103. Lanciani (559) describes Fabretti's interpretation as "verità . . . aritmetica" [mathematical truth]. Deviations from the standard interval of 240 feet were probably the result of unusually winding courses in particular areas. As Lanciani observes (557), *cippi* of only five aqueduct lines have been discovered to date, those of the Anio Vetus, Marcia, the combined Marcia/Tepula/Julia, and the Virgo. *Cippi* appear to have been in use for only some fifty years, having been introduced by Augustus in his rebuilding of the aqueduct system, and while later emperors restored earlier *cippi*, no new such boundary stones dated later than the time of Claudius are attested. For a recently discovered *cippus* of the Aqua Marcia and an updating of Ashby's listings, see S. Gatti, "Un nuovo cippo dell'*Aqua Marcia*," in *Trionfo* II, 93–104.

Fabretti argues here that the use of iugeral *cippi* was abandoned after Claudius's reorganization of the water system because the Roman road system, which roughly paralleled the lines of the conduits outside the city, provided an easier system to mark the distances from the distribution points. He also cites Frontinus's notice that he prepared maps of the aqueduct network to facilitate maintenance and repair of the lines (*Aq.* 17.3); such a system appears to have superseded the iugeral markings and explains Frontinus's practice of citing road distances in miles in his descriptions of the aqueduct courses. Cf. Hodge, 103–4.

Fabretti's observations about the iugeral standard in *cippi* prompt criticism of errors in the work of **Jean-Jacques Chifflet** (1588–1660), who had published a short treatise on the Aqua Virgo in 1657. See Ashby, 167 n. 6 (199 n. 6). However, Fabretti's most severe censure in this passage is reserved for **Pirro Ligorio,** whom he had criticized earlier, in the first dissertation (I.4f–g). Here, Fabretti discusses at length a fabricated inscription published by Ligorio (*CIL* VI, 800*, apparently based on *CIL* VI, 1244), pointing out its errors of fact and historical inconsistencies, and he cites similarly sharp criticism of Ligorio recently published by **Exechiel Spanheim** (1629–1710). For the career of Spanheim, a native of Geneva who studied at Leiden and became a professor at the age of twenty-two years, see Sandys, 2:327. To point out inaccuracies in Ligorio's inscription, Fabretti also cites the work of **Onofrio Panvinio** (1529–68), whose edition of the *Fasti consulares* was published in 1557, and **Adolf Occo** (1514–1606), who published a collection entitled *Imperatorum Romanorum numismata*. See Sandys, 2:145; *EHCA*, 851–52 (R. W. Gaston).

f. Source of the Aqua Augusta

Fabretti next turns to the source of Augustus's supplement to the Aqua Marcia, which he seeks to identify with ancient construction either at springs known as the Lake of S. Lucia (fig. 17, no. 18) or at another spring to the north (fig. 17, no. 17). As a result of his earlier error in fixing the sources of the Aqua Marcia, both of these identifications are incorrect. Modern topographers generally place the source of the Aqua Augusta at the Rosoline springs at kilometer 61.5 of the modern Via Sublacense, some one hundred meters east of the side road to Marano, where remains of two conduits were seen by Ashby. See Fiore Cavaliere and Mari, "Acquisizioni," 464; Ashby, 96, fig. 6 (117–19, fig. 14); Aicher, 161–62.

For an alternate view positing the source of the Aqua Augusta at the modern Mola di Agosta (probably to be identified with springs identified by Fabretti as the "Forma della Mola" [fig. 17, no. 22]), see Panimolle, 86–88. Panimolle places the sources of the Marcia at the Rosoline springs and argues that the sources of the Aqua Augusta are eight hundred paces distant, precisely at the springs below Agosta, and that the name *Agosta* is derived from Augustus's supplemental branch. Panimolle's arguments ignore the detail in Frontinus's notice that the Aqua Augusta ran eight hundred paces to the conduit of the Marcia, not to its source (*Aq.* 12.2), a fact mentioned by Fabretti, albeit in error, in supporting his identification.

g. Origin of the Aqua Marcia

Finally, Fabretti turns to a brief discussion of the origins of the Marcia, refuting the notice of Pliny the Elder (*HN* 31.41) that the Marcia originated in Marsian territory, east of the Fucine Lake. Fabretti quotes Holste to reject the statement of Pliny and sharply criticizes the topographical arguments of **Georg Fabricius,** whom he also censured in the first dissertation (I.4c). Poleni (33 n. 24) comments on Fabretti's criticism in this passage, "Haec quidem videntur duriuscule scripta: attamen fateri debeo, haud video, qua ratione ea Plinii narratio defendi queat" [These words seem a bit harshly written, yet I must confess I do not see how that statement of Pliny can be defended].

5. SETTLING TANKS

a. The "Villa delle Vignacce"

Fabretti now describes in considerable detail several structures closer to Rome that he identifies as *piscinae* (settling tanks) of the Anio Valley aqueducts; the first (fig. 26), which he assigns to the Aqua Marcia, is a construction near the fourth milestone of the Via Latina, an area commonly known today as the "Villa delle Vignacce." In the sixteenth century, Pirro Ligorio had recorded "grandissimi ruini di una villa" [very great ruins of a villa] in the area (*Neap.* xxxiv.179, cited in T. Ashby, "The Classical Topography of the Roman Campagna-III," *BSR* 4 [1907]: 77), but the presentation given here by Fabretti is the first detailed account of a structure subsequently described and cited in Lanciani, 293; T. Ashby, "Classical Topography-III," 77–78; Ashby and Lugli, "La villa dei Flavi cristiani," 183–91; Ashby, 133–34 (166); and, most recently, following restoration work on the nearby Acqua Felice in 1980–81, Pisani Sartorio, "Punto di derivazione dell'acqua Marcia," 55–56. Pisani Sartorio also includes a cross section and reconstruction of its functioning (reproduced in Aicher, 95, fig. 22).

In his account of the structure, Fabretti argues that water tapped from the main conduit of the aqueduct was channeled underground through an intermediary *puteus* (wellhead) to the lower cisterns of the *piscina*, then rose to the upper cisterns through an opening in the middle chamber. However, he is unusually defensive here about his reading of the archaeological evidence, acknowledging that he was unable to explore the underground communication between wellhead and reservoir. His explanation of its operation given here indeed raises more problems than it solves and has been rightly described as "ingenious but erroneous" (I. A. Richmond in Ashby, 133 n. 7 [166 n. 216]).

There have been several later attempts to make sense of the physical remains. Lanciani (293) seemed to accept Fabretti's identification of the complex as the *piscina* of the Aqua Marcia: "Credo che il Fabretti abbia colto nel vero riconoscendo la piscina della marcia in quella" [I believe that Fabretti hit on the truth in his recognition of the settling tank of the Marcia in it]. However, Lanciani also observed that the structure Fabretti describes cannot be the *piscina* described by Frontinus (*Aq.* 19.1); as Fabretti himself indicates, the facing of *opus mixtum* (irregularly faced concrete) in the construction dates from the time of Hadrian, dating also confirmed by the brick stamp cited in this passage. The nearby "Villa delle Vignacce" can also be dated by brick stamps to the same period. In addition, Lanciani noted another seven similar structures in the immediate area, making impossible the firm identification of any one of them with that cited by Frontinus.

Lanciani (293) gave a general account of the structure described by Fabretti, with no explanation of its operation; instead, he raised the possibility that a siphon was perhaps used to transfer the water from the nearby aqueduct conduit: "L'acqua vi giungeva per mezzo di un sifone (?)" [Did the water reach the structure by means of a siphon?]. In 1907, Ashby ("Classical Topography-III," 77–78) presented a more detailed description of the complex, based largely on that of Fabretti, who had seen the site in far better condition in the late seventeenth century; Ashby supplemented this report with more details and new diagrams in 1927, in his joint publication with Lugli ("La villa dei Flavi cristiani," 183–91), and repeated it in abridged form in 1935 (Ashby, 133–34 [166]).

The most recent publication by Pisani Sartorio demonstrates clearly that despite the value of Fabretti's detailed description of the structure as it stood at his time, his account of its operation is in error. Pisani Sartorio ("Punto di derivazione dell'acqua Marcia," 55) offers the following reconstruction of its operation: the water from the acqueduct was drawn through a pipe installed in the conduit wall to the wellhead situated between the aqueduct conduit and reservoir, today at ground level but originally at the level of the conduit itself; from this wellhead, the water was channeled to fill the nearby reservoir, from where it was distributed further in pipes to supply the nearby villa, with a runoff channel provided for excess water. The wellhead itself may have also functioned as a filtering device for the water tapped from the reservoir. The lower chambers, which Fabretti reconstructed as cisterns, appear to have served only as support for the tanks above.

In contrast to his detailed account of the "Villa delle Vignacce" complex, Fabretti describes much more briefly a second cistern of trapezoidal shape nearby (fig. 27), which he identifies as a *piscina* of the Tepula and Julia, arguing that similarities of design in the two structures suggest a parallel functioning. Again, he is defensive about his reading of the archaeological evidence, observing that Frontinus's note (*Aq.* 68.4) that the *piscina* of the Marcia was closer to the city than that of the Tepula and Julia contradicts his identification. Poleni (123 n. 1) noted Fabretti's uncertainty and defended Frontinus's text on this point. Lanciani (298) observes that Fabretti's identification with the *piscina* of the Tepula and Julia cannot be correct, because the structure stands too close to the city, "ad sixtum ab urbe miliarium" [at the sixth milestone from the city]. Ashby (134–35 n. 8 [167 n. 224]) identifies it as a cistern supplying a nearby villa and suggests from its deposit that the water it carried was that of the Aqua Marcia.

To strengthen his arguments on the functioning of the two structures described here, Fabretti also presents a third example as a parallel, a reconstruction of a *piscina* that he identifies as that of the Aqua Virgo (fig. 28), the operation of which he states is similar to that of the *piscinae* on the Via Latina. Fabretti describes this structure, located in Rome in the Vicolo del Bottino (to which it seems to have given its name), as full of mud in the seventeenth century; it appears to have served a branch line of the Aqua Virgo, not the main conduit itself. Because Frontinus reports that the Virgo had no settling tanks (*Aq.* 22.1), the structure must be assigned not to the original aqueduct but to a later branch. Lanciani (336–37) remarks that, in contrast to Fabretti's description here, the *piscina* was still functioning as a "castello di distribuzione" [distribution tank] in his day; Ashby (173–74 [205]) reports that it was destroyed in the construction of a public cable elevator to the Pincian Hill, when one of its chambers became the waiting room. Fabretti's diagram is reproduced by Van Deman (173, fig. 19) and by P. Pace in "Tecniche di conduzione e distribuzione dell'acqua in epoca romana," in *Trionfo*, 141, fig. 3. The hydraulic functioning of the Vicolo del Bottino *piscina* is also illustrated and discussed by Hodge (123–25).

b. The Aqua Claudia/Anio Novus

Fabretti now turns to another reservoir, further distant from Rome, which he identifies as the *piscina* of the Aqua Claudia/Anio Novus, located by Frontinus's treatise "ad septimum ab urbe milliarium" [at the seventh milestone from the city] (*Aq.* 72.3). His description here is general: he cites only the poor state of the remains in the seventeenth century, the size of the construction (which, he plausibly argues, exceeds that of a private cistern), and the double nature of the construction, which he suggests was planned to accommodate both aqueducts.

Fabretti's identification was accepted by Lanciani (357–58), who described "poche vestigia di piscina" [a few traces of a settling tank] in 1881 and quoted Fabretti's account here at length, also reproducing his diagram of the structure (Lanciani, table VII.10). Only a few years later, however, the discovery in 1884 of the *piscina* of the Anio Novus in the Villa Bertone near Capannelle, showed that the structure Fabretti describes here had nothing to do with either the Aqua Claudia or the Aqua Anio Novus; the actual *piscina* was found filled with a huge amount of pebbles from the Simbruine mountains by the aqueduct, so many, in fact that the owner, Cavaliere Bertone, was able to pave a kilometer of roads and make plaster for six or seven buildings on his property. For more detailed discussion, see Ashby, 225–26 (264–65); Hodge, 124; Blackman and Hodge, 38.

c. Measurement of Water

Fabretti's discussion of *piscinae* now introduces another topic, the measurement of water carried out in settling tanks, as cited by Frontinus (*Aq.* 72.3). Fabretti treats this complicated problem in cursory fashion, focusing mainly on criticism of the recent hydraulic studies of **Benedetto Castelli,**

O.S.B. (1578–1643). Castelli, a student of Galileo, had been professor of mathematics at Pisa until 1626, when he was called by Pope Urban VIII to be papal consultant on hydraulics and professor of mathematics at Rome. His treatise *Della misura dell'acque correnti* (Rome, 1628) was frequently reprinted in the seventeenth century (Fabretti appears to have consulted its third edition [Bologna, 1660]) and was translated into English by Thomas Salisbury in 1661; it is considered by many scholars as the beginning of modern hydraulic research. For Castelli's career and other publications, see *DSB*, 3:115–17 (S. Drake).

Fabretti shows little interest in how Frontinus measured the volume of individual aqueducts: for discussion of this problem and the hydraulics involved, see Hodge, 123–24; R. Taylor, "Torrent or Trickle? The Aqua Alsietina, the Naumachia Augusti, and the Transtiberim," *AJA* 101 (1997): 468–71. Instead, Fabretti sharply criticizes Castelli for presenting as his own discovery the principle that velocity and head affect the flow of water through an orifice; Fabretti cites several passages of the *De aquaeductu* to demonstrate that Frontinus had indeed understood this principle of hydraulics.

However, as Poleni, himself an engineer, observed in his commentary on the *De aquaeductu*, Fabretti's criticism of Castelli as presented here is excessive: Frontinus recognizes the effect of head and velocity on the flow of water, as seen in the passages quoted here, but nowhere in the *De aquaeductu* does he give a full demonstration of the principles of hydraulics set forth in Castelli's treatise and the relationship between the size of pipes and flow of water. Poleni (197 n. 8) writes, "Dum autem amore in Frontinum, odio (ut ita dicam) sive ira quadam . . . in Castellium Fabrettus ducitur, multum in hac re a veritate (quod humanum est) aberrat" [While influenced by love for Frontinus or hatred (if I may) or anger for Castelli, Fabretti wanders far from the truth, a human shortcoming].

Frontinus, of course, is not to be faulted for not having written a theoretical work on hydraulics, since that was not his purpose in publishing the *De aquaeductu*. However, to be fair to Castelli, we must recognize that his seventeenth-century hydraulic study approached the problem on a scientific basis and amplified Frontinus's observations, a fact that Fabretti seems deliberately to overlook here. See Blackman and Hodge, 17–18; M. Lewis, "Theoretical Hydraulics, Automata, and Water Clocks," in Wikander, *Handbook*, 347–48.

6. THE BUILDER OF THE AQUA MARCIA

In the final section of this dissertation, Fabretti treats summarily the problem of identifying the Roman who introduced the Aqua Marcia into the city. He first presents a review of the literary evidence (Frontinus, Pliny the Elder, and Plutarch), citing for Frontinus the text of *De aquaeductu* 7.1 before Giovanni Poleni emended "Marcus," the reading of the Monte Cassino manuscript (C), to "Marcius," a correction accepted by all later editors of Frontinus (see Poleni, 27–28 n. 4). Fabretti's acceptance of the attribution of the Marcia to Q. Marcius Rex rests primarily on the authority of Pliny's notice, but he devotes more attention to the famous coin of L. Marcius Philippus that he reproduces in the text (fig. 30); the coin depicts five arches of the aqueduct on its reverse, with the caption "AQVAM" superscribed (British Museum, Department of Coins and Medals, *Coins of the Roman Republic in the British Museum* [London, 1970], 1:485, no. 3890).

In his discussion, Fabretti plausibly argues that Philippus's depiction of Ancus Marcius on the obverse of the coin for reasons of ancestral pride was the inspiration for Pliny's notice (*HN* 31.41) that the aqueduct was originally planned by the fourth king of Rome. He also cites, in passing, the topographical treatise *Roma vetus ac recens* (Rome, 1638) by **Alessandro Donati, S.J.** (1584–1640). A native of Siena, Donati taught rhetoric in Rome. See Sandys, 2:279; *EHCA*, 553 (C. Sperling). This is Fabretti's first mention of Donati, whom he cites several times in the third dissertation (III.3, 5a, 6).

CLOSING

Fabretti concludes this dissertation with an intriguing list of "doubtful" matters that he has chosen to omit from the present discussion: the intake of the Anio Vetus and the course of that aqueduct and the Anio Novus; the Fons Antoninianus, which was tapped as a supplement to the Marcia by Caracalla; the Fons Herculaneus and Fons Albudinus cited by Frontinus (*Aq.* 14.2, 15.4); the capacities of the individual aqueducts listed in the *De aquaeductu*, along with the mathematical errors in the text; and the levels of the lines. He indicates that he will treat these topics elsewhere, a promise never met in his extant publications. This dissertation closes with a deferential salute to Carpegna.

CHAPTER FIVE

Dissertation III on the Waters and Aqueducts of Ancient Rome

To a most noble gentleman and oldest of friends, Giulio dei Conti di Montevecchio, Raffaello Fabretti, son of Gaspare, of Urbino, sends greetings.

1. INTRODUCTION

No one certainly will be more surprised than you, most outstanding Giulio, that I have decided to dedicate to you this third dissertation on the aqueducts. Everyone has come to know our old close association, extended "through a series of so many years as each of us has" [Ov. *Pont.* 4.12.21]—who then acquainted with one of us is ignorant of that? Ours is a friendship "known almost more than you and I are known" [Ov. *Tr.* 3.6.6]. Everyone, then, will judge me to have acted rightly if, through the duty of writing, I am eager to nurture that harmony that, "begun in the green of youth, comes unweakened to hair turning white" [Ov. *Pont.* 4.12.29–30], which has been joined, as it were, by a destined meeting of minds—"each of our stars agrees in an incredible way" [Hor. *Carm.* 2.17.21–22]—and by "common bonds of sacred study" [Ov. *Pont.* 4.13.43]. I shall leave behind whatever glory I have produced for myself as a result, as a testament to those present and those to come.

But, indeed, I seem to see that you, because of this unusual literary subject (as you will call it), in comparison with others, are affected by no small amazement at me and break forth into these words or others quite like them: "Why does that man (curse him) babble at me about aqueducts? I do not abide even the sight of them and would rather hear with ready ears the praises of wine with Homer, the prince and master of our poets. And why, I say, does he recount those restless suburban excursions of his, like a story to a deaf man? I am so devoted to the city and leisure that not at any time have I even sought out country stays at Tusculum, pleasant as they are. And whatever territory lies beyond the city, would it seem to be 'beyond the paths of the year and the sun' [Virg. *Aen.* 6.796]?"

Nevertheless, having spoken first this pardon, in this exchange and perhaps for the first time, I am forced to disagree with you. There is more than one reason at hand why to you most of all, before all my friends, this treatise should be dedicated. Although you often display yourself to us—through poetic boasting, I think—both as most fond of Homer and no less than him devoted to wine, it has nevertheless seemed good to you to pursue lyric poetry and to imitate Pindar rather than Homer. Not wrongly, even to a Pindaric bard, will the ancient abundance of waters entering Rome be described, since Pindar himself proclaimed on the very threshold of his odes that "water is best" [*Ol.* 1.1].

Moreover, when we seek out the fourteen aqueducts existing at the time of Procopius [*Goth.* 1.19.3], ten of them have been lost altogether, and four remain. The Crabra, or Damnata, the dirtiest of all and foul in its very appearance, serves not only for irrigation of gardens, as once before, but for washing off skins from the butcher. The Virgo is almost reduced to nothing and unworthy of its own name, poured out for the most part into the Rivus Herculaneus—fleeing from it once, the Virgo is said to have acquired its name [Plin. *HN* 31.42]—and leaking from damaged conduits. The Traiana is polluted from water recently tapped for it from Lake Sabatini (not to speak of worse sources) and rightly destined, as once before, for driving mills on the Janiculum, according to the same Procopius as our authority [*Goth.* 1.19.8]. Finally, there is the Alexandrina, scorned by the ancient Romans for a long time and very greatly to be scorned in like manner today, since it is polluted by rains and all too often arrives muddy.

In reading about so many losses of aqueducts, who else, therefore, would bear them more steadfastly than you? You have been accustomed either, by Pyrrhic indifference and lack of feeling, to forgo the most important of things or, by continual abstinence from water, to do without a most excellent substance. Your quite sedentary pursuits now (indeed, your interest in sitting as the one and only thing) and your very great ignorance of this *ager suburbanus*, which you have never inspected, either at close hand or at a distance, force me, because I know that you are excellently versed in foreign geography, in like manner to take care to aid your leisure through my constant excursions and to instruct you in this domestic knowledge of Roman lore. Indeed, especially "that holy and venerable name of friendship" [Ov. *Tr.* 1.8.15] also demands, as a bond of human society, that one friend help the other and that one contribute things he has sought for himself to the benefit of others. "But nevertheless let us seek serious things, with play put away" [Hor. *Sat.* 1.1.27].

2. TOPOGRAPHIC MAP

It is our custom to prepare a topographic map with the course of the roads worked out as much as possible and the location of the more well known ancient monuments, for an easier grasp of what is to be discussed. If ever it has been useful elsewhere, here it is certainly most necessary, as you will see [fig. 31]. In this, I demand that you grant me some assumptions now, which I will prove in fuller detail elsewhere. First, the measurements of distances began not from the Golden Milestone at the head of the Forum, as many have thought, but from the ancient location of the gates before Aurelian's extension of the walls. Likewise, the circumference of the republican walls was quite restricted; indeed, it proceeded along the high ground of the hills, the course of the Tiber from the west, and the *agger* of Servius or Tarquinius Superbus from the east. As a result, both the distances of some places that we have learned and the courses of the roads—especially the branching of the Via Latina from the Appia outside the city, according to Strabo [5.3.9], concerning which there was a long discussion in my first dissertation [I.4d]—are successfully defended from different absurdities of modern writers.

3. THE EASTERN GATES OF THE "SERVIAN" WALL AND THEIR ROADS

Just as, in the first dissertation [I.4d], we briefly discussed restoring the Porta Capena to its proper location, we shall shortly attempt the same

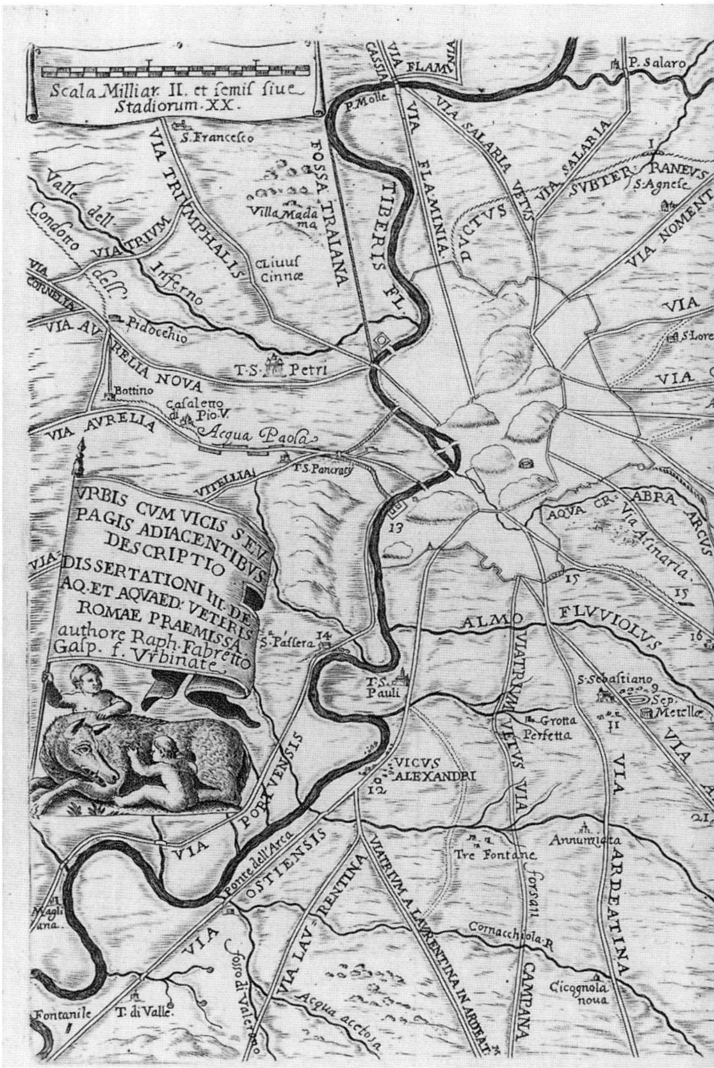

Fig. 31. Topographical map of Latium depicting settlements adjacent to Rome

1. Substructure of the Aqua Virgo, between the Via Salaria and Via Nomentana, below the Church of S. Agnese
2. Another substructure at the bank of the Anio, to the right of the Via Nomentana
3. Another, under the very pavement of the Via Collatina
4. Another on the Via Collatina, on the property Bocca di Leone
5. Ruins of a town or a village on the Via Praenestina
6. Other ruins, on the Via Labicana
7. Other ruins, on the Via Latina, perhaps of the Pagus Lemonius
8. Other ruins of a further village on the same road, on the property Settebassi
9. Others, on the Via Appia, perhaps of the neighborhood of the Camenae
10. Others, of a further village on the same road, on the property *di Statuario*
11. Others, of a village or a neighborhood on the Via Ardeatina
12. The Vicus Alexandri on the Via Ostiensis
13. The shipyard and Emporium on the Tiber, below the Aventine
14. The shrine of SS. Cyro and John on the Via Portuense
15. Remains of the conduit of the Anio Vetus, or the Specus Octavianus
16. The intersection of the Via Asinaria with the Via Latina, between the second and third milestones

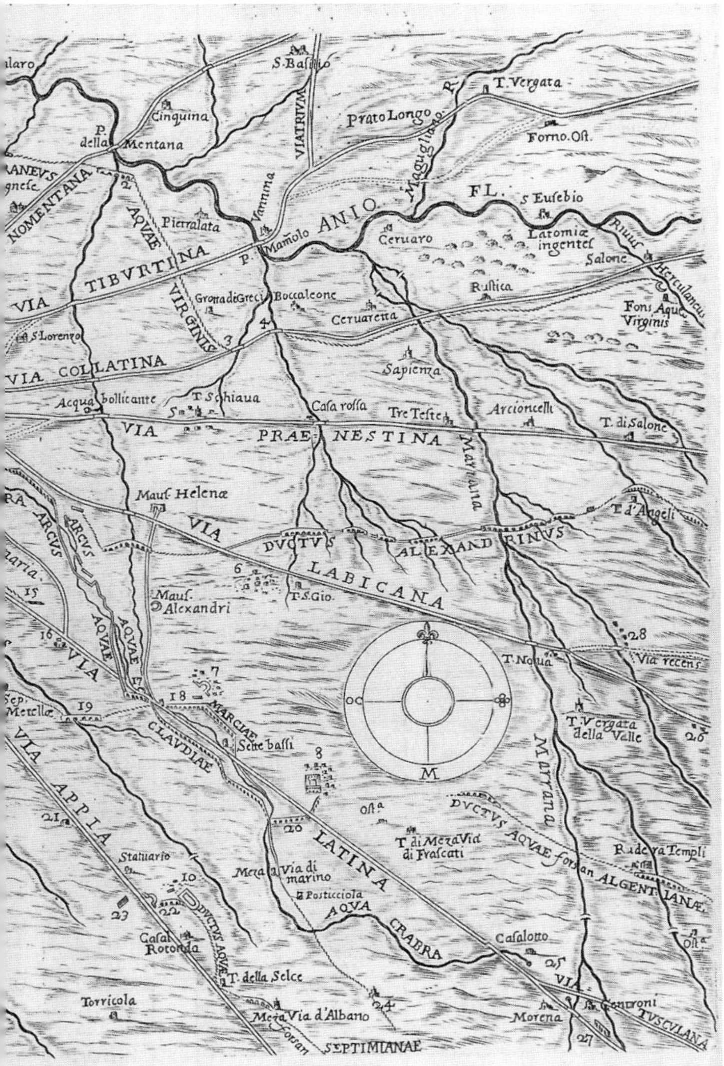

17. The remains of the Temple of Fortuna Muliebris, as we believe, on the Via Latina at the fourth milestone
18. The settling tanks of the Marcia, Julia, and Tepula
19. The conduit led off from the arcade of the Claudia for the use of the village of the Camenae, crossing under the spring of Acqua Santa
20. Another branch from the same arcade for the use of a further village on the Via Latina, in the area of Settebassi
21. A temple on the Via Appia, today the Torrone de'Borgiani
22. Structure in the shape of a theater, on the same road, where the arches of a conduit end (we believe it to be the Septimianus)
23. A very large site for burning dead bodies in the same spot
24. Remains of the settling tanks of the Anio Novus and Claudia
25. Underground construction and emissary of the Aqua Crabra
26. A pair of cisterns on the Via Labicana
27. Traces of a particular conduit on the Via Latina, for the use of a huge building nearby, named *li Centroni*
28. Other remains, conduit for the use of a villa at the seventh milestone of the Via Labicana

thing concerning the Porta Ostiensis or another gate, perhaps with a different name, that was on the same road. Here, with the same conciseness, we shall make an investigation concerning the location of the Porta Collina, Porta Viminalis, and Porta Esquilina that closed the eastern side of Rome, to explain our map no less than to understand ancient writers.

Strabo [5.3.7] is our authority that all three of these gates were located along the *agger:* "Servius completed a wall after the Esquiline and Viminal Hills had been added to the city. Since all these areas could be easily seized from without, they dug a deep trench, and when its dirt was received within, they built an earthwork of about six stades on the inner edge of the trench and set up walls and towers on it, from the Porta Collina up to the Porta Esquilina. The third gate, the Porta Viminalis, is in the middle of the earthwork."

Something similar concerning enclosure of the *agger* by the Porta Collina and Porta Viminalis is found in Dionysius, who indeed omits the Porta Viminalis but does not for that reason exclude it from the place designated by Strabo: "Where, however, the city is most vulnerable to enemy attacks, from the Porta Esquilina up to the Porta Collina, it was fortified by artificial construction, for a trench was dug in front, the minimum width of which is one hundred feet and more, the depth thirty feet, and above the trench rises a wall, supported by a high and broad earthwork, not easily to be shaken by battering rams nor pierced by tunnels. Along this fortification, almost seven stades long and fifty feet wide, the citizens were then assigned by squadrons" and so on [Dion. Hal. *Ant. Rom.* 9.68.2–4].

We would begin to explain the maximum length of seven stades, according to Dionysius, if we start from the Porta Collina, not far from the highest ground of the Quirinal, and proceed along the *agger* already mentioned; sure traces of it remain in the vineyard of the Carthusian fathers and subsequently in the Villa Peretti. It will come to an end in front of the Church of S. Eusebio, more or less, and will lead in a certain way to the Porta Esquilina itself. Antiquarians are still fighting among themselves about this gate and are so disgracefully in a fog about an obvious matter. Between these two end points, if the *agger* is divided again, the Porta Viminalis will obviously be between the two gates the Collina and the Esquilina; for this reason, it will quite correctly be that gate that has sometimes been called "between the earthworks." In like manner, it is also the one "under the middle of the *agger*," as Strabo says [5.3.7], the place where we see even today the *agger* proceeding from the Porta Collina straight up to this point, taking a turn, and being divided in two.

Since we stand on the testimony of ancient and most respected authors and the nature of the place itself, we do not have the leisure to refute here all the whims of modern writers. Instead, we have set forth this argument only in passing, to free our path from brambles and obstacles, such as those authors cited have scattered everywhere. It will therefore be enough for us to put before our eyes and consideration, through this sketch [fig. 32], completed not with finished art but rather simply, how well all the roads, which begin from this direction, correspond to the location we have assigned to the gates and the common course of the other roads that we have measured elsewhere. In this way, each reader, through his or her own reasoning, will understand, from consideration of the topography alone, what most modern writers have been ignorant of. Not paying attention to the roads and the gates from where they go out, they have decided different things, one against another.

With our topographical arrangement then set in place, there appear next to each other, issuing from the Porta Esquilina excellently and with unbending course, each of two roads, namely, the Via Praenestina and the Via Labicana, which went forth from there according to Strabo [5.3.9]. Indeed, although now the Via Praenestina, in its course up to the Porta Maggiore, is confused with the Via Labicana, it is nevertheless possible to see the ancient course of the Praenestina through the gate, now closed, which was left in the extension of the walls by Aurelian for the sake of this road, as well as through traces of the same road from there; none of these is visible from the Porta Maggiore, unless, afterward, we follow the ancient Praenestina by turning to the left. Nardini also describes the rut of this road leading to the closed gate already mentioned, inside the modern walls of the city,[1] and there remains a trace of its beginning, distorted and less than a right angle in each direction, along the north side of the *castellum* of the Aqua Claudia, described in our first dissertation [I.4c]. As a result, the trace will show the most direct course of the Via Labicana from this point to the right and that of the Praenestina from there to the left. On this course, from the *castellum* to the closed gate already mentioned, the Via Praenestina runs into that decagonal structure in the Vigna Coltelli

1. Nardini, 39.

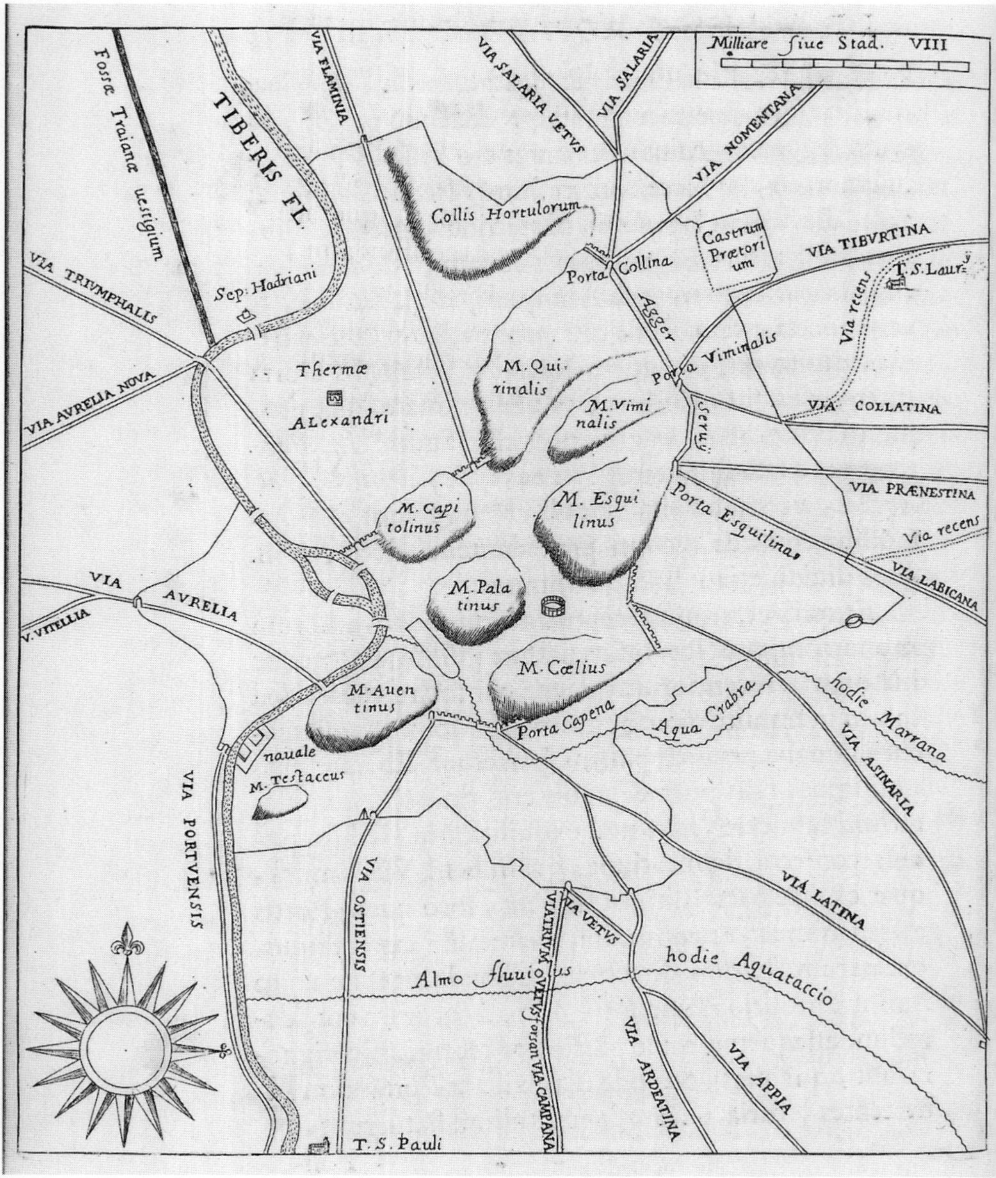

Fig. 32. Plan of ancient Rome, with gates and major roads from the city

that some reasonably have named the Pantheon of Minerva Medica (so called from its round shape) and situated in this fifth region of the Esquiline, according to Sextus Rufus.[2]

Something similar indeed happens along the two roads, the Via Collatina and Via Tiburtina, that depart straight from the Porta Viminalis, or "the gate between the earthworks." By an equal shortcut, because the gate along the Via Tiburtina is closed (indeed, I say that this road appears at the south side of the Castra Praetoria, with very sure traces of its ancient course), through the Arch of S. Lorenzo on the Via Collatina, by a route twisted to the left, they have forced the road to seek the Via Tiburtina below the Basilica of S. Lorenzo. As a result, the Porta Collina will finally send forth its own roads, the Via Nomentana and the Via Salaria, in a straight line, in accordance with the intention of our ancient sources and the pattern of the other roads. Fabricius, however, writes: "The Via Salaria was allotted its name from the gate from which it began."[3] Among ancient authors, you will indeed read of a Via Salaria but not a Porta Salaria and of a Porta Collina but not a Via Collina; for example, Festus [p. 9L] mentions a Porta Agonensis, or Aegonensis (which, for him, is the Porta Collina), and a Via Salaria.

However, the roads had been paved at one time, before the aqueducts were introduced; for this reason, those who either introduced or restored the aqueducts enlarged the conduits by special and noted work when they had come to the roads, both for their adornment and so that the roads would not be narrowed, as we see Claudius did along the Via Labicana in the arch of the modern Porta Maggiore and Augustus along the Via Collatina, with the arch built at oblique angles to align it with the road, as has been said elsewhere [I.4f]. Indeed, modern writers, troubled in vain and disagreeing among themselves about their identification in a matter not at all supported by the truth, have mistakenly accepted these arches as the gates of the ancient city.

Nevertheless, it is necessary to correct the error of Nardini who stands in our way; he posited a second earthwork next to the one already mentioned, as if that one was the work of Servius and as if it is necessary to attribute another to Tarquinius, from a misunderstanding of ancient sources that he cites in support of this opinion, which is new and, I might

2. Sextus Rufus, *De regionibus urbis* (= Graevius, 3:93F).

3. Fabricius, 51 (= Graevius, 3:471A).

say, by leave of so great a man, absurd.[4] I indeed would not believe even Mercury were he to tell me that more than one earthwork ever stood toward the east, or against Gabii (two things that are the same, although Nardini tries to separate them). If inquiry is made of Livy [1.14.3] and Strabo [5.3.7], whom Nardini cites as sources, they will say nothing further than that Servius brought to completion a wall from this direction, up to the Quirinal, which Ancus Marcius had omitted, as Strabo says, and that Servius surrounded the city completely with an earthwork, trenches, and a wall, as Livy says. Moreover, the evidence of Strabo will deny that the Porta Esquilina in particular, from which two famous roads went out, at his time "was closed as unnecessary," as Nardini, twice falsely deduces from him, as if standing on a very sure matter.[5]

Aside from the fact that Dionysius says nothing about Servius as the builder of any earthwork, he might also complain that they have barbarously translated him when he states correctly in Greek, "Tarquinius was active and employed a large number of workmen in strengthening those parts of the city walls that looked toward Gabii, by widening the ditch, raising the walls, and placing the towers at shorter intervals" [Dion. Hal. *Ant. Rom.* 4.54.2]. These words are as follows in Gelenius's Latin translation, with emendations by Sylburg: "Tarquinius, with a large number of workmen employed, fortified the part of the city that turned toward Gabii, with the trench widened"—understand the trench that Servius had dug—"and the walls built higher"—as at one time Servius had decided—"and with towers erected at more frequent intervals along that part."[6]

Nardini did not act with good faith: he distorted the true meaning of the words "widening the ditch" and "raising the walls" (which imply the completion of someone else's work, already standing) to invent, rashly and with no authority, a construction from scratch.[7] I understated the case when I said "with no authority," since, on the contrary, both Dionysius and Strabo openly oppose Nardini's statement. Dionysius, as is understood from his other words quoted earlier [Dion. Hal. *Ant. Rom.* 9.68.2–3], describes a complete fortification of a triple nature around the city of Rome (such as was found in the consulship of Lucius Aebutius and Publius

4. Nardini, 157.

5. Nardini, 39–40.

6. S. Gelen, trans., *Dionysii Halicarnassei antiquitatum sive originum Romanarum* (Basel, 1549); F. Sylburg, ed., *Dionysii Halicarnassei opera omnia* (Leipzig, 1774–77).

7. Nardini, 157.

Servilius—indeed, in the eightieth year after the expulsion of the kings), where, "situated on elevations and steep cliffs, it [Rome] was so protected by nature itself that it needed little fortification"; where "it was defended by the Tiber, not otherwise than by a wall"; and finally, pertinent to our discussion, where "the city is most vulnerable," that is, "from the east," according to Pliny [*HN* 1.9]. Nevertheless, the *agger* had not occupied this part and region, except from the Porta Esquilina to the Porta Collina. Moreover, Strabo, after he reviews the wall completed from all sides and the single earthwork at the least secure part, finally concludes, "Of this sort is the fortification of Rome" [5.3.7], so that he excludes every other addition of fortification.

Nonetheless, this common desire in many writers of this age—that of enlarging the walls of ancient Rome—has forced them to make up these things and others like them. Yet, as I have said many times and as you have learned from Dionysius, cited earlier, the Tiber, the raised elevations, the high ground of the hills, and an earthwork used to enclose the city. Dionysius wanted this to be understood concerning the time when he was writing, as is evident from these words: "The walls of Rome are difficult to find, because of the structures joined to them everywhere, yet they keep traces of their ancient structure in many places. However, if someone wished to measure their circuit and compare it with that by which the Athenian city is contained, the circuit of Rome will not appear much larger" [Dion. Hal. *Ant. Rom.* 4.13.4–5]. Lipsius and Clüver (still not agreeing with this comparison) and all those who follow them—seeming to diminish the size of Rome, or, rather, that of the ancient city, and to not accept the things that Dionysius had also said elsewhere about the restricted course of its walls, by thinning them out as reason demanded—have enlarged Athens itself beyond limit and have taken into the comparison the Long Walls, Munychia, Astu, the port of Piraeus, and the Phalaric Wall, even though they are parts separated from Athens eighty years before.[8] On this matter, look at Fr. Donati, who fully, no less than ingeniously, refutes Lipsius and Clüver.[9] We mention this author more willingly because he has not only enriched our studies by his very great learning but also added weight and glory to them by his example, when he shows that they are matters not unworthy of a most pious and serious elder. This is something I also wish

8. Lipsius, 108–10 (= Lipsius, *Opera omnia*, 3:423–24); Clüver, 509.

9. Donati, 58 (= Graevius, 3:598A–602D).

earnestly to say about Fr. Athanasius Kircher (although there is sometimes disagreement between us).

We will discuss fully elsewhere [III.5c] how, when the walls had been reduced into a narrow space, Pliny's first measurement of the circumference of the city (which we will present in its appropriate place) could climb to thirteen miles and beyond. If others have dared to reduce Pliny's seventy miles to seven, perhaps we also will not be ashamed to reduce these thirteen miles to eight ("XIII" to "VIII," with a single letter slightly changed), since today we intend in this introduction and short examination to report nothing further than the correct location of the ancient gates and the courses of the roads from them. Having accomplished this, now we turn our pen to the subject set before us.

4. THE NUMBER OF THE AQUEDUCTS

a. Frontinus's Evidence and Pliny

Not only has Dionysius most deservedly put the aqueducts among the three things to be admired in the city, when he says, "To me clearly, among these things from which the greatness of the Roman Empire is most evident, there seem most magnificent the aqueducts, the paved roads, and the sewers" [Dion. Hal. *Ant. Rom.* 3.67.5], mentioning the same things that Strabo most admires in his *Geography* [5.3.8]. The aqueducts have also provided the first and special indication and evidence of Roman greatness, such as Pliny has seriously and truthfully said: "If anyone will have considered more carefully the vast supply of water arriving for public use, for baths, pools, homes, canals, suburban properties, villas and space, the arches constructed, the mountains tunneled through, and the valleys leveled, he will confess that there has never been anything more wonderful in the whole world" [*HN* 36.123].

In agreement with this is the following passage of Frontinus: "We turn to the maintenance of the lines, just as we have promised, a matter worthy of earnest concern, since they are a most eminent testimony to the greatness of the Roman Empire" [*Aq.* 119.1]. Likewise, Cassiodorus says about Rome, "So many rivers have been brought there, on structures built something like mountains, that you would believe that their channels were natural masses of rock, because so great a force of the river could be strongly sustained during so many centuries" [*Var*. 7.6.2]. Rutilius has surpassed all of this, with poetic exaggeration.

> Why should I speak of the channels hanging from their airy arches, where scarcely Iris would support the rain-bearing waters? Rather, you would say that these mountains have grown into the stars; let Greece praise so great a work of giants. [Rut. Namat. 1.97–100]

Concerning the number of the aqueducts from the time of Augustus up to Frontinus, there is very little doubt, since Frontinus himself reviews nine and describes their source, capacity, and length of conduit, separately and in order. Pliny does not contradict him, handing down in the chapter cited earlier that "seven rivers were channeled into the sewers by Agrippa in his aedileship." Nardini has wrongly charged this so that he may provide an even worse answer, that the Julia and Tepula, because they arrived joined together, were counted as one and that the Alsietina was excluded because it was on the other side of the Tiber.[10] The simple and natural solution of this imaginary discrepancy between Pliny and Frontinus consists in only the difference of chronology: when Pliny reviews the work of Agrippa, he correctly reports that only seven aqueducts were introduced into the sewers by him, since the Claudia and Anio Novus did not yet exist but were added to the others a long time afterward. If, indeed, Nardini's frivolous reason, "because before Frontinus they were mixed together," has to be applied some place, why should not the Claudia with the Anio Novus have been considered only as one in the same way? According to Frontinus, the aqueducts "were distributed outside the city, each one from its own conduit, and inside the city were mixed together" [*Aq*. 72.6] (he repeats the same thing in almost similar words at *Aq*. 91.3).

b. The Evidence of the Regionary Catalogs and Procopius

In the centuries after Frontinus, however, as new sources of water were channeled into old conduits, so also an addition of new aqueducts was made to the city and those buildings contiguous with it. But we find up to

10. Nardini, 508.

this point that no one has written how many lines were added beyond the old ones, and we look in vain in Publius Victor and the *Notitia Imperii*, since, indeed, these authorities differ considerably among themselves. However, they have spoken in the same way about the principal conduits as they have about water sources, more than one of which some conduits received. As a result, a remarkable confusion has arisen among modern writers.

Indeed, in his epilogue, Victor numbers the water sources in this way:[11]

Twenty Waters

1. Appia
2. Marcia
3. Virgo
4. Claudia
5. Herculanea
6. Tepula
7. Damnata
8. Traiana
9. Annia
10. Alsia, Alsietina, or Augusta
11. Caerulea
12. Julia
13. Algentiana
14. Ciminia
15. Sabatina
16. Aurelia
17. Septimiana
18. Severiana
19. Antoniniana
20. Alexandrina

The *Notitia Imperii*, however, arranges its water sources in a different way, as follows [Jordan, 2:569–70; VZ, 1:185–86]

Nineteen Water Sources

1. Traiana
2. Annia

11. Publius Victor, *De regionibus Romae liber aureus* (= VZ, 1:256; Graevius, 3:112E).

3. Alsia
4. Claudia
5. Marcia
6. Herculea
7. Julia
8. Augustea
9. Appia
10. Alsietina
11. Setina
12. Ciminia
13. Aurelia
14. Damnata
15. Virgo
16. Tepula
17. Severiana
18. Antoniniana
19. Alexandrina

Nevertheless, Procopius, a most careful author in other respects and a diligent investigator of Roman lore, seems to stretch forth a thread in this labyrinth, with these words: "Indeed, there are fourteen aqueducts in number at Rome, constructed by men of old in baked brick, having that width and depth that they allow a man to ride in them mounted on a horse" [*Goth.* 1.19.13]. As other authors generally do, we will grant credibility to him when he speaks directly about the conduits, but not concerning the water sources. Resting on this stable foundation, as it were, let us strive as best we can to identify the aqueducts themselves.

It would be superfluous to repeat here the nine well-known aqueducts cited by Frontinus and examined elsewhere by us (namely, in the first dissertation, on the Aqua Alexandrina) and to "recook cabbage." Instead, we shall only connect to each one of them the water sources reported by Publius Victor and the *Notitia*, after a discussion of the fourteen conduits supplied by Procopius.

Beyond the nine conduits that I say are described in Frontinus—indeed, those of the Appia, the Anio (which afterward was called the Vetus, for distinguishing it from the other), the Marcia, the Julia, the Tepula, the Alsietina, the Virgo, the Claudia, and the second Anio (which earned the name *Novus*)—there is another, that is, [the Aqua Crabra].

c. The Aqua Crabra

We can seek the tenth aqueduct from Frontinus, and it will be that of the Aqua Crabra, which "used to flow beyond the intake of the Julia," as Frontinus himself says [*Aq.* 9.4], that is, to the right of the Via Latina, where the Julia also began, according to *Aq.* 9.1. Although he adds in the same place, "this water was not tapped by Agrippa," that is, excluded from the conduit of the Julia, "since he had condemned its use or because he believed it should be left for the property owners of Tusculum" [*Aq.* 9.5], it is nonetheless clear that the Crabra indeed would have been not further received afterward into the channel of the Julia but brought to Rome in its own channel, which it occupies even today. Indeed, there remain traces of the antiquity of this conduit in the ancient construction of its cataract, or emissary, at the ninth milestone of the Via Latina, below certain large ruins called Centroni. Here, after that volume of water carried to Rome by the artificial channel already mentioned has been received in an underground conduit, the remainder is carried through the Roman Campagna into the Anio River. In addition, this same opening of the channel gives off the smell of ancient magnificence; here, it goes underground for almost four stades, so that after overcoming the ridgeline in between, it can be easily carried into the city from that point on a downward course. It is self-evident that the channel of the Aqua Crabra is today supplemented by the Aqua Julia and Aqua Tepula, which flow into the same Alban valley from their broken conduits, a matter that, however, Martinelli has disclosed as if by a certain secret warning of a friend.[12]

Just as at one time, in the territory of Tusculum, "all the villas of that region," according to Frontinus [*Aq.* 9.5], "used to receive" the Aqua Crabra "distributed in turns, according to fixed days and amounts, but not with the same restraint," so perhaps by the same rule was it distributed in the city. This fragment of an inscription [fig. 33], seen in the garden of S. Maria on the Aventine, can be understood as concerning this distribution and a similar means of allotment through division by measures and unequal lengths of time (*CIL* VI, 1261):

> for Thyrsis, freedman of Augustus, two pipes from the second to the . . . hour, on the fourth day before . . .

---

12. F. Martinelli, *Roma ex ethnica sacra exposita* (Rome, 1653), 31.

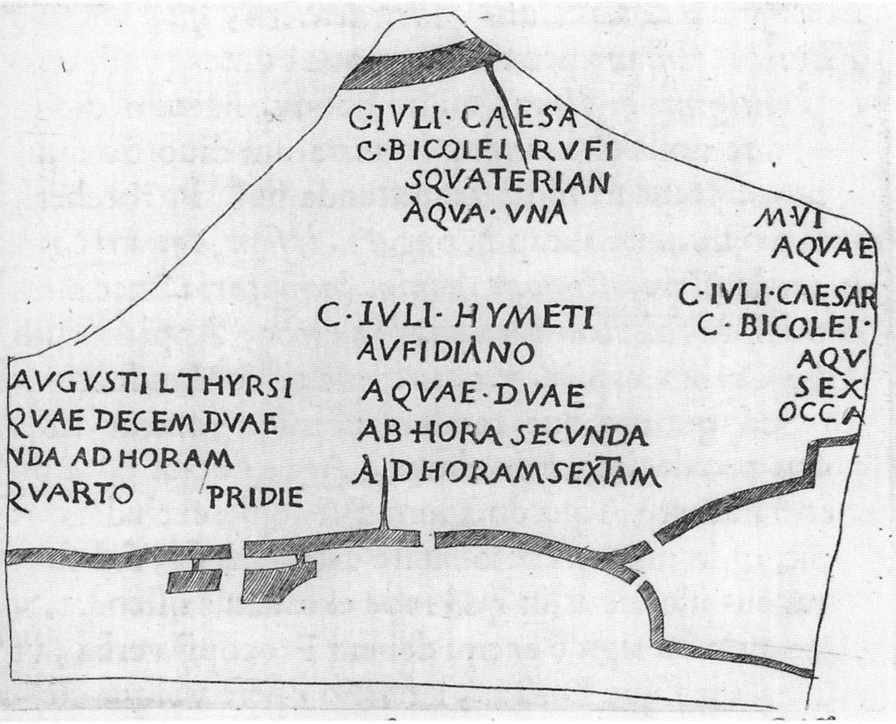

Fig. 33. Inscription depicting staggered distribution from water source [*CIL* VI, 1261]

> for the freeman of C. Julius Caesar, C. Bicoleus Rufus Squaterianus, one pipe . . .
>
> to the Aufidianum of C. Julius Hymetus, two pipes from the second to the sixth hour . . .
>
> to Vibius . . . pipes, to C. Bicoleus, freedman of C. Julius Caesar, . . . pipes from the sixth hour until sunset . . .

The same passage of Frontinus just cited [*Aq.* 9.5] is proof that this aqueduct was afterward called Damnata. When Frontinus mentions water that was "untapped," "not approved of," and "excluded," he seems to have given the opportunity of coining the name *Damnata*. Both Victor and the

Notitia for that reason omit the name *Crabra,* because they have kept the name *Damnata,* similar in actuality. From the origin described earlier and the downward course of the Aqua Crabra, or Damnata, it becomes obvious that it is different from the Almo, which is mixed in with the Tiber between the Porta Ostiensis and the Basilica of S. Paolo. Aside from other testimony of pagan writers, the following statement of Gregory I in his *Registrum epistolarum* is proof: "In addition, the two properties lying next to the same grant, between the Tiber and the porticus of that church, for those coming from the gate of the city on the right side, which the River Almo divides. . . ."[13] Indeed, the springs of the Almo, quite nearby—Ovid describes it as "very brief in its course" [*Met.* 14.329]—are separated from the distant sources of the Crabra by a long, intervening strip of land. As a result, Fr. Kircher's identification of the two can scarcely be admitted: "the Aqua Crabra itself in the city is indicated by the name *Almo.*"[14] Nor can another similar error of Fabricius, who ignorantly confuses the Appia with the channel of the Crabra.[15]

But by whatever name that aqueduct that we posit for the tenth is finally called, it is not to be excluded from the number of Procopius's aqueducts because its channel was not constructed of fired brickwork but is shown to be underground, dug out in the very rock. Indeed, Procopius's words ought not to be taken so literally that we understand that all the aqueducts and their entire conduits were constructed everywhere of brickwork: the channels of both the Aqua Alsietina and the Aqua Virgo were opened for the most part in the living tufa, and all the construction of both the Marcia and the Claudia above ground was built exclusively of *opus quadratum* [squared stone]. Indeed, it is enough, for the credibility of Procopius's words, to verify these things in most cases, although not in all.

Similar is that [other] characteristic that Procopius attributes to all these aqueducts, namely, a width and depth that "allow a man to ride in them mounted on a horse" [*Goth.* 1.19.13]. The channels of some, at a depth of six or seven feet, perhaps barely admitted a man and a horse (provided he led the horse restrained by a bridle); the channels of the Julia and Tepula, two feet wide and three and a half feet deep as Pliny describes

13. P. Ewald and L. M. Hartmann, eds., *Gregorii I papae Registrum epistolarum* (Munich, 1978), 2:434 ("Epist. 14.14").

14. Kircher, 62.

15. Fabricius, 186 (= Graevius, 3:521B).

them [*HN* 7.26], were not even large enough for a pygmy mounted on the back of a ram or goat.

Clearly, however, I am amazed that someone would have been found who, in his commentary on Frontinus,[16] has unhappily confused this volume of the aqueducts with respect to height, or (as Procopius calls it) "depth," with that elevation that Frontinus promises to report in the introduction of his work. No other "height" of aqueducts is to be spoken of there, except what Frontinus [*Aq.* 18] calls by his own word the "leveling" [*libra*] of each line. Indeed, except for the measurement of the conduit of the Aqua Appia, which Frontinus [*Aq.* 65.3] says he found to be five feet in height and one and a half feet in width, there is not even a trace in his treatise about the height and width of the others.

Listen to a monstrosity of an interpretation: "Afterward, with respect to the height of each one and the system of measurements, ancient writers hardly agree with themselves on the height of the underground courses and the arches. Procopius, however, defines the height as sufficiently great that a horseman could pass through those arches easily. Frontinus, below, says that they were raised to a height of 109 feet. Strabo of Amaseia, book V, [writes] 'The sewers, vaulted with close fitting stone, have left room for wagons loaded with hay.'"[17] Keuchen, people will say that your thinking is not consistent when you make Frontinus, who refers to the height of arches, to be in total agreement with Procopius, who refers to the depth of channels, and—what provokes the greatest laughter—with Strabo, who talks about the width of sewers.

I am not unaware that I should say this interpretation should be first attributed to Dempster,[18] from whom Keuchen has taken it in almost as many words. Nevertheless, I think that Keuchen has behaved worse and is obviously guilty of ignorance in a matter that gives special proof of his handling of it and of theft at the same time.

But since we have now strayed too long, let us return to the remaining aqueducts that are missing from Procopius's number. The Aqua Traiana will therefore provide the eleventh aqueduct for us, just as the Aqua

16. R. Keuchen, ed., *Sexti Julii Frontini viri consularis quae extant* (Amsterdam, 1661).

17. Keuchen, *Sexti Julii Frontini*, 371.

18. T. Dempster, *Antiquitatum romanarum corpus absolutissimum in quo praeter ea, quae Johannes Rosinus delineaverat, infinita supplentur, mutantur, adduntur* (Cologne, 1620), 137 ("ad caput xiv Paralipomena. De aquaeductuum forma mirabili et iure concesso," 136–39).

Alexandrina, about which we have spoken in its own dissertation, will provide the twelfth.

d. The Aqua Septimiana

For the thirteenth aqueduct, there comes before us that one included on our map [fig. 31] under the name *Aqua Septimiana;* to the left of the Via Appia, before the seventh milestone from the city, after a construction on arcade of more or less 630 paces, it seems to have carried its water from the Alban hills to the Via Appia itself, beyond the location known as Casal Rotondo, near the tower now called the Torre della Selce. Its conduit is raised above that of the Claudia by the entire height of its arches where they are higher, that is, by an elevation of about twenty-five feet (there was no need for more exact measurement because of the clear evidence of its appearance).

However, I would not deny that this aqueduct could have received the Anio Novus; although the Anio Novus is carried near the city on the arches of the Claudia, it surpasses it by a moderate height, as we have shown in the cross section in our first dissertation [fig. 7]. Nevertheless, before the Anio Novus was received into its settling tank, I think that it always maintained the same parallel course, as I have found, by constant observation, that in many places far from the city, it is carried forty feet and more above the Claudia. I will pursue this more fully sometime in a separate dissertation, *On the Course of the Aqueducts*.

As a result, I do not indeed believe—but, as I have said, I do not consider impossible—that this aqueduct was tapped from the conduit of the Anio Novus. This same aqueduct, after it goes under that long ridge of the Via Appia that runs all the way from here to the tomb of Cecilia Metella, used to be distributed at first for the use of a certain huge settlement, or neighborhood, which is, as it were, the image of a large town, to the right for those approaching the city, in a place called Statuario and marked by no. 10 on our map [fig. 31]. Then again, from here, turning back to the left onto the Appia—its conduit divided into two channels, which are evident up to this day in a structure in the shape of a theater at no. 22—it continued toward the city and the boundaries of Region I. Indeed, far below remain traces of this same conduit (for it could not have been a new source of water), between the third and fourth milestone for those approaching on the left, near the walls of an ancient temple at no. 21, commonly known as Torrone de'Borgiani.

Fr. Kircher, in his *Latium*, says unacceptable things about this very aqueduct after Casal Rotondo, stating that its water was "drawn" all the way "from the Lucrine Lake" or from "the Caerulean spring tapped by Appius Claudius Crassus and augmented by Claudius Caesar."[19] My deep respect for a man so distinguished and well known for so many labors causes me to not refute these things but only touch on them in passing. What, indeed, could the Aqua Appia have had in common with the Claudia? Although Eutropius [2.9.3] called (incorrectly) the Appia itself the Claudia, as if introduced by Appius Claudius Caecus, nevertheless, far more absurdly, by reversing the names like this, will we say that Appius Claudius "brought the water of the Caerulean spring by various wanderings of circles," with an anachronism not, indeed, of a few months but of more than 350 years. Who will have said that water came down to Rome from the Lucrine Bay, as though from a higher elevation? No one will have except perhaps Ligorio, who brings the Alsietina to Rome from Alsium (a maritime city of Etruria),[20] or someone who would not be afraid to speak of a watery spring below Tivoli "much deeper than the Roman city,"[21] although the long and moreover winding descent of the Anio toward the city, into which the spring flows down, is roaring back against it. Love of truth and authority of the speaker compel these clearly troublesome criticisms, although from one unwilling, first so that they not seem to be believed by us, next that they not lead away off the path others who will defer more than is right to the efforts of a great man and his personal inspection emphasized so many times. "Indeed, the more that those who publish their concerns are crammed full with reputation of name," to use the words of Salmasius against Scaliger, in a similar matter, "the more wrong they do, with greater destruction to the republic of letters, whenever it has been their practice to do wrong."

e. The Aqua Algentiana

Finally, another aqueduct will complete the number we want, one called by us perhaps with the name *Aqua Algentiana*, which first emerges under

19. Kircher, 31.
20. Ligorio, *Paradosse*, 38.
21. Kircher, 199.

the Tusculan hills at approximately the ninth milestone from the city; again runs on many arches through another valley, under the tower *di meza Via di Frascati* [at the halfway point to Frascati]; and without a doubt heads in the direction of Rome. However, I could not find even the smallest trace of it beyond here.

I am indeed of the opinion that each of these two conduits [the Aqua Septimiana and Aqua Algentiana] brought water to Rome, but I do not claim that they would have been brought inside the walls of the city, especially since that earlier one, which descends over the ridge of the Via Appia, could not have overcome the quite deep valley of the Almo except on a very long arcaded structure. At least some ruins, from a lengthy and strong construction, would have remained of it.

5. THE LIMITS OF THE ANCIENT CITY

Why, you will ask, were these two later aqueducts counted among the fourteen aqueducts of Procopius, if, as we confess, they did not enter the city? To this objection, therefore, we will here, where the occasion especially demands it, pursue at greater length the solution that we prefaced in a few words in our first dissertation [I.5].

The words *Roma* and *urbs* seem to mean one and the same thing, but two very different things are nevertheless indicated by them. The term *urbs* includes whatever is contained within the walls and has been marked off by the plow, or *urbum*, as Pomponius writes [*Dig*. 50.16.239.6]. We follow him rather than Isidore [*Etym*. 15.2.3], who derives the word from *orbs*. The term *Roma*, however, also embraces the built-up area of *aedificia con-tinentia* (continuous buildings), as is established in many laws of the same title, namely, the *lex urbis* [*Dig*. 50.16.2], the *lex Alphenus* [*Dig*. 50.16.87], the *lex aedificia* [*Dig*. 50.16.139], the *lex qui in continentibus* [*Dig*. 50.16.147], the *lex mille passus* [*Dig*. 50.16.154], and the *lex collegarum* [*Dig*. 50.16.173.1].

Now it is worth our effort to define how far this built-up area extended and to interpret its expanse, for an appropriate understanding of both the laws cited and of other ancient authors. Lipsius so loosens the boundaries of Rome that he has included in it cities and towns situated around it—

Ostia, Aricia, Ocriculum, and others.[22] In this, he speaks for the expansion of Roman greatness elegantly, to be sure, but against the mind of our jurisconsults and contrary to law. It will be far from every correct and probable interpretation that the jurisdiction of the urban prefect, granted within one hundred miles from the law on his duties [*Dig.* 1.12.4]—reckoned from the continuous buildings, not from the Golden Milestone of the City, as is held in the *lex mille passus* [*Dig.* 50.16.154]—would be extended almost by half, if the distance would begin to be reckoned from Ocriculum (which is thirty-nine miles away from the city). Moreover, if Lipsius is correct, a man whom the earth of Aricia or Tibur first took up in its embrace would be said to have been born at Rome, according to the *lex qui in continentibus* [*Dig.* 50.16.147]. Finally, that which is "up to the continuous buildings of the city," although it is distant from the city by three days' journey, such as the independent city Ocriculum, would be considered not to be away from the city, according to the *lex collegarum* [*Dig.* 50.16.173].

In addition, in the same chapter, Lipsius expands Rome, "extending without a perimeter into the suburbs," into a measureless space. I do not know why, shortly before, he contracts the city no further than the seventh milestone, by distorting the words of Pliny [*HN* 3.66–67].[23]

However, to our proposed "limit of expanding buildings" already set forth, we believe no ancient authorities are in opposition. On the contrary, there is outstanding support for it, since the very appearance of the topography supports our theory, to the extent that we may reconstruct it because of the losses over a very long time and the hostile madness that has devastated everything by fire and sword.

a. Settlement on the Via Ostiensis

Along the Via Ostiensis (to take our beginning from this point), before it comes to the intersection of the Via Laurentina, on the very bank of the Tiber, there stand huge ruins of a settlement or neighborhood at no. 12 [fig. 31]. We have both its name and its distance from the city in Ammianus [17.4.14]: "It was brought"—he is speaking about the obelisk brought from Egypt by order of Constantius—"into the Vicus Alexandri, distant

22. Lipsius, 113 (= Lipsius, *Opera omnia*, 3:426).

23. Lipsius, 109–11 (= Lipsius, *Opera omnia*, 3:424–25).

from the city at the third milestone. From there, put on windlasses and drawn more slowly through the Porta Ostiensis and the Piscina Publica, it was brought in the Circus Maximus." Indeed, anyone who will have looked at the place will certainly come to the conclusion that tombs or other buildings did not stand forth at the edge of the road here, as everywhere else, but, rather, that homes were arranged far and wide on either side in the manner of a town, houses that the Tiber, deviating from its ancient course, presently cuts through, showing part of the settlement separated from Latium on the Etruscan bank on the right.

Moreover, we should note here the position of the Porta Ostiensis in the circuit of the walls of ancient Rome. It could not be elsewhere than beneath the high ground of the Aventine, on its south side, between the Churches of S. Prisca and S. Saba. Only from this direction is there a way open through Region XII of the Piscina Publica very close by, toward the Circus, which bordered on the same region to the left. For if Porta Ostiensis were beneath the Aventine on its north side, at the place of the modern Salinae—although what antiquarians call the Via Ostiensis is commonly said first to have gone forth from the Porta Trigemina, which they locate there—the transport of the obelisk would have been made through Region XIII (the Aventine) or Region XI (the Circus Maximus), in which the Porta Trigemina was located, according to Victor and Rufus, and not through Region XII (the Piscina Publica). It would also not have run into these things on its journey but would have remained far and outside the area of the entire Circus, in the direction of the Porta Capena. Even if we exclude this account of Ammianus, the ancient Porta Ostiensis or the beginning of the Via Ostiensis cannot be imagined elsewhere; in this way, the gate may look directly out on its road and not present a monstrous angle where it begins, contrary to that which we observe in all other roads, especially in an open and unencumbered area.

In addition, the location of the Septizodium, the remains of which used to be visible in the memory of our fathers between the Caelian and the Palatine, seems to contribute a substantial amount to this argument. Indeed, Septimius Severus (as Spartianus says about him), "when he built it, thought of nothing else than that his construction would meet those coming from Africa" [SHA *Sev.* 24.3]. Yet for those coming from Africa a bit behind the entrance of the gate, from which the Via Ostiensis began (if it should be situated here), the Septizodium of necessity met also those not thinking about it. If the gate, however, had been between the Aventine and the Tiber, it would have to have been sought by a long way around and including a guide for the journey, and the reason Spartianus gives for it would have been altogether meaningless.

There is, moreover, a ready explanation why I have named the Via Ostiensis as the appropriate and particular route for those arriving from Africa: indeed, this road, being straight, was the shortest of all for those journeying after they had put to shore at the mouth of the Tiber, namely, at the sixteenth milestone, according to the Antonine Itinerary [*It. Ant.* 301.6] and the Peutinger Table.[24] Reality itself, with measurement made by me more than once and in a consistent way, has also confirmed this. The Via Portuensis, opened on the right bank of the Tiber across the river, three miles longer than the Ostiensis in its curvings—for excellently does the Antonine Itinerary [300.8] count nineteen miles from Rome to Portus—served only for the conveyance of goods and for drawing cargo ships by cattle up the river, as is understood from Procopius [*Goth.* 1.26.10–13]. Procopius's particular remark concerning this road and the very sure traces of the road itself, observed by me and faithfully rendered on the topographic map [fig. 31], show that it clearly diverged from the modern course of the Tiber, where the river withdraws from the fixed and permanent road (indeed, where its course once was) and where, besides, it had to be channeled to avoid dangers to modern transport (which are extremely great at this Vicus Alexandri, notorious for frequent shipwrecks).

(1) *The Emporium*

Nardini shows excellently, in my opinion, that the Porta Trigemina is not at all synonymous with the Porta Ostiensis, as they commonly hold.[25] Uselessly, however, he adds from Victor and Rufus (who say no such thing) that it remained useless, for that was common to all the other ancient gates included within Aurelian's walls. Instead, a persuasive argument that the Porta Trigemina was the same as the Porta Navalis is presented by the narrowness of its location between the Aventine and the Tiber, enough for only one gate. Fr. Alessandro Donati proves forcefully that the Navalia were "at the Aventine" and "the first walls of the city,"[26] against the opinion of Ligorio, Panvinio, and others.[27] It is also appropriate for the Empo-

24. *Tab. Peut.* IV.5 (cf. Bosio, *La Tabula Peutingeriana*, 84).
25. Nardini, 50.
26. Donati, 252–53 (= Graevius, 3:788E–789C).
27. O. Panvinio, *Descriptio urbis Romae* (= Graevius, 3:377A).

rium to have been connected to the Navalia. Indeed, after you have gone past the high ground of the Aventine, you will notice that very many ruins along the Tiber, which up to this time have remained confused and scattered (and for this reason unobserved by others), constitute a single structure, which can represent nothing else than this Emporium that we are searching for. There are several reasons: first, the great number of warehouses and porticoes; next, their gradual descent toward the Tiber, which seems to have served for unloading goods from ships to the warehouses; then, the very broad open area everywhere around it for exposing these same goods, whether for manufacture, reworking, or the loading of ships; and finally, the fold and curve of the riverbank, seen and noted by Flaminio Vacca, in that space that lies between the last and lowest arches and the Tiber, for establishment of a port.

Vacca describes it like this in his letter to Simonetti cited elsewhere by us [I.6]: "I remember that, last year, near the place called La Cesarina, because it belongs to the Casa Cesarina"—likewise today—"there were found certain yellow columns, which, transported on the Tiber, had been unloaded in that area above the bank; there are also seen there many unfinished pieces of different material that were found in that spot in times past, but they are full of ugly spots and surrounded by very hard quartzes, and that is the reason why they remained in that place. Above ground are seen walls in the manner of warehouses, and in the bank of the Tiber is seen the curve of the port."[28]

Within the area of these ruins, I think there existed that inscription with the navigational law "Whatever is carried for owner's use does not owe duty" [*CIL* VI, 8594] that Fulvio reports as found on the bank of the Tiber below the Aventine.[29] Certainly, a base two feet in height still exists there, uncovered in earlier years within the boundaries of our Emporium, with the following inscription.

> Anteros Caes(ar), warehouseman of the third cohort, gave and dedicated freely as sacred to Silvanus. [*CIL* VI, 588]

On its right side is sculpted a patera and on the left a wolf, sacred to this divinity, as to Mars among the Romans. There, also, stood that altar found in the vineyard once belonging to Marcello di Capozucchi:

---

28. Cf. Montfaucon, *Diarium italicum*, 167–68.
29. Fulvio, 322–23.

Sacred to the conserving fortune of the Horrea Galbiana, M. Lorinus Fortunatus the master gave and dedicated from his own money. [*CIL* VI, 236]

Finally, another small base is kept in the same place in a vineyard bordering on this Emporium to the south. Whether it was brought back here from the home of Pomponio Leto, where Mazochi observed it, or whether it is another from the same vineyard, it should be referred perhaps to the public warehouses of this Emporium, from the following double inscription with emblems (the first inscription occurring on the front of the base, the second on the back) and the *magistri* cited in the inscriptions.

L. Dunius Apella, C. Annius Tyrannus, first masters, gave and dedicated it to the fortune of the warehouse.
(patera and pitcher)
(on the right side, a club; on the left, a globe)

C. Annius Tyrannus the master and L. Dunius Apella gave and dedicated it to the fortune of the warehouse.
(two cornucopias with fruits; a plowshare placed crosswise) [*CIL* VI, 188]

Therefore, I have decided not to present the construction—notable and unusual in its great antiquity—in a single and simple sketch but to give a detailed explanation of it in sections, shown in ground plan [fig. 34] and cross section [figs. 35–36]. After this, I shall return to investigating the settlements along the military and principal routes that enclosed the continuous buildings of Rome.

The entire area of the porticoes and the warehouses is of stone construction, without any mixture of brickwork. Indeed, the dividing walls are of *opus incertum* [concrete faced irregularly with stone], from which their antiquity is detected, but the arches and supports of the arches are of cut stone, of most careful craftsmanship.

b. Other Settlements on Main Roads

The Via Ardeatina shows ruins of its own settlement (indeed, not certain and extensive, yet not to be scorned) in that place where it begins to diverge from the Appia (joined to the Appia, it proceeded almost up to that spot) on a modern road that connects the Appia with the Ardeatina below the tomb of Cecilia Metella. The mass of buildings virtually touches

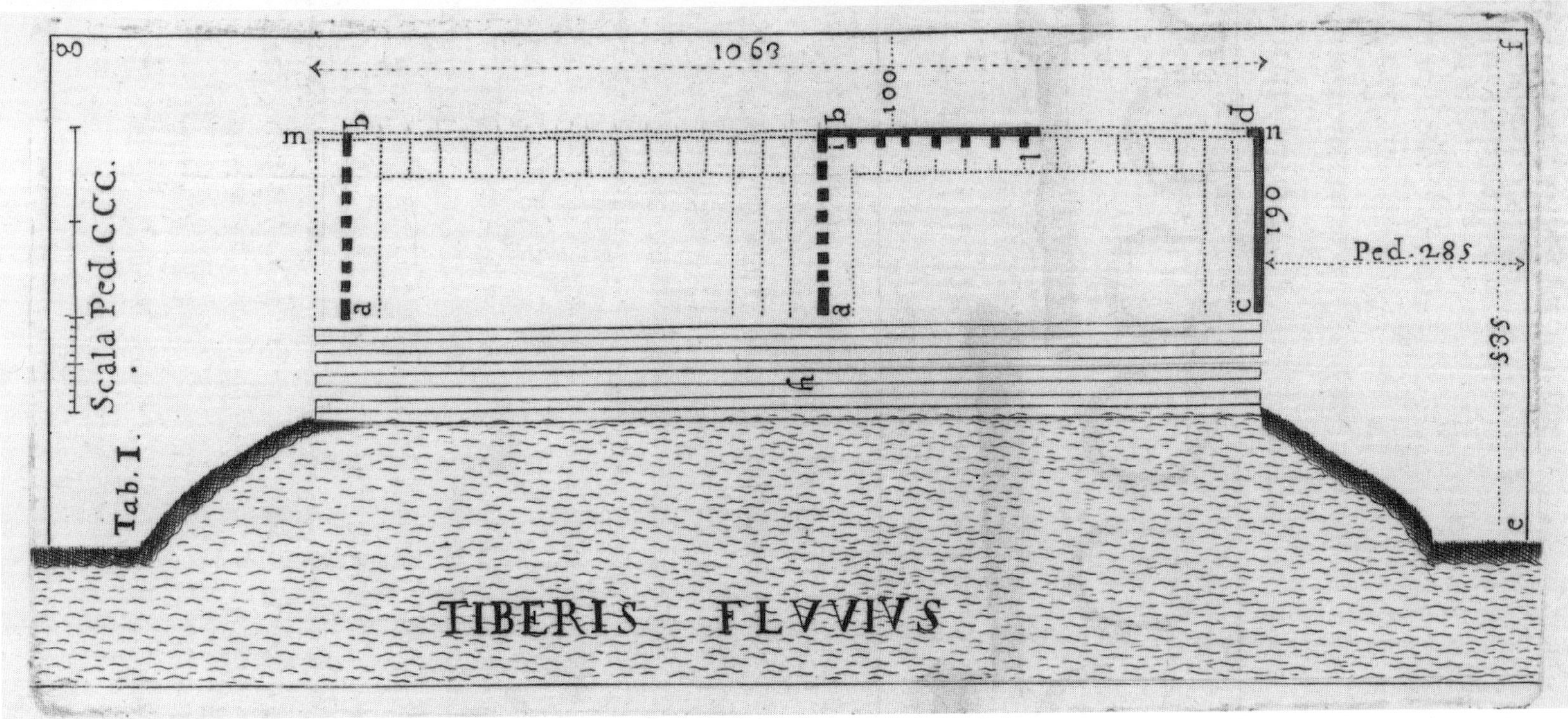

Fig. 34. Emporium structure: plan

- ab. Two walls, equal in length, structure, and number and downward movement of their arches, showing a double curvature at each face of the arch, from which we believe there were porticoes with a descent to the Tiber
- cd. Another wall, corresponding to the preceding in structure and length, but lacking arches, since it is external
- efg. Substructure of the space or exterior area bounded by a brick wall
- h. Steps by which we conjecture from Livy [41.27.8] that there was an ascent from the Tiber into the warehouse
- il. Remains of chambers or storage spaces into which the longer side *mn* seems to have been divided

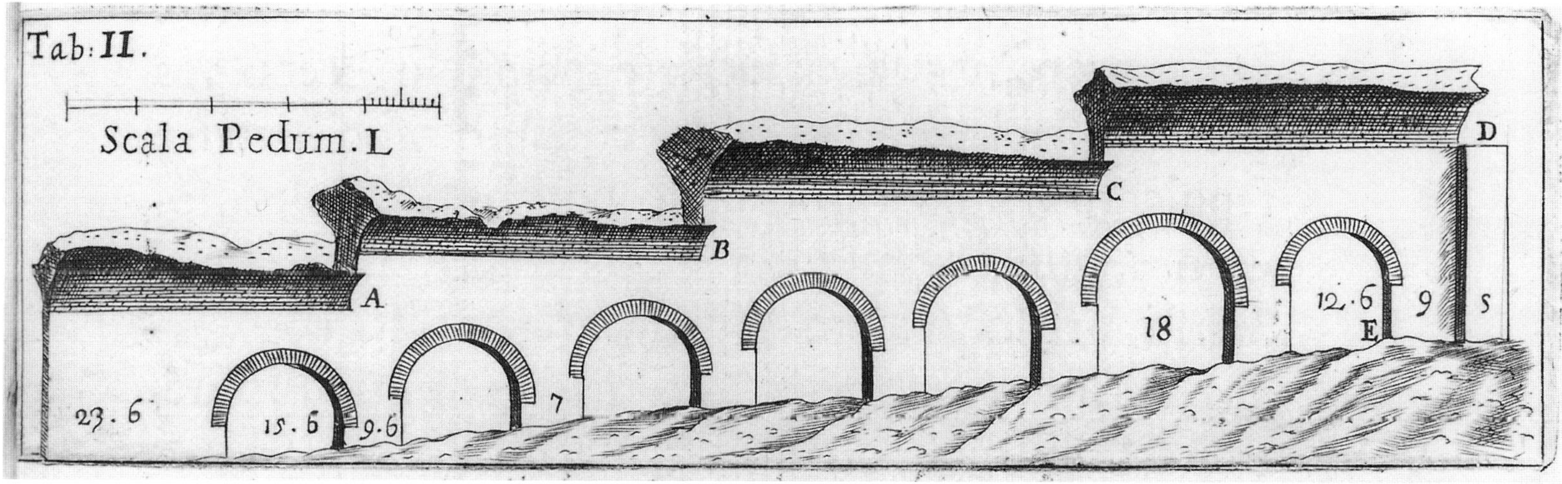

Fig. 35. Porticus: cross section. The ascent of the porticoes—from which, according to the instruction of Vitruvius [5.12.1], there ought to be an entrance to emporia—from bottom to top is twenty feet: from A to *B*, six feet: from *B* to C, eight feet: and from C to *D*, six feet. The place is filled with earth and debris.

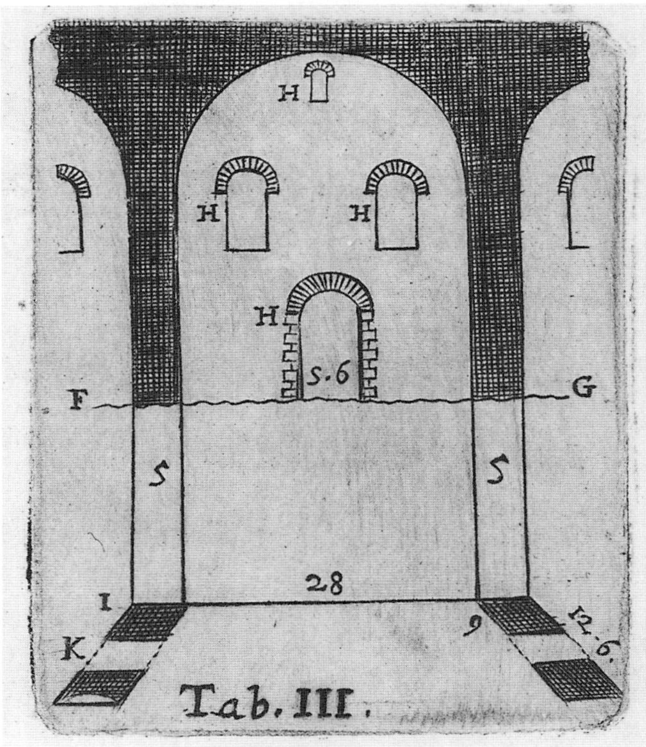

Fig. 36. Cross section of vaulted chamber. The part under the line *FG* is buried under earth and its own debris. There remain traces of doors between one and the other warehouse, from side *IK*, of the same length as the other between *E* and *D* of the earlier table [fig. 35], that is, nine feet.

H. Four windows on the exterior side toward Monte Testaccio, on which excavation shows there were lattices, two inches wide and deep, distant by a space of one foot from the exterior facade

on others that lie scattered to the left, away from the Via Appia, near the Church of S. Urbano. Between them was a circus, which, from coins and its more recent structure, those more informed judge was that of not Caracalla but Gallienus.

The Appia, as the most famous of the other roads, at first included that district—its name was perhaps that of the Camoenae, if we believe Nardini[30]—and afterward gained another one most splendid, beyond the fifth

30. Nardini, 81–82.

milestone to the left, at a place to which the name *Statuarii* is still attached, perhaps from the large number of statues and adornments. This district is distinguished before the others for its own praetorium, temple, amphitheater, and circus, as well as other most notable elements.

Another district closer to the city along the same road (we shall call it Ad Camoenas) competed with that one in the rest of its adornment and, in rivalry with it, had its own conduit, which took water from the arcade of the Claudia. Indeed, there still remain traces of the structure from the wellhead at the start of the conduit (no. 19 of our map [fig. 31]), filled with incrustation everywhere, through which the water descended to this new conduit. It is conspicuous even now from the arches that cross the valley near the source of the Aqua Salutaris in a course of four hundred feet and from the channel of the same aqueduct, again within the boundaries of the district itself, even now in the Vigna Cortesia.

The Via Latina also shows a double settlement, the first a bit beyond the fourth milestone, at the settling tanks of the Aquae Marcia, Julia, and Tepula, as we believe (which perhaps was the Pagus Lemonius); the second, the largest of all, extended ten stades beyond, in the area of Settebassi. To the west of this is shown the villa of some emperor (for the ruins are not unworthy of such an identification), with a station for a garrison and quarters of the praetorians, as well as porticoes, baths, and other adornments. For the use of this, its own conduit, consisting of an arcade of six hundred paces, was brought from the arches of the Anio Novus.

Along the Via Praenestina, around the circular temple that contemporaries call the Tor de'Schiavi, there remain very many traces of a most elaborate settlement.

To the left of the Via Praenestina was the Via Collatina, according to Frontinus [*Aq.* 5.7]—not, indeed, as Holste notes on Clüver,[31] starting from the Praenestina at the fifth milestone or after the *Tor de'Schiavi* already mentioned, where there is a modern side road. The Via Collatina maintained its own paved course from the city all the way to Collatia, as I wish you to believe now on behalf of my good faith. Holste indeed observed neither its course nor its terminus; if he had observed it, he, like us, would have located Collatia at the ruins *del Castellaccio*, a mile away from the Via Praenestina (which was reported as the ancient Castrum Osae or Losae more than once in the *Bullarium casinense* for the census of

31. Holste, 102–3.

the venerable monastery of S. Paolo),[32] instead of identifying them as Gabii, as he did, rashly and inconsistently.[33] Both Strabo [5.3.10] and Dionysius [Dion. Hal. *Ant. Rom.* 4.53.1], as cited by Clüver,[34] expressly prove that Gabii stood on the Via Praenestina itself, and Holste admits the same thing, that Gabii was on the Via Praenestina, in his annotations on Ortelius.[35]

Holste must be corrected, moreover, when he believes that the *castello* of S. Giuliano was on the Osa stream and accepts it as our Castrum Osae;[36] more accurately, the ruined walls and the name *S. Giuliano* remain around one and a half miles beyond, on the other side of the Castrum di Castiglione, in the direction of Tivoli, and at a still greater distance from the Via Praenestina, as I shall show in a more appropriate place.

For this reason, nevertheless, no settlement appears along the Collatina, either because it is to be associated with the settlement already mentioned on the Praenestina, to which it was very close at that point, or because the Via Collatina was to be considered not one of the main roads but a secondary route. The road was only eight feet four inches in width, not equaling the customary measurement of consular roads, generally fourteen feet four inches. As a result, Pliny [*HN* 31.25], in showing the source of the Aqua Virgo, which arises along that road, cited the Via Praenestina, as the more famous and truly military route, and a side road of two miles from its eighth milestone. You will see from our map [fig. 31] that this is nevertheless the same as the eighth milestone of the Via Collatina (about which [see] Frontin. *Aq.* 10.5). How little, however, has Georg Fabricius, in his *Roma*, understood our topography (that I may again raise a cry against that notorious writer) when he says that this Via Collatina "leads to Collatia, outside the gate of the same name, and after a brief distance runs into the Salaria"![37] There are two whole roads, the Tiburtina and the Nomentana, in between.

The Via Tiburtina and the Nomentana went forth along the sides of the Castra Praetoria; although the gates themselves from which they

---

32. Margarini, *Bullarium casinense*, vol. 1 (Venice, 1650), 26 (constitutio 22), 32 (constitutio 25), 35 (constitutio 30); vol. 2 (Todi, 1670), 139 (constitutio 150), 218 (constitutio 211), 282 (constitutio 262).

33. Holste, 199.

34. Clüver, 954.

35. Holste, 35.

36. Holste, 131, 199.

37. Fabricius, 54 (= Graevius, 3:476F).

issued are today closed, ancient ruins of each road do indeed survive. Each road had this very sort of a district or settlement as its terminus, according to Pliny, in words to be cited shortly.

Along the Via Salaria, I do not doubt that its own settlement existed, but I confess that I have looked for it in vain, since perhaps, as Lucan laments about Troy, "All of Pergamum is covered with brush, and even the ruins have perished" [9.968–69].

The Campus Martius, with the Campus Minor, expanded in its entirety and intact to the Via Flaminia and the Via Triumphalis, closed the circuit of our Rome there outside the walls.

c. Measurements from the Golden Milestone

Pliny states the following, first about the size of the city strictly speaking; then about the measurement of distance, still not understood, from the Forum to individual gates; and finally about that even more obscure measurement to the edge of the continuous buildings: "The walls in their course surrounded an area at the time of the principate and censorship of the Vespasians, in the 826th year after the foundation of Rome, of thirteen miles and two hundred paces, having embraced seven hills. The city itself is divided into fourteen regions, with 265 crossroads of the Lares. If a straight line is drawn from the milestone set up at the head of the Forum to the individual gates, . . . the result is a total of thirty miles and 765 paces in a straight line. But the total length of all the streets from the same milestone through the districts to the farthest edge of the buildings with the Castra Praetoria . . . results in a little more than seventy miles" [*HN* 3.66–67].

Putting aside for now the first two measurements of distance, the third, from the Golden Milestone to the farthest edge of the buildings and the Castra Praetoria, through the settlements of all the roads, corresponds excellently to our reconstruction, as you can recognize from the following summary set forth roughly and established for a simple grasp of the matter.

From the Golden Milestone via the Via Ostiensis, including the Vicus Alexandri, and return by the same route to the Forum, seventy-five hundred paces.

From the [Golden] Milestone via the Via Ardeatina, including its settlement, and return by the same route to the Forum, seventy-five hundred paces.

From the [Golden] Milestone via the Via Appia, including the settlement of the Camoenae, and return by the same route, seventy-five hundred paces.
From the [Golden] Milestone via the Via Latina, including its settlement (the Pagus Lemonius?), and return by the same route, eleven thousand paces.
From the [Golden] Milestone via the Via Labicana, including its settlement, and return by the same route, eight thousand paces.
From the [Golden] Milestone via the Via Praenestina, including its settlement, and return by the same route, seven thousand paces.
From the [Golden] Milestone via the Via Tiburtina, including the Castra Praetoria, and return by the same route, three thousand paces.
From the [Golden] Milestone via the Via Nomentana, including the Castra Praetoria, and return by the same route, three thousand paces.
From the [Golden] Milestone via the Via Salaria, including a settlement of uncertain location, and return by the same route, seven thousand paces.
From the [Golden] Milestone via the Via Flaminia, including the Campus Martius, and return by the same route, six thousand paces.
From the [Golden] Milestone via the Via Triumphalis, including the Campus Minor, and again by the same route, three thousand paces.
Total 70,500 paces.

In interpreting Pliny's words, we have deliberately omitted those most distant settlements along the Via Latina and the Appia (nos. 8 and 10 [fig. 31]), since, indeed, we believe that they were built after Pliny's time and interrupted the series of other settlements arranged generally in a circle. Indeed, two factors here suggest the work of an earlier time: the aqueducts, which we see were constructed for the supply of each settlement, and the special praetorium of the settlement on the Via Latina. Here, I observed first the reworking of walls among the ancient monuments, with the inner surface straight but the outer tilted back slightly, which they call *a scarpa* today (although I am not unaware that this technique, which Vitruvius [6.11] calls "structural leaning," was of older invention). Nevertheless, it is certain that these settlements existed at the time of Procopius—what indeed would have stimulated so splendid a mass of structures after the abandonment of the Roman state and the transfer of the seat of empire to the East?—and that they received the water delivered by two later aqueducts. We may therefore conclude that the aqueducts from this point entered what was called "Rome" (to the extent that they did not also proceed farther).

These observations on Pliny will certainly seem presumptuous or bold to some, but they come closer perhaps to his idea and offer with themselves proof noted by no one, although very certain, from the topography itself.

Moreover, the words of Dionysius describe the circuit of these continuous buildings as follows: "The places inhabited around the city, many they are and great, are unfortified and especially subject to incursions of enemies. If anyone wishes to ascertain the size of Rome, with his eyes turned to these things, he must be greatly deceived. And he will surely not be able to learn how far the city extends and where it stops; in such a way does the area of the entire city cohere in an unbroken bond, furnishing to those looking at it the idea of a city stretched into infinite distance" [Dion. Hal. *Ant. Rom.* 4.13.4].

Whoever is forced to recognize the boundary of the Ambarvalian ceremonies and the *ager Romanus* itself within the fifth or sixth milestone—according to Strabo [5.3.2], "at the place known as Festi"—will charge that these words of Dionysius are no doubt presented with exaggeration, just as other words of Pliny, "the expanding buildings adding many cities" [*HN* 3.67], can scarcely be extended beyond the limits we have designated. But beyond our own extension, which we have given to the name *Rome*, Nardini seems to stretch what is called the *urbs* itself in a certain manner.[38] He has wished to include in the territory of the first region, described by Rufus and Victor and with its circuit of thirteen thousand paces unchanged, the Temple of Fortuna Muliebris, which was distant from the city by twenty thousand feet in a straight line (indeed, at the fourth milestone on the Via Latina), and the Fossae Cluiliae, at the fifth milestone on the Via Appia, "whose name, with physical evidence," at the time of Livy (for so the historian says [1.23.3]) "disappeared through antiquity"—Nardini thus imitates carelessly the license of Panvinio (to be sure, in not so unrestrained a fashion).[39]

Well, then (so that my discourse, having digressed, may return to its point of departure), from the correctness of the Latin and the support itself

38. Nardini, 73–74.
39. Panvinio, *Descriptio urbis Romae* (= Graevius, 3:284).

of the jurisconsults, it is not inappropriate to call by the name *Rome* that area outside the city and within which some of the aqueducts listed by Procopius delivered their water. Let us add as a crowning touch that Frontinus also, like us, recognized Rome outside the city: although he states as preface in his introduction that he will describe "the aqueducts that flow into the city" [*Aq.* 3.1], he nevertheless includes among them the Alsietina, which he later announces "was consumed entirely outside the city" [*Aq.* 85], not in the Campus Martius, as Ligorio, that most daring stick, understood in his *Paradosse:* "The Alsietina spring, taking itself from the Lacus Alsietinus over the Via Claudia, supplied a fountain in the Campus Martius, as Frontinus shows."[40] I correct him with the comments of Celso Cittadini, saying the same things in Italian, "He makes authors that he has never read say in his own way what the Latin does not mean. Frontinus did not mention the Campus Martius but states that the Alsietina terminated at the *naumachia* [naval amphitheater], meaning the *naumachia* that was at San Cosimato, not the one below the Pincian."[41]

6. THE WATER SOURCES LISTED IN THE REGIONARY CATALOGS

There remains now the task (that we promised we would undertake) of directing the flow of all the sources of water listed by Victor and the *Notitia* to be absorbed into those fourteen aqueducts of Procopius. Indeed, for the first four, the Appia, the Marcia, the Virgo, and the Claudia, its own conduit has been assigned to each.

The fifth, the Herculanea, was received into the conduit of the Anio Novus, according to Frontinus [*Aq.* 15.4]: "There is joined to it [the Anio Novus] the Rivus Herculaneus, which arises on the same road at the thirty-eighth milestone, from the region of the sources of the Claudia across the river and the road . . . , but when mixed, it loses the charm of its own limpidity." Accordingly, it was perhaps listed by Victor after the Claudia in place of the Anio Novus, which does not appear in his summary.

Also called the Herculaneus was that branch in the city through which

40. Ligorio, *Paradosse,* 39.

41. Cited in Martinelli, *Roma ex ethnica sacra,* 429.

part of the Aqua Marcia distributed itself below the Gardens of Pallas throughout the Caelian, as Frontinus says [*Aq.* 19.8–9]. Yet its name was not taken from that Herculanean branch joined to it in Tiburtine territory (for how could that happen?), as Fr. Donati mistakenly states.[42] As a result, this branch of the Marcia could have been called in the city by a name sought at closer distance and not from its source. But if, indeed, among these more thorny topics, it is pleasing to delight one's mind by Ligorio's persuasiveness, listen to him teaching like this in his *Collectanea de aquaeductibus:* "The Aqua Herculanea. It was tapped in the aqueduct of the Aqua Virgo by the emperor Claudius and was the reason that the Virgo lost its reputation, as a virgin does when corrupted by a man. As a result, as Pliny says, the masculine conduit was removed by oracle from the Aqua Virgo to return it to its purity."[43] Laugh now (if bad bile does not affect you more, as I feel sometimes happens to me in the case of this fool) at the rash confidence of the man and his lack of concern toward critics.

The sixth, the Tepula, and the seventh, the Damnata, had their own conduits and names.

The eighth, the Traiana, as we proved in the first dissertation [I.4g], was brought from Sabatine territory to the top of the Janiculum and the region of the Transtiber. Fulvio, who wishes the Aqua Virgo to have been called the Traiana on the occasion of its restoration and based on that coin with the fountain that we considered to be understood in connection with it, is therefore wrong.[44] Nor, indeed, does a reference to the Aqua Traiana inscribed as follows on a lead pipe found on the Aventine near the Porta S. Paolo contradict our opinion:

> Aqua Traiana. Quintus Anicius Antoninianus, the son of Quintus, curator of the Thermae Varianae.[45]

It seems far more easy and more probable that the water crossed from the Janiculum very close by the Aventine over the Pons Aemilius and arrived through pipes from the more distant bank to the nearer one, with the advantage of linked conduits, just as, from the other direction, aqueducts were customarily brought from Rome across the Tiber, according to

42. Donati, 298 (= Graevius, 3:837C).
43. Ligorio, *Collectanea*, Ottob. lat. 3365, fol. 5v.
44. Fulvio, 185.
45. Gruter, clxxxii.7 (= Lanciani, 511, no. 81).

the testimony of Frontinus [*Aq.* 11.2]. As a result, I may say with Statius, "to cross rivers with a bold pipe . . ." [*Silv.* 1.3.67], rather than fantasize, apart from the authority of any ancient writer, that the Marcia was restored to the Aventine by Trajan. Indeed, Frontinus's words "to these hills, more than one aqueduct has been brought back, in particular, the Marcia, carried on a larger structure from the Caelian to the Aventine" [*Aq.* 87.4] are declared, with these immediately following, "The queen and mistress of the world feels this concern of its most devoted emperor Nerva day by day . . ." [*Aq.* 88.1].

By these words, Trajan's father, Nerva, not Trajan himself, is indicated as builder of the extension. Besides, they cannot be a reference to Trajan, since at the time Frontinus wrote, that is, "at the beginnings of his administration," as he says in his introduction [*Aq.* 1.1], or when "Nerva the emperor"—to whom he dedicates his work—"and Virginius Rufus were consuls for the third time" [*Aq.* 102.17], Trajan had not yet fully gained imperial power and was far from bringing in an aqueduct and striking a coin distinguished by his fifth consulship to commemorate the deed (as we have noted elsewhere, namely, in our first dissertation [I.4g]). Even if this were true, this accomplishment would not have deserved to be marked by a coin with depiction of a fountain, nor would the name *Marcia* have to be erased from a pipe so that *Traiana* might be substituted; especially to distinguish aqueducts, it was customary that their names be marked on the pipes, as we see the abbreviation "AQV. MAR." and likewise "ANIO. VET." indicated on other pipes.[46]

This Aqua Traiana (as I have briefly touched on already, citing Procopius) is said to have been intended for mills that were once on the slope of the Janiculum, a function for which we see that it has finally returned after a long interval. In such a way, there is "nothing new under the sun, nor can anyone say, behold, this is new," as the Scriptures maintain [Ecclesiastes 1.9–10]. So, indeed, Procopius states about this: "From the region of the Transtiber, a great hill is conspicuous, and there all the mills have been constructed; indeed, the great force of the water is brought through its channel to the top of the hills, from where it falls on the slope with a mighty force" [*Goth.* 1.19.8]. Some people should therefore cease to wonder that this most ancient function of the Aqua Traiana has been resumed and to find fault with it (as we have heard they are doing). To me, it has

46. See Gruter, clxxxii.8–9.

rather always been in my prayers to see the Tiber freed from blockage by river mills. Belisarius indeed devised these mills, when necessity dictated, after the aqueducts had been cut by the Goths in the siege of the city. As Procopius adds [*Goth.* 1.19.19–20], why, once the siege was ending, should we allow these structures to endure, to our loss and danger, with the greatest ugliness brought about by the obstruction of the river basin, no less inconvenience, since commerce between the upper and lower river has been interrupted? Not only are we taught by private losses that the expenses of river mills of this sort outweigh their advantages, since to replace those mills that are destroyed or carried off headlong by the river in flood (as not rarely happens), others are not sufficient. It is also clear from mathematical demonstrations that public losses from flooding of the city are increased as a result (I do not deny that other reasons are also factors here). When the course of the water has been slowed down by those huge mills by which the river banks are obstructed (Holste calls them "stationary piers" in a specific dissertation with this title),[47] the result is that the volume of water is increased, since water acquires greater speed when blockages against it are removed. As a result, its elevation drops, as Fr. Castelli concludes in his treatise *On the Measurement of Running Water*.[48]

In the case of these mills, the following worse thing has happened: their occupation of the banks of the Tiber has been a mistake by new discovery and not from imitation of Belisarius. Indeed, Belisarius blocked the river bank not with "stationary piers" but, rather, with a harmless structure (to describe it with Procopius's words): "In front of the bridge"—beneath the Janiculum—"with ropes fitted and stretched most strongly from each bank of the river, he tied two boats, with a two-foot space left between each, where the water was rushing with greater force from the arch of the bridge. Then, with two millstones installed in each boat, he installed in the middle a machine by which mills are accustomed to be turned. He attached other small boats in a line, according to the plan of those that were behind, and installed machines in the same manner, all of which, in a row, driven by the force of the water flowing forward, turned the mill set next to them and milled as much as the city required."

For the ninth, the Annia, or, as the revised Victor has it, the Amnia, we agree with Panciroli that the name could be said to have been derived

47. L. Holste, *Dissertatio de pila stassilari* (= Graevius, 4:1803–6).
48. Castelli, *Della misura*, 37 (corollary 16, app. 9).

from *Ania* or *Anio*, which was taken from the river itself;[49] we should not say instead that Victor omitted an aqueduct in existence and put in its place another uncertain one by mistake. Now let us show praise for this ingenious solution of Panciroli through comparison with a much worse and monstrous interpretation: "The Aqua Annia was brought from the Aqua Traiana in different places of Rome by Lucius Annius Verus the censor, in the principate of Antoninus Pius and Marcus Aurelius." Already, without me, you have recognized that this is Ligorio;[50] who else would know how to include so many foolish ideas in a very few words?

The tenth, the Alsia, or the Alsietina, which is also the Augusta, is the very Alsietina to which Frontinus also attributed the name *Augusta*, stating, "What reason moved Augustus, an emperor with the greatest foresight, to introduce the Aqua Alsietina, which is also named the Augusta, I do not know; its water enjoys no popularity and indeed is hardly wholesome" [*Aq.* 11.1].

The name *Augusta* is also connected with another source, by which Augustus (as stated in his *Res Gestae* [20.2]) "doubled the aqueduct that is called the Marcia, with a new source tapped for its conduit," which was afterward tapped in part for the Marcia, in part for the Claudia, as we have observed from Frontinus [*Aq.* 14.3] in our second dissertation. Dio Cassius writes that this name was also given to the Aqua Virgo by Agrippa: "Agrippa brought the aqueduct that was called the Virgo into the city at his own expense and named it the Augusta" [54.11.6]. This, however, is not so certain; indeed, it may be stated more securely that Dio mistook one aqueduct for a second—namely, the Virgo for the Appia, which had its source in the same property of Lucullus—when the writer first and alone said that the Virgo acquired the name *Augusta*. Frontinus also would not have been silent about this; indeed, he indicates instead the reason for Dio's mistake. Frontinus says that a branch of the aqueduct, which was called the Augusta, was tapped as a supplement for the Appia, as we see in *Aq.* 5.6–8, although with some lacuna, which is made clear from the things he adds later [*Aq.* 65.1–4].

As a result, so as not to disagree with Frontinus, closer in time and more diligent in the account of his official duty and a more faithful writer, let us give the name *Augusta* back to the branch of the Appia, which, "on

49. G. Panciroli, *De quattuordecim regionibus urbis Romae earumdemque aedificiis tam publicis quam privatis libellus* (= Graevius, 3:446E).

50. Ligorio, *Collectanea*, Ottob. lat. 3365, fol. 5v.

the Via Praenestina at the sixth milestone, on a side road to the left" (as Frontinus declares in *Aq.* 5.6–8), "received a source." [About this Frontinus says that] "its conduit up to the Gemelli (which is a place below Spes Vetus), in an underground line, completes 6,380 paces."

Nardini correctly observes that the name of the Alsietina was discovered to have been incorrectly added to this branch of the Augusta from a gloss that crept into the text of Frontinus.[51] But he does not correct Frontinus with equal success when he states that the Aqua Appia could not have been brought "intra Spem Veterem," that is, to the area of the Porta Maggiore. For if he is moved by the argument (indeed, he claims none) that the Appia, as it were, would not have been able to rise to that place, the point is meaningless, since Frontinus [*Aq.* 5.6] indicates the place where the conduit crossed, not its elevation, and states that it was underground everywhere (except for the section at the Porta Capena [*Aq.* 5.5]). It is not at all absurd to keep in mind that the same conduit was sunk underground, in part at a deeper level under the hills, in part at a more shallow level under level ground.

Angeloni wrongly applied this same name, *Augusta,* to the Marcia: "Agrippa, reintroducing into the city, at his own expense, the Aqua Marcia, already destroyed, the best and the purest that there is today, called it the Augusta."[52] But he does so from authorities unknown to us and even to himself, unless we might say that he would have been mistaken in his reading of "Marcia" instead of "Virgo." He could have called the Virgo the "Augusta," having followed Dio. However, even this prop does not sufficiently strengthen Angeloni's opinion; according to Frontinus [*Aq.* 10.1], Agrippa "brought the Virgo that he had tapped to Rome" and did not indeed reintroduce it a second time, as Angeloni's words indicate.

The eleventh, the Caerulea, is part of the Claudia, which "took its beginning from two most abundant and beautiful springs, the Caeruleus and Curtius," as Frontinus says [*Aq.* 14.1].

The twelfth, the Julia, gave a name to its own aqueduct, from Frontinus [*Aq.* 9] afterward distinguished by the title *Venocis* by Ligorio, who transferred the name from the man discovering the aqueduct to the aqueduct—indeed, not the same one.[53] As Frontinus says [*Aq.* 5.2], "Indeed, to Gaius Fabius, the name *Venox* was given, on account of his discovery of the *venae*

51. Nardini, 503.

52. Angeloni, 11.

53. Ligorio, *Collectanea,* Ottob. lat. 3365, fol. 5v.

of the Aqua Appia." Since, properly speaking, *Venox* otherwise means "searcher of the source," attribution of such a name to the water source itself being searched for and sought out seems to be truly Ligorian.

The name of the thirteenth, the Algentiana, will be able to match that conduit on the Via Tusculana, because it most directly faces the Algidus, as our map shows [fig. 37]. Certainly, to no other aqueduct is it to be equally well applied, unless our identification is deceptive.

Mount Algidus, we argue, was that one that the monastery of S. Silvestro and Rocca Priora occupy, stretching with more than one peak behind Tusculum to the left of the Via Latina, toward Praeneste and the southeast. That intervening valley, in which the ruins of Castra Molaria are seen, used to separate both Algidus and Tusculum (which was on the same ridge) from the Alban Mount. The Via Latina, rising indeed on the right—I have discovered traces of it more than once—was brought to the sides of the Tusculan Mount and that of the Algidus and the valley already mentioned, as is clear from the following passage of Strabo [5.3.12], not cited by Clüver among so many statements of this writer and others that he amasses:[54] "Between the mountains against which Tibur and Praeneste are established, there stretches forth another mountainous and lofty ridge, leaving behind the valley between Algidus and the Alban Mount; in that place, Tusculum was situated."

This also is very well confirmed by Livy's description of Hannibal's march [26.9.11–12]: "Hannibal, after the territory of Fregellae had been laid waste in hostile fashion, because the bridges had been cut, came into Labican territory through that of Frusinum, Ferentinum, and Anagnia; from here, he sought Tusculum by way of the Algidus." If Hannibal sought Tusculum through the Algidus from the territory of Labicum (this is, with Holste, next to the town Colonna),[55] then the Algidus, Tusculum, and Labicum, with all its territory, are on the same left side of the Via Latina, as we have claimed. For those who claim that the Algidus is to the right of the Via Latina, with that valley intervening, on the same and unbroken high ground with the Alban Mount, today called Monte Cavo (as Fr. Kircher does in his *Latium*),[56] Hannibal's march is conceived as not straight, like that of a man hurrying, but in a circle, like that of one overflowing with leisure.

54. Clüver, 472.
55. Holste, 194.
56. Kircher, 37, 70–72.

But the excellent Fr. Kircher will perhaps introduce another error in his defense of this one, namely, the location of Labicum at the town of Valmontone (as he promises he will demonstrate shortly thereafter), from which a means of crossing was given from his Mount Algidus to Tusculum. Nevertheless, Kircher cannot make use of this assumption, since, later (forgetful of his promise), he orders "Labicum to remain for pressing reasons" on the Via Praenestina and in the place where the town Zagarolo now stands.[57] Even if we admitted this and transferred Labicum from the right of its own Via Labicana to the right of the Via Praenestina, Hannibal's march will still have been to the left of the Via Latina over the Algidus; from this, we have perceived correctly from Strabo that the ridge of the Algidus was continuous with the high ground of Tusculum but not with that of the Alban Mount.

So that we may sample something about the true position of Labicum, beyond the authority of Holste, who places it at the town Colonna, I will introduce here, from my daybooks for the work I am planning about the *ager suburbanus*, the text of the following inscription recently discovered and fixed today in the wall of the Vigna Lazarini; from it, not only is the position of the ancient town obvious, but, moreover, it is clear why Antoninus, in his itinerary [*It. Ant.* 304.7], and the Peutinger Table[58] mark as "ad Quintanas" the place at the same distance of fifteen miles that Strabo attributes to Labicum. From this little-known marble—

> To the spirits of the dead, to Parthenius, treasurer of the republic of the Lavicani Quintanenses [*CIL* XIV, 2770]

—the "Lavicani" are the same as the "Quintanenses," so that Fr. Athanasius Kircher may decide on a fixed and lasting home for them at long last.[59]

What, finally, will we do about Gronovius? In his third epistolary dissertation to Goefius,[60] boasting that he is emending the passage of Livy already cited [26.9.11–12]—indeed, intact and correct—he throws every-

57. Kircher, 120–21.

58. *Tab. Peut.* V.1 (cf. Bosio, *La Tabula Peutingeriana*, segment 5.1).

59. Kircher, 120–21.

60. *Epistula tertia Wilhelmo Goefio*, in J. Gronovius, *Epistolae in quibus multa Titi Livii loca geographica emendantur et illustrantur* (Amsterdam, 1678), 21–28, reprinted in A. Drakenborch, ed., *T. Livi Patavinii Historiarum libri, qui supersunt, omnes* (Stuttgart, 1828), 15:246–54.

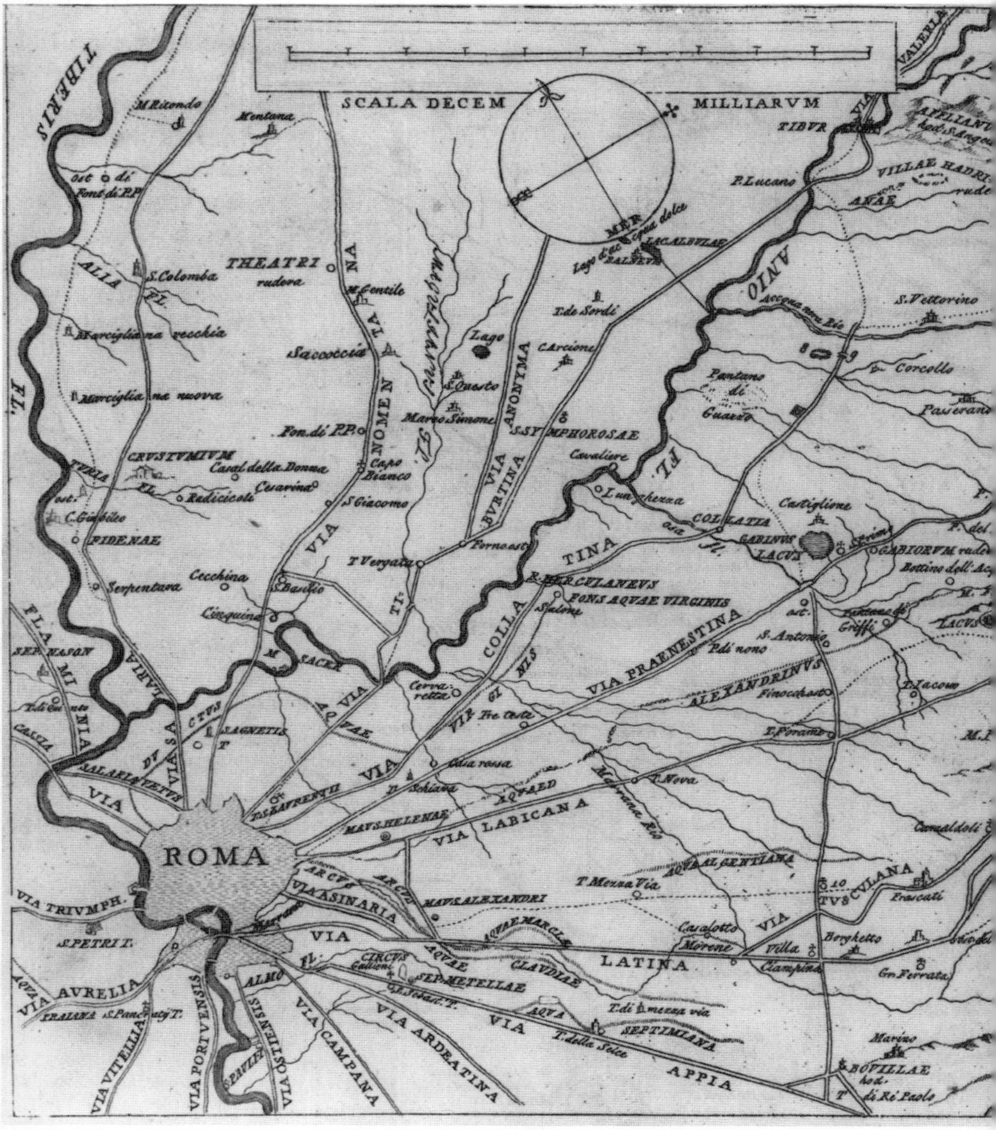

Fig. 37. Topographical map of Latium showing the course of the Via Latina, with the Algidus and Praenestine mountains. (Courtesy of Biblioteca Apostolica Vaticana, Vatican City.)

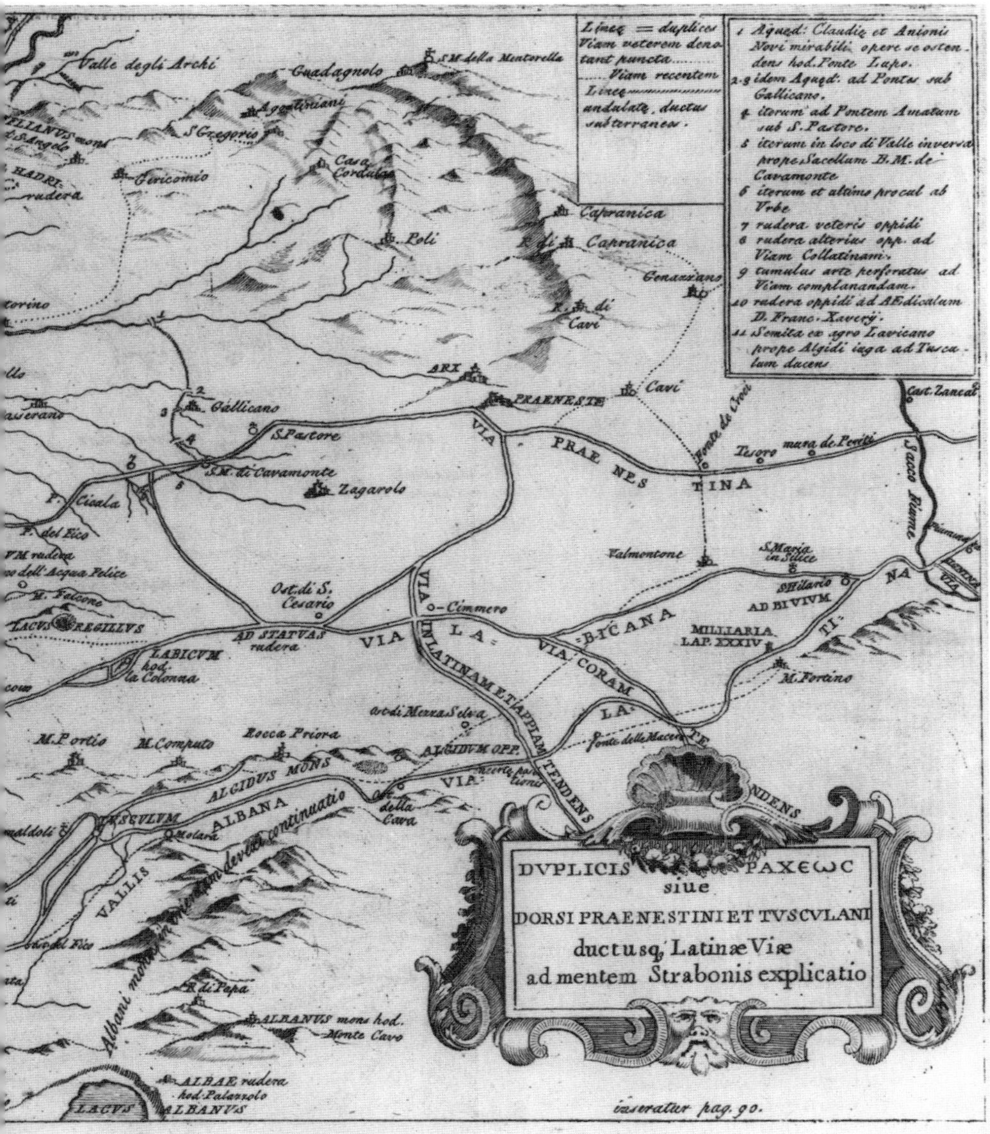
DVPLICIS PAXEWC siue
DORSI PRAENESTINI ET TVSCVLANI
ductusq. Latinæ Viæ
ad mentem Strabonis explicatio
Lineæ = duplices Viam veterem denotant puncta ...... Viam recentem Lineæ andulatæ ductus subterraneos.
1 Aquæd. Claudiæ et Anionis Novi mirabili opere se ostendens hod. Ponte Lupo.
2.3 idem Aquæd. ad Ponte sub Gallicano.
4 iterum ad Pontem Amatum sub S. Pastore.
5 iterum in loco di Valle inversa prope Sacellum B. M. de Cavamonte
6 iterum et altimo procul ab Urbe
7 rudera veteris oppidi
8 rudera alterius opp. ad Viam Collatinam.
9 tumulus arte perforatus ad Viam complanandam.
10 rudera oppidi ad AEdiculam D. Franc. Xaverij.
11 Semita ex agro Lavicano prope Algidi iuga ad Tusculam ducens
Valle degli Archi
Guadagnolo
S. M. della Mentorella
S. Gregorio
Gericomio
Casa Cordula
Poli
Capranica
R. di Capranica
Genazzano
R. di Cavi
ARX
PRAENESTE
Cavi
Gallicano
S. Pastore
S. M. di Cavamonte
Zagarolo
Cicala
VIA PRAENESTINA
Tesoro
mura de Periti
Fonte de Croci
Sacco Fiume
Cast. Zancati
Valmontone
S. Maria in Silice
S. Hilario
AD BIVIVM
MILLIARIA LAP. XXXIV
M. Fortino
Ost. di S. Cesario
Cimmero
AD STATVAS rudera
VIA LABICANA
VIA CORAM TENDENS
VIA IN LATINAM ET APPIAM TENDENS
LACVS REGILLVS
LABICVM hod. la Colonna
M. Falcone
Ost. di Mezza Selva
M. Portio
M. Computo
Rocca Priora
ALGIDVM OPP.
ALGIDVS MONS
VIA ALBANA
TVSCVLVM
Molara
della Cava
VALLIS
Albani montis in orientem devexi continuatio
R. di Papa
ALBANVS mons hod. Monte Cavo
ALBAE rudera hod. Palazzolo
LACVS ALBANVS
inseratur pag. 90.

thing into marvelous confusion so that he may then substitute "Pedum" for "Algidus,"[61] with not only Livy but Silius Italicus contradicting him and with no similarity at all. Oh how plentiful a crop of jokes and witticisms would that Marcus Meibom, wounded by so many reproaches, gather from here! With them, he would attack his antagonist in turn and fart in the face of "Pedum," so inappropriate a word! Indeed, refraining from such things, we will warn you and others in brief fashion that Gronovius is mistaken on three very serious points.

First, [Gronovius is mistaken] in his strict acceptance of the word *Algidus* for a town by that name lying at the base of its mountain on the Via Latina.[62] Livy, in all the passages cited by Gronovius, permits this to be understood not about a town of Algidus itself but about the mountain or the ridge of the same place, which has a not inconsiderable extension. Therefore, not at all carelessly does Silius [*Pun.* 12.536–37] refer to the Algidus with a plural noun in the description of Hannibal's march: "nor do pleasant Algida hold him back." Indeed, so many mountains could seem to the poet Silius (by an especially poetic figure of speech) as tops of a mountain, even though a single one.

Gronovius is mistaken a second time when he denies that between the Alban Mount on this side and the ridge of Tusculum and the Algidus on that side, a deep valley is opened up, extended for many miles.[63] He does this in error, according to others, as a result of what he believes is Strabo's judgment, but it is actually Strabo badly interpreted. From our point of view, however, he acts very foolishly, since we know these places better than Gronovius knows his own home. In the valley, by a gigantic device and in a continuous line, from the city Tusculum to the Alban Mount, Gronovius constructs a stupendous earthwork of twenty-five stades at least. Yet according to Strabo himself, it is understood, in no obscure fashion, from these words next to those already cited [5.3.12], that each height is separated by an intervening valley: "next to this"—[Tusculum]—"there lie regions sloping toward the Alban Mount."

Finally, Gronovius is mistaken when he dryly and grammatically interprets the passage of Strabo (discussed by us in our first dissertation [I.4d]) concerning the branch road of the Via Latina—"The Via Latina takes its

61. Gronovius, *Epistolae,* 28 (= Drakenborch, *T. Livi,* 15:254).
62. Gronovius, *Epistolae,* 25 (= Drakenborch, *T. Livi,* 15:250–51).
63. Gronovius, *Epistolae,* 26 (= Drakenborch, *T. Livi,* 15:252).

beginning from the Appia, bending off to its right near Rome, and above the height of Tusculum, it passes between the town Tusculum and the Alban Mount" [5.3.9]—reading this as a crossing of the high ground of Tusculum to suit his own thesis.[64] With the aid of the topography itself, we shall demonstrate to the believers of this sort what we are setting forth with pragmatic truthfulness concerning that lofty climb of the Via Latina at the sides of the Tusculan height that we have discussed.

I would not know how to conceive even in my mind how we might include among the Roman aqueducts the fourteenth, the Aqua Ciminia of Publius Victor, except by combining it with the Sabatina. I have seen that this pleased the distinguished Fulvio Orsini, in the margin of the Aldine book published at Venice in the year 1518, access to which the abbot Michelangelo Riccio, distinguished for his learning and his mode of life, made available to me; here, Riccio noted "Ciminia, or Sabatina" in Fulvio's own handwriting. It is not that I do not recognize that the Lacus Ciminius is different from Sabate, both in Strabo [5.2.9] and in Silius [*Pun.* 8.490–91], in passages cited by Clüver.[65] Instead, either the proximity of the places deceived Victor or the author of the *Notitia*, in an age not yet given to scholarship, or the breadth of the Ciminian glade, very greatly expanded by Livy [9.36.1], also included the Lacus Sabatinus.

Panciroli, in his notes on Publius Victor, derives the name *Ciminia* "from the Ciminian forest near Tusculan territory, concerning which Pliny writes in book 9."[66] As a result, our hopes were raised that it could have been the same as the Algentiana. We believe, however, that the most famous man was mistaken because of the lack of punctuation of his text. Because the dividing punctuation had been removed, Panciroli joined by chance different things that Pliny keeps separate in different places. These are the words of Pliny in that passage where he reports different wonders of the lands [*HN* 2.211]: "At the Arae Murtiae in Veii, and near Tusculan territory, and in the Ciminian forest, there are places in which things fixed in the earth are not drawn forth."

The fifteenth, the Sabatina, we believe to have been that water that, according to Frontinus [*Aq.* 71.1], brought from the Lacus Sabatini, was tapped into the Aqua Alsietina in the vicinity of Careiae. The text is as follows, with a slight emendation: "The capacity of the source of the Alsi-

64. Gronovius, *Epistolae*, 26 (= Drakenborch, *T. Livi*, 15:251).

65. Clüver, 331.

66. Panciroli, *De quattuordecim regionibus* (= Graevius, 3:448F).

etina neither is listed in the record books nor could be determined with accuracy under present circumstances, since from the Lacus Alsietinus and then from the Lacus Sabatinus in the vicinity of Careiae, as much as the watermen have arranged, it has no more than two thousand *quinariae*."

The sixteenth, the Aurelia, must be said to be the same as the Traiana, since the Traiana was brought to Rome for a long distance along the Via Aurelia, as we have already shown.

The seventeenth, the Septimiana, Nardini thinks was drawn from some principal conduit reworked for new purposes and given its name.[67] But we exclude this sort of minute listing of aqueducts by Victor as secondary and spillover (so to speak). All the other lines either had their own aqueducts or, having received an appropriate name of their own, were a substantial part of them. Indeed, someone who had thought to leave his own name on water drawn from a nearby *castellum* would rightly have incurred the censure of vanity.

The Thermae Severianae, named from Septimius Severus (as Spartianus reports in his *Life*), Victor and the *Notitia* in agreement describe as being in the first region of the Porta Capena, which stretched far outside the city, including the Almo stream and the settlement of the Camenae. From this, we may conjecture that our thirteenth aqueduct [III.4d], which was brought on the back of the Appia toward this region, was perhaps called the Septimiana, which we are seeking. This is more probable because we find that great conduit without a name and without a builder from those previously cited.

The eighteenth, the Severiana (after we have seen Victor overflowing with use of synonyms), could indicate by another name the same aqueduct of either Septimius Severus or Severus Alexander.

That the nineteenth, the Aqua Antoniniana, was added to the conduit of the Marcia is shown by the inscription [*CIL* VI, 1245]—which we cited in the first dissertation [I.4a]—at the arch under which the Via Collatina once certainly passed, today serving the modern Via Tiburtina (for the ancient road runs along the south side of the Castra Praetoria, as we have said).

Fr. Athanasius Kircher, as an aside in his *Latium*, makes Marcus Aurelius Antoninus, the son of Pius, the builder of the conduit and author of this inscription.[68] He also cites an inscription of Antoninus Pius concern-

67. Nardini, 510.
68. Kircher, 208–9.

ing the rebuilt conduits of the Claudia and Anio Novus at the Porta Maggiore.[69] But beware of each statement, for the third inscription at the Porta Maggiore, which Kircher wishes to be understood as Antonine, is that of Titus Vespasianus, inscribed with this message:

> The emperor Titus Caesar, son of a god, Vespasianus Augustus, pontifex maximus, in the tenth year of tribunician power, imperator seventeen times, father of his country, censor, consul seven times, at his own expense saw to the reintroduction with a new conduit of the Curtian and Caerulean aqueducts brought by the divine Claudius and afterward restored to the city by his father, the divine Vespasian, since, from their source, they had collapsed from their foundations because of age. [*CIL* VI, 1258]

The other inscription on the arch on the Via Collatina [*CIL* VI, 1245] certainly pertains to Caracalla, the son of Severus. To be sure, the name *Marcus Aurelius Antoninus* is common to both [emperors], along with the added titles *Pius Felix Augustus* and *Parthenicus Maximus*, but the title *Britannicus Maximus* fits Caracalla alone. With such similar titles, Caracalla is also honored in the following inscription.

> The emperor Caesar Marcus Aurelius Antoninus Pius Felix Augustus Parthicus Maximus Britannicus, pontifex maximus, father of his country, consul three times, consul designate four times, rebuilt the road cut off by flood. [*CIL* X, 6876]

Moreover, the designation here of a fourth consulship is not appropriate for Marcus Aurelius, for he is found as only consul for the third time in that year in which his death occurred or when he is inscribed also with a third consulship on coins cited by Occo, under "tribunician power for the thirty-fourth year."[70] The designation could not yet be assumed for the following year, since this office was not taken except at the end, with the year waning, but Marcus Aurelius himself died in the month of March, as is known from excerpts of Dio Cassius [71.1] and all the *Fasti*. In short, so common is this, from the great number of coins with the title *Britannicus*, more than forty of which you will see cited by Occo,[71] that, among schol-

69. Kircher, 31.
70. Occo, *Numismata*, 227–28.
71. Occo, *Numismata*, 284–97.

ars, Caracalla is accustomed to be marked by the name *Britannicus* to distinguish him from the son of Pius and from Elagabalus. If you indeed should demand both how this Aqua Antoniniana could be brought from the channel of the Appia and how it arrived over an arcade at the hippodrome of Caracalla on the Via Appia and from there again along the Via Latina to the Thermae Antoninianae, as the distinguished Ligorio fearlessly affirms,[72] "you would do nothing more than to make an effort to be mad with reason" [Ter. *Eun.* 62–63].

Concerning the twentieth, the Aqua Alexandrina, you may read thoroughly our entire [first] dissertation, which, unless we are mistaken, proves that this is the same source of water brought to Rome with its own conduit and notable arcade by the emperor Severus Alexander and afterward delivered to Rome by Pope Sixtus V, although by a different route.

Of the following four, which are found in the new edition of Victor, the twenty-first, the Anio Novus, and the twenty-second, the Anio Vetus, are among the nine aqueducts listed by Frontinus. The twenty-third, the Albudina, was part of the Claudia, as is apparent from the following words of Frontinus [*Aq.* 14.2]: "The Claudia also receives that spring that is called the Albudine, of such good quality that it serves as a supplement to the Marcia whenever there is need, with the result that it changes nothing of its quality by its addition." The twenty-fourth, the Crabra, last of all, was the same as the Damnata, as we have already argued at length [III.4c].

The *Notitia Imperii* presents nothing new beyond what has already been stated, except the name of some Aqua Setia, which Nardini, by sufficiently good conjecture, believes came about by the error of some unskilled scribe, when the name of the Alsietina was divided, by replacing "Alsietina" with "Alsia" and "Setina."[73] Another name, *Aufeia,* was an old name of the Marcia, according to Pliny [*HN* 31.41]: "The Marcia was once called the Aufeia and its source itself the Piconia."

7. UNDERGROUND CONDUITS IN ROME

There will indeed not go unmentioned some underground water conduits that are believed to remain from antiquity, one showing itself between the Church of S. Anastasia in Circo and the Church of S. Giorgio in Velabro

72. Ligorio, *Collectanea,* Ottob. lat. 3365, fol. 6r.

73. Nardini, 510.

and flowing at once into the channel of the Cloaca Maxima very close by there; the second under the Palazzo Grimani on the Quirinal; finally, the third, flowing not far away, in the well of the Pharmacopolium below the statue of King Mithridates, on the street of the Blessed Virgin of Constantinople, which they say is the same as that which is seen at the Piazza dell'Olmo under the Dye Factory and is carried away by an uncertain course toward the West and the Tiber.

We think, however, that all of them, as a spurious work of Roman magnificence in these conduits, should be removed from the family of aqueducts about which we are speaking, first because of the moderate and uneven supply of water with which they are provided. It is unlikely to think that this water was brought from a distance, since we find that the watery and mossy soil around the city and in the city itself is not lacking in natural springs. Indeed, it can be no wonder that level places, especially under the hills, spring forth with different water sources, even when springs are idle, since falling rainwater has been filtered through and is seeping frequently in this wet season. Another reason for excluding them is the humble and poor construction, in comparison with that we see in our aqueducts. As a result, it is better that you allow these channels to creep unseen with their inglorious moisture than that "you give the honor of this name" [Hor. *Sat.* 1.4.44] to them.

If, indeed, they boast anything from antiquity, they are accordingly to be linked not to aqueducts proper but to urban springs, once irrigating the city or suburbs, to which we saw Galen made mention in the passage cited in our earlier dissertation [II.4e]. Among their number was the Petronia Amnis in the Campus Martius, about which Festus reports as follows [p. 296L]: "The Petronia is a stream flowing into the Tiber, which the magistrates cross, having taken the auspices, when they wish to conduct some business in the Campus; this sort of auspices is called perennial. Ancient sources, moreover, used to treat *amnis* as feminine." Likewise, we find the spring of Mercury near the Porta Capena (which is under [the Church of] S. Anatasia, very close to its course, from our account of the Porta Capena [I.4d]), cited by Ovid [*Fasti* 5.673]: "There is the spring of Mercury, close to the Porta Capena. If you wish to believe those who know, it has divine power." Finally, there is the spring of Juturna, that of Picus and Faunus, the spring of the Lupercal, and others that Nardini lists.[74]

Our delightful Ligorio (to close happily the dissertation and, at the

74. Nardini, 502.

same time, the book with him), not having read Festus or contradicting him, wished to give as follows the course and etymology of the Petronia Amnis: "It is the name of a spring and small river that runs in the Tiber, and it is renowned as that pretty and clear stream that comes from the roots of the Palatine Hill and is called the spring of S. Giorgio, from the nearby church that was built there. It was named Petronia from issuing under the rock."[75] In another place, as he overflowed with little inventions, "Ligorio thought it was that spring of the Euripus, which flowed in the Circus Maximus," if we believe Holste, making reference to him in his annotations to Ortelius's *Geographical Thesaurus*[76] (for where Ligorio says this, I considered it hardly worthwhile to investigate in the such great mishmash of his books)—as if this little spring could suffice to fill a ditch ten feet deep and wide, almost a mile in length, as Dionysius describes it [Dion. Hal. *Ant. Rom.* 3.68.2]! We read in Frontinus [*Aq.* 84.2] that 460 *quinariae* of the Aqua Virgo were supplied to the Euripus of Agrippa, to which the aqueduct itself gave its name.

Nardini reports, from the same Ligorio, that the water of the fountain of Piazza Mattei, distinguished by bronze statues of boys, tortoises, and dolphins, was once brought by Augustus for the use of a second Euripus at the Circus Flaminius.[77] But Nardini's mistake is clear. By whatever stake you wish, I would contend that Ligorio was referring not to the water of this fountain (which in its highest jet is that of Sixtus, or the Acqua Felice, with that of the Aqua Virgo in the four lower pipes issuing from the mouth of the dolphins) but to another spring that we have said was under the Dye Factory, in the Palazzo Mattei from the other side to the west, with only the next street separating it. Indeed, because of this channel, believed (not incorrectly) to be ancient, it is probable that Ligorio also concocted his own Euripus. For why would Nardini's observation that the Circus Flaminius indeed was surrounded by no Euripus stand in his way? According to the old adage, Hippocleides does not care about these things [Hdt. 6.129]. Ligorio, who announced that he was always self-sufficient, is so untroubled and unconcerned about the truth, I say, that we must exercise the highest scruples in dealing with him.

Nardini was childishly and disgracefully ignorant that the fountain of

75. Ligorio, *Collectanea*, Ottob. lat. 3372, fol. 119v.
76. Holste, 101.
77. Nardini, 324.

Piazza Mattei, thirty-six feet higher than this fullers' water, supplied not waters unknown but those of the Aqua Virgo and Acqua Felice, as we have said. While he believes he is criticizing someone else, he stirs up trouble and disputes for himself. Again, he seems to have fallen back into the same shadows, still doubting whether the water of the fountain of Piazza Mattei should be called the Petronia,[78] an idea that ought not to have fallen into the mind of a sensible man, especially one open to investigations of these matters.

Wearied by extensive reading, you will sing at me with your Martial, "Oh, enough is enough now, oh little book!" [Mart. 4.89.1]. It is time, therefore, to close the channels, and with all the water drained, which we have taken for abundant speaking, at long last let us be silent by law of the court. To you, my friend, a fond farewell. I wrote this from my museum, on September 27, 1679.

Commentary

Fabretti's third dissertation was published, like the first two, without section headings, but it is here organized as follows:

1. Introduction
2. Topographic Map
3. The Eastern Gates of the "Servian" Wall and Their Roads
4. The Number of the Aqueducts
   a. Frontinus's Evidence and Pliny
   b. The Evidence of the Regionary Catalogs and Procopius
   c. The Aqua Crabra
   d. The Aqua Septimiana
   e. The Aqua Algentiana
5. The Limits of the Ancient City
   a. Settlement on the Via Ostiensis
      (1) The Emporium

78. Nardini, 357.

b. Other Settlements on Main Roads
c. Measurements from the Golden Milestone
6. The Water Sources Listed in the Regionary Catalogs
7. Underground Conduits in Rome

OPENING

Giulio dei Conti di Montevecchio. Giulio dei Conti was *scriptor in Graeca lingua* [Greek secretary] of the Vatican Library and one of the learned friends of Giovanni Ciampini. See Ashby, 2 n. 4 (20 n. 9). The commentator of the Barbiellini edition writes of Giulio (127 n. a): "Romae diu, multumque vixit, litteratis viris carus, et iis maxime, qui nocturnis horis frequentes aderant apud clarissimum Ciampinum, cuius domi omnium fere musarum chorus consedisse videbatur. Poeticam facultatem sedula cura excoluit; quod testantur aliqua typis edita, ac longe plura, quae MSC. apud haeredes reperiuntur" [He lived much at Rome over a long period, dear to its literati, and especially to those present in great numbers in the evenings at the home of the famous Ciampini, at whose home the chorus of almost all the Muses seems to have gathered. He cultivated his poetic talent with constant care, as his published works and far more things found in manuscripts belonging to his heirs bear witness].

1. INTRODUCTION

The literary interests of Fabretti's addressee no doubt inspired Fabretti's many quotations of Latin poetry in his introduction, particularly those from Ovid and Horace.

In this introduction, Fabretti lists the four aqueducts functioning in Rome at his time, citing them not by their modern names but by those of the ancient conduits with which they are associated: the Crabra (which, as in his first dissertation [I.2], Fabretti mistakenly names the twelfth-century papal Marrana Mariana); the Virgo, or Acqua Vergine, restored by Luca Peto in the mid-sixteenth century; the Traiana, reworked by Pope Paul V as the Acqua Paola; and the Alexandrina, or Acqua Felice of Sixtus V, which tapped the same sources as the ancient Aqua Alexandrina.

None of Fabretti's references is favorable: the Marrana Mariana, Acqua

Vergine, and Acqua Paola are described as polluted; the Vergine's conduits are cited as leaking, with its flow reduced; and the Acqua Felice is described as frequently muddy. Given the relatively recent date of introduction for the last three lines, Fabretti's criticisms appear exaggerated for rhetorical effect, and his citation of Pliny's far-fetched explanation of the name of the Aqua Virgo seems to be little more than a joke to amuse Giulio, whom Fabretti describes as devoted to wine and indifferent to the city's water supply. The final quote from Horace's *Satires* confirms the conclusion that we are not to take this discussion too seriously.

2. TOPOGRAPHIC MAP

As in his first two dissertations, Fabretti begins the third with a detailed map depicting the topography of the area to be discussed. This map, entitled "Plan of the Ancient City with Adjacent Neighborhoods or Settlements," presents the Roman roads issuing from the city, with the settlements and monuments, focusing primarily on the eastern *suburbium*. Fabretti's numerical listing of monuments and ruins is again selective, omitting many topographical details; his numbering corresponds to descriptions and identifications of modern topographers as follows:

1. Aqua Virgo substructure: Ashby, 172–73 (203); Van Deman, 171.
2. Aqua Virgo conduit: Ashby, 172 (203–4); Van Deman, 171–72, textcut 18.
3. Aqua Virgo conduit: Ashby, 172 (203–4); Van Deman, 171.
4. Aqua Virgo: Ashby, 172 n. 1 (203 n. 40); Van Deman, 171.
5. "Villa dei Gordiani": Tomassetti, 3:546–53.
6. Centocelle: Tomassetti, 3:478–80.
7. "Villa delle Vignacce": Tomassetti, 4:81–82.
8. Imperial complex at Settebassi: Tomassetti, 4:102–5.
9. Imperial complex of Maxentius: G. Pisani Sartorio, *La villa Massenzio sulla via Appia* (Rome, 1976); A. Frazer, "The Iconography of the Emperor Maxentius' Buildings in Via Appia," *Art Bulletin* 48 (1966): 385–92.
10. Imperial Villa of the Quintilii: L. Quilici, "La villa dei Quintilii sulla Via Appia," *Bollettino Unione Storia ed Arte* 1–2 (1974): 15–26.
11. Neighborhood of Tor Marciana, perhaps to be identified with the

ancient Capitonis, located at the third milestone of the Via Ardeatina: Tomassetti, 2:488–89.

12. Vicus Alexandri: Tomassetti, 5:154–57.
13. Emporium (Fabretti's identification of the Navalia here is problematic): Nash, 1:380–86; Richardson, 143–44; *LTUR*, 3:139–40 (C. Mocchegiani Carpano).
14. Area of S. Passera on the Via Portuense: Tomassetti, 6:348–54.
15. Specus Octavianus: Ashby, 87 (107).
16. Via Asinaria at the Via Latina: Tomassetti, 4:47–48; Ashby, "Classical Topography-III," 43.
17. Temple of Fortuna Muliebris: Tomassetti, 4:81–82; Quilici, *Via Latina*, 62–65.
18. Settling tanks of the Marcia/Tepula/Julia: Ashby, 133–35 (160); Pisani Sartorio, "Punto di derivazione dell'acqua Marcia."
19. Branch line of the Claudia: Ashby, 234–35 (272–74); Lanciani, 277.
20. "Settebassi" branch of the Anio Novus: Ashby, 228 (266–67); Van Deman, 322; Aicher, 102.
21. Tomb of S. Urbano: Tomassetti, 2:124–25; on the name *Torre dei Borgiani*, see Tomassetti, 2:97.
22. Imperial villa of the Quintilii: Quilici, "La villa dei Quintilii," 15–26.
23. *Ustrinum* (crematorium): Tomassetti, 2:129–30.
24. Settling tanks of the Claudia/Anio Novus: Ashby, 225 n. 6 (264 n. 143).
25. Emissary of the Marrana Mariana: Ashby, 222–23 n. 5 (262 n. 130).
26. Via Labicana cisterns (also shown in fig. 1).
27. Villa dei Centroni aqueduct: Ashby, 222 (261 n. 121); G. M. de Rossi, *Bovillae*, FI 1.15 (Florence, 1979), 155–56 (nos. 109–10).
28. Another branch conduit along the Via Labicana (also shown, but not identified, in fig. 1).

This map is in many respects a more detailed enlargement of much of the area shown on the topographical map of the first dissertation (fig. 1), with many of the same topographical features depicted: the tributaries of the Tiber and Anio Rivers, the courses of the major Roman roads from the city, and important topographical landmarks along them. Also included are the courses of the Aqua Marcia/Tepula/Julia and the Claudia/Anio Novus from Capannelle into Rome, the line of the Aqua Alexandrina, and remains of three aqueducts to be discussed in this dissertation, those which Fabretti identifies as the Septimiana, Crabra, and Algentiana. Although Fabretti does not discuss the Aqua Virgo in any detail in this dissertation, sections of that aqueduct and its course are also depicted. Surprisingly, the Aqua Alsietina and Traiana are not shown west of the city, although the course of the papal Acqua Paola does appear. The remains of cisterns and an unidentified branch of an aqueduct along the Via Labicana (fig. 31, nos. 26, 28) do not figure in Fabretti's discussion of the *ager Romanus* but might have been included as evidence of the existence of an ancient settlement along that road, discussed in III.5b. They are, however, some distance east of the *pagus* (to be identified with Centocelle) indicated on the same road (fig. 31, no. 6).

Fabretti's presentation of Rome itself on the map is also selective, showing only those features to be discussed in III.3: the topography of the seven hills, the Emporium along the Tiber below the Aventine, and the circuits of the republican "Servian" Wall and Aurelian Wall.

3. THE EASTERN GATES OF THE "SERVIAN" WALL AND THEIR ROADS

Fabretti first addresses a topographical issue not directly germane to the subject of this dissertation but important in arguments to be presented later concerning the size and extent of the ancient city (III.5).

Fabretti's locations of the three gates of the eastern republican wall along the *agger*, the **Porta Collina, Porta Viminalis,** and **Porta Esquilina,** agree with those posited by modern topographers; see G. Säflund, *Le mura di Roma repubblicana,* 43–44, 63–66, 74–75; *LTUR,* 3:319–24 (M. Andreussi); Richardson, 262–63. Fabretti's presentation of the circuits of the "Servian" Wall and Aurelian Wall in the map accompanying his discussion (fig. 32) also generally coincides with those in modern studies.

However, the map and arguments presented here focus primarily not on the walls and locations of the gates but rather on the courses of the major Roman roads issuing from them: the **Via Praenestina** and **Via Labicana,** the **Via Collatina** and **Via Tiburtina,** and the **Via Nomentana** and **Via Salaria.** Here, Fabretti's map becomes more problematic, and the arguments and evidence presented become more complex.

The Via Salaria and Via Nomentana did indeed issue from the Porta Collina, as Fabretti argues, the Nomentana branching off from the Salaria just outside the gate. See Ashby, *Campagna*, 82; *LTUR*, 3:326 (F. Coarelli); Richardson, 417–19. Fabretti's map, which shows the Via Salaria Vetus diverging from the main road outside the Porta Nomentana, is generally correct on the course of these roads. Cf. S. Quilici Gigli, *La Via Salaria da Roma a Passo Corese*, Passeggiate nel Lazio 3 (Rome, 1977), 11–12. Fabretti's criticism of **Georg Fabricius** in this section is somewhat imprecise: in chapter 4 of his *Descriptio urbis Romae* ("De portis veteris novaeque urbis" = Graevius, 3:476F) Fabricius does not specifically name a "Porta Salaria" but does cite a gate that he says "nomen a deportando sale adepta est" [got its name from the transport of salt].

The courses of the Via Tiburtina and Via Collatina are more uncertain. Topographers generally agree that the Via Collatina, the less important of the two roads, diverged from the Via Tiburtina outside the Porta Tiburtina (Porta S. Lorenzo) of the Aurelian Wall: Ashby (*Campagna*, 143–45) traced its course outside the city along the route of the Aqua Virgo, as Fabretti depicts it in his topographical map (fig. 31). The more important Via Tiburtina, having diverged from the Via Praenestina outside the Porta Esquilina of the "Servian" Wall, appears to have run to the later Porta Tiburtina (Porta S. Lorenzo) in the Aurelian Wall; Augustus's monumentalization of the conduits of the Aquae Marcia/Tepula/Julia carried above that gate points to its significance as a major thoroughfare at the end of the first century B.C. See Ashby, *Campagna*, 94; *LTUR*, 3:312–13 (G. Pisani Sartorio); Richardson, 419. Fabretti, however, reconstructs the course of the Via Tiburtina as issuing from the Porta Viminalis and running to the so-called Porta Chiusa (or Porta Clausa), a postern gate southeast of the Castra Praetoria in the circuit of the Aurelian Wall. This is highly unlikely, given the prominence of the road that ran under the modern Porta S. Lorenzo and Augustus's monumentalization of the aqueduct conduits above it. On the Porta Chiusa, see Richmond, 181–84; Nash, 2:208–9; *LTUR*, 3:303 (G. Pisani Sartorio); Richardson, 302.

The courses of the Via Praenestina and Via Labicana, which Fabretti discusses at greatest length, are the most problematic of all. Fabretti cites Strabo 5.3.9 and archaeological evidence to argue that the Via Labicana diverged from the Via Praenestina just outside the Porta Esquilina to run to the Porta Praenestina of the Aurelian Wall (the modern Porta Maggiore) and that the ancient Via Praenestina itself ran to the north, crossing the circuit of the Aurelian Wall midway between the Porta Maggiore and Porta S. Lorenzo. Strabo's description of the course of the Via Collatina and the point of its divergence is too vague to be decisive: Strabo states only that the Via Collatina leaves the Via Praenestina and Campus Esquilinus to the left, to run 120 stades to Collatia (5.3.9). However, the archaeological evidence cited here merits closer scrutiny.

Fabretti argues that a gate (closed at this time) in the Aurelian Wall, between the Porta Maggiore and Porta S. Lorenzo, accommodated the Via Praenestina. A postern gate in the area has indeed been documented by modern topographers; see Lanciani, *FUR*, 24; Richmond, 231–32. Richmond names it "the postern of the Licinian Gardens" and also observes that the gate may "have served a road passing through the Horti Liciniani, in front of the *nymphaeum* known as the temple of Minerva Medica," a monument that Fabretti specifically mentions in this discussion. Fabretti also cites traces of an ancient road described by Famiano Nardini (perhaps to be identified with the road labeled "Via Strata" in Lanciani *FUR*, 24), as well as the orientation of the north and south sides of the "Trophies of Marius" *castellum* just outside the Porta Esquilina, as evidence of the course of the Via Praenestina.

Fabretti's argument from the "Trophies of Marius" is particularly interesting, since the orientation of the third century A.D. *nymphaeum* might well have been determined by surrounding street patterns; certainly, the modern Via di Porta Maggiore, which follows the line of the southern side of the *castellum*, appears to indicate the course of an ancient route directly to the Porta Maggiore itself. The northern orientation of the *castellum* is, however, more problematic: a road running at that angle would pass far north of the Minerva Medica *nymphaeum* and would meet the circuit of the Aurelian Wall, not at the postern of the Licinian Gardens, but much further to the northwest, near the second curtain of the wall south of the Porta S. Lorenzo. Fabretti's argument is therefore not persuasive.

However, Fabretti's methodology, that of arguing from the orientation of the "Trophies of Marius" *nymphaeum* to reconstruct the course of the roads it fronts, demonstrates ingenious foresight, even if his conclusions are incorrect. Fabretti was not (and could not have been) aware of the most telling archaeological evidence for the courses of the Via Labicana and Via Praenestina, that of the plan and orientation of the tomb of Eurysaces just outside the Porta Maggiore. This evidence was unknown to Fabretti because in the seventeenth century, the tomb was still encased in a semicircular tower of the gate complex, not cleared until 1838. The trapezoidal form of this tomb of the late first century B.C. and its orientation indicate clearly what Fabretti sought to argue from the evidence of the "Trophies of Marius," the bifurcation of the two roads the tomb fronts one kilometer from the Porta Esquilina. The two routes in question must be the Via Praenestina and Via Labicana.

Strabo's comment (5.3.9) does not conflict with this interpretation of the evidence. Strabo merely locates the point of divergence in the Campus Esquilinus, the size and extent of which are unknown. This campus was certainly bounded by the Esquiline necropolis to the south but may well have extended as far east as the circuit of the Aurelian Wall or even beyond it. See *LTUR*, 1:218–19 (F. Coarelli); Richardson, 64–65.

One last piece of evidence omitted by Fabretti must also be cited here: the ancient name *Porta Praenestina* for a gate in the Aurelian Wall clearly indicates that the gate was identified with the course of the Via Praenestina and the destination of Praeneste reached by the road passing through it. Fabretti's arguments about the courses of the Via Tiburtina and Via Praenestina are therefore incorrect; his methodology, however, was sound. We are left with the impression that he might well have argued for a different (and presumably correct) course for the Via Praenestina if he had been aware of the evidence available today.

Minerva Medica. The name of a temple listed in the regionary catalogs as in Region V (Esquiliae). Fabretti follows the common attribution of the *nymphaeum* on the Via G. Giolitti between the Via Labicana and the Aurelian Walls to the temple complex. See *LTUR*, 3:255–56 (C. Carlucci); Richardson, 269–70.

Sextus Rufus. Fabretti's citation of the regionary catalogs under this name reflects an erroneous attribution of the listings to an imaginary late-third-century *vir consularis* (former consul), an attribution that originated in Flavio Biondo's *De Roma instaurata* (Venice, 1510) and that later led to the *Sexti Rufi V. C. de Regionibus Urbis Romae Libellus* published by Onofrio Panvinio as part of his *Reipublicae Romanae commentariorum libri tres* (Venice, 1558). Panvinio's text of Rufus, however, was in actuality a compilation of Pirro Ligorio, a fact unknown to Fabretti, who attacks Ligorio frequently elsewhere. For discussion, see VZ, 1:200–206; Jordan, 2:300–302.

Nardini's double *agger*. Although Fabretti cites Famiano Nardini to support his own arguments for the course of the Via Praenestina in this section, he vigorously attacks Nardini for having posited a double *agger* along the eastern circuit of the "Servian" Wall, the first constructed by Servius Tullius and a second by Tarquinius Superbus. This error, he argues, is based on a distorted reading of Strabo and Dionysius's *Antiquitates Romanae*. For the text of Dionysius, Fabretti cites the sixteenth-century translation of **Gelenius** (Siegmund Ghelen, 1497–1554) and the later edition by **Friedrich Sylburg** (1536–96).

The circuit of the walls. Fabretti's final subject in this section is criticism of **Justus Lipsius** and **Philip Clüver** for positing extensive walls for Athens to make sense of Dionysius of Halicarnassus's comparison of the size of Rome and Athens at *Antiquitates Romanae* 4.13.4–5. Fabretti's criticism of these two scholars seems gratuitous, perhaps inserted only to introduce his praise of the Jesuits **Alessandro Donati** and **Athanasius Kircher;** however, Fabretti severely criticizes Kircher elsewhere, in the first and second dissertations (I.2, II.4b) and later in this one (III.4c–d, 6).

4. THE NUMBER OF THE AQUEDUCTS

Because Fabretti's introduction to Giulio dei Conti did not directly indicate the specific subject of this dissertation (in contrast to the introductions of the first two dissertations), the author begins by citing famous passages of Dionysius, Strabo, Pliny, Frontinus, Cassiodorus, and Rutilius Numantianus on the achievement of the Roman aqueduct system. Lipsius (157–61) cites the same passages in his chapter on the aqueducts.

a. Frontinus's Evidence and Pliny

The discrepancy between Frontinus's listing of nine aqueducts (*Aq.* 4) and Pliny's cite of seven (*HN* 36.123) permits Fabretti to attack Nardini once more; otherwise, his citation of Pliny appears gratuitous.

b. The Evidence of the Regionary Catalogs and Procopius

As claimed here, Fabretti appears to have been the first to confront the discrepancies between Frontinus's listing of nine aqueducts (*Aq.* 4), the citation of fourteen aqueducts by Procopius, the much longer listings of water sources found in the fourth-century regionary catalogs, and the list given by Publius Victor. As in his first dissertation (I.3), Fabretti treats Victor as an authentic source—indeed, presenting his evidence before that of the *Notitia Imperii*.

Fabretti does not specify the edition of the *Notitia* he cites here, but it appears to have been that of **Guido Panciroli** (1523–99), whose *De quattuordecim regionibus urbis Romae earumdemque aedificiis tam publicis quam privatis libellus* was first published in Rome in 1602 and was reprinted several times in the seventeenth century; Panciroli's text also appears in Graevius, 3:383–456. Fabretti cites Panciroli by name in his discussion of the Aqua Annia and Aqua Ciminia later in this dissertation (III.6).

Procopius's citation of fourteen aqueducts (*Goth.* 1.19.3) provides a starting point for the discussion. Fabretti logically begins with the smaller number from Procopius because it is directly concerned with conduits, not water sources, and can be more easily reconciled with Frontinus than can the longer lists of the regionary catalogs.

c. The Aqua Crabra

Fabretti's identification of this conduit as Rome's tenth aqueduct is based primarily on the evidence of Frontinus's *De aquaeductu* 9.5, which Fabretti quotes at length, although glossing the text of Frontinus to support his argument; the commentator of the Barbiellini edition writes (143 n. c): "Cuius loci Frontiniani verba Fabretti nonnihil detorsit, quamquam sensus est idem prorsus" [Fabretti twists somewhat the words of this passage of Frontinus, although the sense is indeed the same]. The major change is Fabretti's citation of the phrase "hanc Aquam ab Agrippa emissam" (which he glosses with the phrase "a ductu Iuliae exclusam") for the manuscript reading "hanc Agrippa omisit."

Fabretti seems unaware, however, that in this notice on the Aqua Crabra, Frontinus may have been describing a natural stream, not an aqueduct proper. For a summary of the evidence, see Hodge, 448 n. 17.

Fabretti accepts Frontinus's description of the Crabra as an aqueduct "in agro Tusculano" (in Tusculan territory [*Aq.* 9.5]), arguing that it ran underground near the Villa dei Centroni at the ninth milestone of the Via Latina, then emerged to be channeled into an artificial aboveground canal for delivery to Rome. Much of this argument is highly problematic, the result of a faulty understanding of the evidence. As the commentary to the first dissertation (chap. 3) indicates, the water brought to Rome by the ancient conduit that Fabretti describes here was that of the twelfth-century Marrana Mariana, introduced by Pope Calixtus II to supply the area around S. Giovanni in Laterano. Calixtus used part of the *specus* of the Aqua Claudia below the Villa dei Centroni for his conduit then channeled the water in an open canal for delivery into Rome. The channel can be seen today between the arches of the Claudia/Anio Novus and the substructure of the Aqua Marcia in the area of Roma Vecchia; for a description, see Aicher, 97.

Fabretti is certainly correct in recognizing the emissary as an ancient construction (see Ashby, 222 n. 5 [262 n. 131]), but he is wrong in identifying its water as that of the Aqua Crabra; Callixtus's aqueduct was in fact fed by sources that had supplied the Aqua Julia. It is also surprising that Fabretti pays absolutely no attention to conduit levels in his discussion here.

Fabretti's additional statement that the Crabra receives water from the same sources supplying the ancient Tepula and Julia introduces criticism of the author of a contemporary guidebook, **Fioravante Martinelli,** whose *Roma ex ethnica sacra exposita* was published in 1653.

The **Villa dei Centroni,** which made extensive use of water in its architecture, was certainly supplied by an aqueduct linked to a castellum some seven hundred meters southeast of the complex; the source of its water, however, is unknown. For a description, see de Rossi, *Bovillae*, 70–98 (nos. 101, 109–10); Quilici, *Via Latina*, 133–37, table IV; Ashby, *Campagna*, 159–60.

Fabretti goes on to argue that the rationing scheme described by Frontinus was perhaps followed as well in distribution of the Crabra within the city, citing the evidence of an inscription found on the Aventine (*CIL* VI, 1261), without directly connecting it with the Crabra itself. The inscription, which depicts what appears to be a plan of an aqueduct, with reservoirs and sluices marked, along with names of property owners and a schedule of distribution, does not identify its conduit. Mommsen (*CIL* VI, 1261) attributed the inscription to the Aqua Crabra, but the aqueduct it depicts may well have been one in a nonurban setting; see Bruun, 87 n. 48, for a description of the plan and its listings.

In fairness to Fabretti, it should be noted that he cites and illustrates the inscription only to document a rationing scheme comparable to that described by Frontinus in *De aquaeductu* 9.5; Fabretti does not specifically identify the aqueduct shown as the Crabra. Although he was in error in identifying what was the papal Marrana Mariana as the Aqua Crabra itself, Fabretti's citation of the inscription to illustrate the staggered system of distribution described by Frontinus in *De aquaeductu* 9.5 was remarkably astute, perhaps our earliest documented attempt to link epigraphical and literary evidence on this point. Poleni (41 n. 16) comments on Fabretti's observation here about possible distribution by allotments of time. That such arrangements were used in ancient Rome is now confirmed by the discovery of an inscription indicating staggered distribution of the Aqua Alsietina (*CIL* VI, 31556). See Ashby, 183–84 n. 1 (214 n. 9).

Fabretti's next argument is to identify the Aqua Crabra with the Aqua Damnata listed in the regionary catalogs. This identification, supported only by his gloss of the textual evidence of Frontinus's *De aquaeductu* 9.5 mentioned earlier in the commentary on this section ("emissam, improbatam, atque exclusam"), is not persuasive, especially since the water of the Aqua Crabra itself was not brought to Rome. No topographer to date, however, has made convincing sense of the name *Aqua Damnata*. A recently discovered citation in a thirteenth-century Hebrew translation of a commentary by Galen suggests that the Aqua Damnata may well have been a mineral spring with therapeutic properties in the *ager Romanus* rather than an aqueduct proper; see. A. Wasserstein, "Aqua Damnata," *Hermes* 103 (1975): 382–83; *LTUR*, 1:65 (D. Palombi); Richardson, 17.

To close this part of his discussion, Fabretti devotes considerable attention to another topic not in dispute, the course and length of the Almo (or modern Caffarella), a natural tributary of the Tiber (shown as the Almo Fluviolus in figs. 31–32); this introduces criticism of Kircher and Georg Fabricius for their errors about the Crabra. Finally, Fabretti returns to the passage of Procopius cited earlier, to argue that the details it gives about the brick construction and dimensions of Rome's aqueduct channels are not to be taken literally; this comment introduces a sharp attack on two other scholars, **Robert Keuchen** (1636–73) and **Thomas Dempster** (ca. 1579–1625).

Keuchen, whose edition of Frontinus appeared in Amsterdam in 1661, is severely criticized for having confused *libratio* (the term Frontinus uses for the elevation of the aqueducts) with the dimensions of the conduits themselves. Poleni (6–7 n. 8) comments on Fabretti's criticism here, "Quam turpiter autem hallucinatus sit in sua Nota adcitata verba Keuchenius, qui hanc altitudinem, ceu *altitudinem subterraneorum meatuum et fornicum* sumere visus est, si scire cupit, is consulat Fabrettum, a quo satis iam, immo fortasse nimis, Keuchenius vapulat" [If someone wants to know how disgracefully Keuchen, who took "elevation" here as the height of the underground channels and arches, wandered in his misinterpretation of the words cited in his note, he should consult Fabretti, by whom Keuchen is thrashed sufficiently—indeed, perhaps too much].

Keuchen's source for this misinterpretation was the *Antiquitatum romanarum corpus absolutissimum* of Dempster, a Scottish scholar who studied in Europe and eventually became a professor at Bologna. Most famous for his study *De Etruria Regali* (Florence, 1723–24), a comprehensive account of the Etruscans published a century after his death, Dempster reprinted, with corrections and additions, the *Antiquitates romanae* of Johann Rossfeld, or Rosinus, in 1585. For accounts of his life and career, see R. Leighton and C. Castelino, "Thomas Dempster and Ancient Etruria: A Review of the Autobiography and *De Etruria Regali*," *BSR* 58 (1990): 337–52; *EHCA*, 357–58 (C. L. Sowder).

d. The Aqua Septimiana

After references to the Aqua Traiana and Aqua Alexandrina, which he designates the eleventh and twelfth aqueducts of the fourteen cited by Procopius, Fabretti identifies as the thirteenth aqueduct what is actually a branch of the Aqua Claudia or Anio Novus supplying the Villa dei Quintilii on the Via Appia. Its arcade is visible today from the Via Appia Nuova near the Gran Raccordo Anulare; nothing, however, is known of its course southeast of the modern Via Appia Nuova, and its point of divergence from the main conduit of the Claudia cannot be determined. After reaching the high ground on which the Via Appia Antica runs, at the area of the Torre di Selce, which Fabretti cites, this branch line runs underground to the villa complex, which seems to have been its terminus. Ashby (223–24 n. 4 [263 n. 137]) writes: "[Fabretti] wrongly says that it goes on to Rome, having seen what he believes to be traces of it between the third and fourth mile on the west of the Via Appia near the tomb of S. Urbano. . . . I have never seen anything of the kind there myself." Parker (*Aqueducts*, 133) also connected reservoirs in the valley of the Caffarella with this aqueduct, which he identified as the Aurelia; but Parker's observations are too vague to be significant and have been challenged by Ashby (235 [272–74]). For the state of the aqueduct today near the Villa dei Quintilii, see P. Meogrossi, "Villa dei Quintilii: Il restauro dell'acquedotto," in *Trionfo* II, 211–15; Aicher, 103–4.

Although Fabretti assigns this aqueduct to the Aqua Claudia, he acknowledges that because of its high level, it may well have been a branch tapped from the conduit of the Anio Novus, which ran at a higher elevation. Modern identification of the branch seems to be in question: Ashby (223 [262]) assigns it to the Claudia from deposit found in the villa itself; Meogrossi ("Villa dei Quintilii," 211) associates it with the Anio Novus. In contrast to his interest in levels in the first dissertation, Fabretti is surprisingly vague here, both about the elevation of the branch—indeed, stating that exact measurements are unnecessary—and about the course of the aqueduct itself north of the Villa dei Quintilii. As Ashby (224 n. 4 [263 n. 137]) observes, Fabretti later, in his discussion of the Aqua Algentiana (III.4e), qualifies the statements made here about delivery of the Septimiana to Rome, given the absence of any substructure or bridge carrying an aqueduct line along the Via Appia across the valley of the Caffarella.

Fabretti devotes much of his discussion here to sharp criticism of Kircher and Ligorio, ending with a comment from **Salmasius** (Claude de Saumaise, 1588–1653) concerning Scaliger. The reference, however, is cryptic; Salmasius was distinguished for his learned commentaries on the Scriptores Historiae Augustae and his *Plinianae Exercitationes* (Heidelberg, 1629), but no specific source for this quote can be identified. The commentator of the Barbiellini edition writes (149 n. a): "In catalogo librorum, qui Claudium Salmasium auctorem agnoscunt, quemque catalogum lucubrationi de illius laudibus, et vita subiiciebat Antonius Clementius, nullus est, qui inscribitur adversus Scaligerum. Quocirca sentiebam, vel Fabretti memoria, qua multum poterat, esse deceptum, vel haec forte a Salmasio aliquo ex multis eius scriptis obiter esse" [In the listing of books with Salmasius as author and in the listing Antonio Clement added to his eulogy of his life, there is no title listed against Scaliger. I feel, therefore, that Fabretti was deceived by his great memory or that perhaps these things were said in passing by Salmasius somewhere in his many writings].

e. The Aqua Algentiana

Fabretti's fourteenth aqueduct in Procopius's citation is to be identified with remains of a conduit seen in the Fosso di S. Mauro northeast of the ninth kilometer of the modern Via Tuscolana, described by L. Quilici (FI 1.10, 820 [no. 711], figs. 1854–57) and Ashby (126–27 [157–58]; cf. Ashby, "Classical Topography-III," 140–41. The aqueduct appears to have been a branch of another conduit supplying a suburban villa rather than a principal aqueduct itself; remains of a large villa were indeed found southwest of it. In contrast to his discussion of the tenth through the thirteenth aqueducts, Fabretti appears very tentative about this last line, saying almost nothing about it and expressing uncertainty about the name itself and whether the aqueduct even reached the city of Rome.

5. THE LIMITS OF THE ANCIENT CITY

Fabretti's uncertainty about the aqueducts he calls the Septimiana and Algentiana now introduces a new problem (to which he had alluded in the first dissertation, at I.5), the limits of the ancient city of Rome. He begins this discussion by citing laws implying that Rome extended far beyond the circuit of the "Servian" Wall and the later Aurelian Wall, as far as its built-up area of continuous buildings. This point then permits him to introduce criticism of inconsistencies in Justus Lipsius's arguments on the size and extent of the ancient city; Fabretti cites two passages of Lipsius's treatise *De magnitudine Romana*.

In chapter 3 of his second book ("Magnitudo suburbiorum Romae"), Lipsius included Tibur, Ocriculum, Ostia, and Aricia, towns some distance from Rome, in defining the overall area of the city, but in the chapter immediately preceding ("Diuturna opera, et caput omnium Roma"), he had argued that the figures in Pliny's famous passage on the size of Rome, indicating the distances from the Golden Milestone to the individual gates, the *extrema tectorum*, and the circuit of the city's wall (*HN* 3.66–67, quoted and discussed by Fabretti in III.5c), should be emended to reflect a circuit of twenty-three miles and a distance of seven miles from the Golden Milestone to the edge of the built-up area.

Fabretti's discussion of the extent of the ancient city touches on an important problem still unresolved and probably insolvable: the boundaries of ancient Rome were themselves ambiguous and subject to change, according to purpose and definition. For the concept of the *suburbium* itself and its relation to the city, see E. Champlin, "The *Suburbium* of Rome," *AJAH* 7 (1982): 97–117; on the city's expansion into its surrounding territory during the empire, see E. Frézouls, "Rome ville ouverte: Réflexions sur les problémes de l'expansion urbaine d'Auguste à Aurélian," in *L'urbs*, 373–92.

Fabretti adopts a compromise position on the issue, rejecting Lipsius's arguments for a wide expansion of the city, as well as views of other topographers (whom he does not identify) restricting the boundaries of Rome within the circuit of the republican or imperial walls or its *pomerium*, or sacred boundary. While Fabretti's arguments that the area of the ancient city extended as far as its continuous buildings certainly echo those advanced by modern scholars (on this physical expansion of ancient Rome, see L. Quilici, "La Campagna romana come suburbio di Roma," *PP* 29 [1974]: 410–38), he introduces the point here primarily to support later arguments that those aqueducts that did not run all the way to the city itself are still to be considered part of Rome's system and are therefore to be included in both the number cited by Procopius and the number expressed in the regionary catalogs.

a. Settlement on the Via Ostiensis

To demonstrate the extension of the populated areas of ancient Rome beyond the circuit of the walls, Fabretti now presents a lengthy discussion of settlements along the major roads issuing from the city. The first road to be treated is the Via Ostiensis, chosen no doubt because it permits Fabretti to include digressions on several problems of urban topography, the identification and location of the Porta Ostiensis, as well as structures he identifies as the Emporium and Navalia.

The settlement on the Via Ostiensis is that of the **Vicus Alexandri,** located at its third milestone, before the divergence of the Via Laurentina cited by Fabretti. For a description of the topographical and archaeological evidence for the Vicus Alexandri, see R. Lanciani, "Miscellanea topografica," *BC* 19 (1891): 217–22; Tomassetti, 5:154–57. Tomassetti cites the same passage of Ammianus Marcellinus appearing here but rejects Fabretti's statement that a deviation of the Tiber after antiquity divided the original settlement into two parts. Fabretti's topographical map (fig. 31) appears to indicate remains of a settlement on the west bank of the Tiber directly across from the Vicus Alexandri itself.

The Vicus Alexandri appears to have been a settlement that grew up on either side of the Via Ostiensis, probably to be connected with shipping operations along the river. Ammianus's notice that the obelisk brought by Constantius was off-loaded at the Vicus Alexandri for transport by land into the city through the Porta Ostiensis seems significant here for understanding the location and growth of the settlement.

The **Porta Ostiensis** (or the modern Porta S. Paolo of the Aurelian Wall) is Fabretti's first topic of digression. While the location and history of the gate are firmly established today (see Richmond, 109–21; *LTUR*, 3:307–8 [G. Pisani Sartorio]), many antiquarians from the fourteenth century on, including Flavio Biondo and Alessandro Donati, had identified the Porta S. Paolo with the Porta Trigemina of the "Servian" Wall, arguing that the Via Ostiensis issued from that gate; for example, Giovanni Battista Falda's 1676 map (Frutaz, 3: table 363) labels the gate "Porta s. Paolo o Trigemina." For discussion, see Tomassetti, 5:18–19.

Fabretti refutes this identification from the evidence of Ammianus mentioned earlier, demonstrating that the route of Constantius's obelisk through the Porta Ostiensis could not have been through Region XII (the Piscina Publica) unless the obelisk had entered the city through a gate in the area of the Porta S. Paolo. To strengthen his argument, he also cites a passage of the *Historia Augusta* concerning the Septizodium—a monumental facade constructed by Septimius Severus in the southeast corner of the Palatine—to demonstrate the importance of the approach to the Palatine from the Porta Ostiensis. The Septizodium itself, as Fabretti remarks, had stood until the late sixteenth century, when its demolition was ordered by Pope Sixtus V. See Richardson, 350.

In contrast to his earlier interest in posterns of the Aurelian Wall in his discussion of the Via Tiburtina and Via Praenestina (III.3), Fabretti here seems to accept the Porta S. Paolo as the sole Porta Ostiensis, disregarding the evidence of the small postern (commonly named the Porta Ostiensis West) that appears to have served the Emporium along the Tiber. Because this Porta Ostiensis West had been closed at the time of Maxentius, the gate certainly would not have figured in the transport of Constantius's obelisk. For the Porta Ostiensis West, see Richmond, 219–21; *LTUR*, 3:308 (G. Pisani Sartorio). Fabretti does argue sensibly from the physical remains of the Via Ostiensis itself that the Porta Ostiensis must have occupied the position of the modern Porta S. Paolo.

(1) The Emporium

Fabretti's discussion of the Porta Ostiensis and Porta Trigemina now permits a further digression on structures he identifies as the **Navalia** and

Emporium. Fabretti supports Alessandro Donati's identification of the Porta Navalis with the Porta Trigemina, as well as Donati's location of the Navalia near the Aventine. Donati (252–53 [= Graevius 3:788–89]) had cited the difficulties of positing a location for the ship sheds further up the Tiber in the Campus Martius: "Neque enim Navalia fuerunt in Campo Martio, qui campis illis [Prata Quinctia] est obiectus; etsi non omnibus, et ultra, ad pontem usque Milvium protensis. Quo modo enim si ibi fuissent, intersepto pontibus fluvio, onerariae magnaeque naves praealtis malis, velisque diffusis eo deduci potuissent . . . Ergo enim ad Aventinum, et prima urbis moenia Navale, quo comodus esset traiectus a mari; ut hodie Tiberinus ille tractus navibus opportunam stationem praebet. Porro ager ille quattuor iugerum, seu prata Quinctia, ubi horti sunt ad Aedem S. Francesci, trans tiberim fuere. Neque negotium facessit Plinius scribens: *Cincinnato aranti quatuor sua iugera in Vaticano*. Ut enim Vaticani montes a ponte Milvio ad Ianiculum pertinent: ita et campi. Qui enim trans Tiberim intra illos montes iacent, vaticani sunt" [Indeed, the Navalia were not in the Campus Martius, which is set against the Prata Quinctia, although not all of them were extended as far as the Pons Milvius and beyond. For if they had been there, since the river was restricted by bridges, how could large merchant vessels with tall masts and full sails have been brought to that point? . . . The Navalia were therefore at the Aventine and the first walls of the city, where there might be an easy passage from the sea, as today that stretch of the Tiber offers a suitable moorage for ships. Furthermore, that field of four *iugera,* or the Prata Quinctia, were across the Tiber, where there are the gardens at the Church of S. Francesco. And Pliny does not create a difficulty when he writes, "to Cincinnatus plowing his four *iugera* in the Vatican," for just as the Vatican mountains run from the Pons Milvius to the Janiculum, so, too, do the fields; those that lie below the mountains across the Tiber are Vatican].

Location of the Navalia remains a problem today, the result of confusing topographical evidence. See, for example, Richardson, 266; *LTUR*, 3:339–40 (F. Coarelli). Onofrio Panvinio, whom Fabretti cites in this passage, had indicated that the Navalia were in Region XIV of the ancient city (see Graevius, 3:377A). Ligorio's argument for the location of the ship sheds is unknown; the commentator of the Barbiellini edition writes (154 n. b): "Quae de situ Navalis Ligorius disputat, ea forte inter illius MSC. adhuc delitescunt" [What Ligorio argued about the location of the Navalia perhaps still lies hidden in his manuscripts].

Fabretti devotes much more attention in this passage to a plan and reconstruction of the adjacent vaulted structure (commonly identified as the "Porticus Aemilia") situated along the Tiber in the Testaccio area of modern Rome. Fabretti's description and diagrams are our earliest detailed account of an area not systematically excavated and studied until the mid–nineteenth century and now very much obliterated by buildings of the modern city. For discussions of the complex and nearby structures, see *LTUR*, 2:221–23 (C. Mocchegiani Carpino); Richardson, 143–44; Nash, 1:380–86, 2:238–40; S. L. Tuck, "A New Identification for the 'Porticus Aemilia,'" *JRA* 13 (2000): 175–82.

Fabretti's description of the area begins with a quotation of a letter from **Flaminio Vacca** to Anastasio Simonetti of Perugia (which Fabretti also cited in his account of the Monte del Grano in his first dissertation [I.6]), then moves to a presentation of inscriptional evidence found on the site. Vacca describes in particular the large assortment of marbles off-loaded in the vicinity, from which the area derived the modern name *Marmorata*. See Richardson, 244.

Other scholars are also cited in this discussion. **Pomponio Leto** (Julius Pomponius Laetus, 1425–98) was a leading humanist of the fifteenth century who tilled his land according to the instructions of Varro and Columella and whose vineyard on the Quirinal was frequented by his students: see Sandys, 2:92–93; *EHCA*, 678–79 (I. Rowland). **Jacopo Mazochi** published the first printed collection of inscriptions in Rome in 1521; see J. E. Sandys, *Latin Epigraphy* (Cambridge, 1919), 24–25.

b. Other Settlements on Main Roads

Fabretti's discussion of settlements along major arteries now becomes much briefer, organized by individual roads from the Via Ostiensis, in counterclockwise fashion. He first discusses the **Via Ardeatina,** which he describes (and depicts on his topographical map [fig. 31]) as branching off from the Via Appia close to the Almo (or modern Caffarella) stream. Some topographers have posited a separate course for the first stage of the Ardeatina, starting from the Porta Ardeatina of the Aurelian Wall and running due south from it, although there are no remains of the bridge by which it crossed the Almo: see *LTUR*, 3:300–301 (G. Pisani Sartorio); Tomassetti, 2:485–86. However, Fabretti's reconstruction appears to be supported by Ashby (*Campagna*, 207): "The Via Ardeatina at present diverges to the right from the Via Appia at the Church of *Domine quo vadis?*, and I think we must suppose that it always did so." Cf. L. Quilici, "La posterula di Vigna Casali nella pianificazione urbanistica dell'Aventino e sul possibile prospetto del Tempio di Diana," in *L'urbs*, 713–45, linking the Porta Ardeatina with an ancient street running from the gate to the Via Appia Antica north of the Almo, as Fabretti indicates in figure 32.

Fabretti's settlement along the Ardeatina as described here and depicted on figure 31 (no. 9) is to be identified with Maxentius's circus/palace complex along the Via Appia Antica, along with other ancient structures connected with the estate of Herodes Atticus extending east to the Church of S. Urbano off the modern Via Appia Pignatelli. For a general description, see Quilici, *Via Appia*, 34–43; Pisani Sartorio, *La villa Massenzio*. Fabretti describes this complex and its structures as the first settlement along the **Via Appia,** identifying the area as Ad Camoenas, with the name given to it by Famiano Nardini. The name itself is incorrect, since the Vicus Camenarum is rather to be located in the immediate vicinity of the Porta Capena, according to our evidence; see Richardson, 63–64, 421. Fabretti cites a second settlement, Statuarii, at the fifth milestone of the Appia, to be identified with the imperial complex of the Villa dei Quintilii, also shown in figure 31 (no. 10).

Fabretti also describes a branch of the Aqua Claudia crossing the valley of the Caffarella on an arcade of four hundred paces (fig. 31, no. 19), now completely lost. However, as Ashby, (234–35 [272–73]) observes, the arches do not appear on the famous maps of Ameti (1693) or Cingolani (1704). Lanciani (277) attributed them to a channel supplying the so-called *nymphaeum* of Egeria. This arcade cited by Fabretti cannot be attested and therefore remains one of the mysteries of Roman aqueduct hunting.

The first **Via Latina** settlement, which Fabretti tentatively names the Pagus Lemonius (fig. 31, no. 17), is to be identified with the area of the Via del Quadraro and nearby "Villa delle Vignacce." See Quilici, *Via Latina*, 62–65. The second is the imperial complex at Settebassi (fig. 31, no. 20), on which see N. Lupu, "La villa dei Sette Bassi sulla via Latina," *Ephemeris Daco-romana* 7 (1937): 117–88. For the branch line of the Anio Novus supplying Settebassi cited by Fabretti, see Ashby, 228 (266–67); Van Deman, 322; Aicher, 102.

Fabretti's **Via Praenestina** settlement (fig. 31, no. 5) is the complex commonly known as the Villa dei Gordiani. Fabretti says extremely little about it but moves directly to a much longer discussion of the **Via Collatina,** where he does not cite particular remains of a settlement but instead attacks the topographical work of Lucas Holste and Georg Fabricius.

Fabretti had strongly criticized Holste on his reading of the topography of the Via Praenestina at the end of the first dissertation (I.7); in like spirit, he now censures him for arguing that the Via Collatina diverged from the Via Praenestina in the vicinity of the Villa dei Gordiani. Although some modern topographers reconstruct the beginning of the Via Collatina from that point (see Coarelli, *Dintorni di Roma,* 129), the ancient Via Collatina probably followed a course separate from the modern road until its fifth kilometer: see Ashby, *Campagna,* 143; Tomassetti, 3:554; Richardson, 415. Fabretti is therefore correct on this point.

Fabretti is wrong, however, in his attempt to identify Castellacio dell'Osa as the ancient Collatia, more accurately located by modern topographers on the site of the modern Lunghezza. See FI 1.10, 199–237; Ashby, *Campagna,* 145; Tomassetti, 3:569–75. The *castello* of S. Giuliano cited in this discussion no longer exists, its name surviving only in the name of a trench just east of the site of Gabii on the Via Praenestina; see Tomassetti, 3:570. As in his first dissertation, Fabretti's censure of Holste here is harsh, but he reserves his greatest scorn for Fabricius, whom he quotes to demonstrate a faulty knowledge of the Roman Campagna and to introduce the next two roads.

Settlements along the **Via Tiburtina** and **Via Nomentana** are also not discussed. Fabretti simply repeats arguments made earlier about the course of each road near the Castra Praetoria and promises that there will be further explanation later in the dissertation. Likewise, Fabretti is unable to cite particular settlements along the **Via Salaria, Via Flaminia,** and **Via Triumphalis;** for the last two roads, he indicates only their courses through the Campus Martius and an area that he identifies, without explanation, as the Campus Minor. However, the name and location of the Campus Minor are problematic: see *LTUR,* 1:224 (T. P. Wiseman); Richardson, 67. Indeed, Fabretti seems ready to end this part of his discussion as quickly as possible and move to the next topic, an attempt to make sense of Pliny's famous description of the size of ancient Rome.

c. Measurements from the Golden Milestone

Fabretti now tackles one of the most problematic passages in our literary sources on the size of the ancient city, Pliny's demonstration in his *Natural History* of the size of Rome through measurement of distances from the Golden Milestone in the Roman Forum to the gates of the city and of the lengths of the streets from the same point to the edges of the built-up area (*HN* 3.66–67). Fabretti's citation is partial, omitting Pliny's comments on the number of the gates themselves, primarily because the question was not directly germane to his immediate argument (and certainly would have complicated his earlier discussion of the gates along the eastern *agger* in III.3). The figures Fabretti cites for the sum of the distances from the Golden Milestone to the gates (thirty miles and 765 paces) and from the milestone to the edge of the built-up area (seventy miles) also differ from those accepted by modern editors of the *Historia Naturalis:* while Fabretti's figures are not without manuscript authority, Mayhoff's Teubner edition (C. *Plini Secundi Naturalis historiae libri XXXVII* [Leipzig, 1933], 1:257–58) gives twenty miles and 765 paces, and sixty miles, respectively, for these distances.

The discrepancy between the distances from the Golden Milestone to the gates is not important, because it does not figure in Fabretti's argument. However, his reading of seventy miles as the sum of the distances from the milestone to the edge of continuous inhabitation along the principal roads becomes quite significant, because Fabretti takes it as the foundation for presenting a rough calculation of distances from the milestone to individual settlements along the routes, resulting in a total of 70,500 paces.

Fabretti's argumentation is ingenious but flawed. To produce this figure, roughly equivalent to the seventy miles given by Pliny (itself open to question in the manuscript tradition), Fabretti bases his case on three premises. The first is that the Via Tiburtina and Via Nomentana both led to the Castra Praetoria, issuing from the city at that point, not through the gates in the later Aurelian Wall that have been traditionally assigned to them. This assumption rests on Fabretti's earlier arguments concerning the course of the roads in III.3. The second premise is that there was once an ancient settlement (for which Fabretti can cite no physical evidence remaining in the seventeenth century) on the Via Salaria, approximately three and a half miles from the Golden Milestone. The third premise is that the Via Triumphalis began at the "Servian" Wall and ran across the Campus Martius (which Fabretti identifies as the Campus Minor) to the area of the Pons Neronis, as indicated in figure 32, a distance included in Fabretti's calculations. However, the topographical evidence for the course of the Via Triumphalis makes this extremely doubtful; see Richardson, 419–20.

Fabretti appears unusually defensive about his attempts to confirm the distances given in Pliny's passage, conceding that some readers may find them "insolentia" [presumption] and "audacia" [boldness]. By no means is his argument persuasive, but this discussion is a highly interesting attempt to make sense of a notoriously difficult passage, a convincing interpretation of which continues to elude topographers to this day.

Fabretti's main purpose was to demonstrate that ancient Rome extended to the edge of the built-up area of continuous inhabitation; as a result, aqueducts delivering water to such areas outside the later circuit of the imperial wall are here considered among the fourteen lines cited by Procopius. Fabretti concludes this section with criticism of Famiano Nardini and Onofrio Panvinio, for having exaggerated the size of the ancient city, and with an even more pointed attack on Ligorio, for incorrect statements about the Aqua Alsietina. Panvinio, in his *Descriptio urbis Romae*, had indeed argued for expanding the boundaries of Region I far beyond the line of the Aurelian Wall: "Haec a Porta proxima denominata est Capena, quae omnia loca, quae tunc circa eam partem tam in urbe quam extra, usque as basilicam S. Sebastiani obtinebat" [This region, which held all the places in that area both in the city and outside it, up to the Basilica of S. Sebastiano, took its name from the Porta Capena very close by] (quoted in Graevius, 3:350). Panvinio's extension of Region I far outside the city was no doubt influenced by the common identification of the Porta Capena itself with the Porta S. Sebastiano, a point refuted by Fabretti in his first dissertation (I.4d).

Ligorio's statement about distribution of the Aqua Alsietina to the Campus Martius was based on confusion of the Naumachia of Augustus in the Transtiber with another in the northern Campus Martius, possibly a construction of Domitian; see Richardson, 265–66. In correcting him, Fabretti cites Frontinus's *De aquaeductu* 85, stating that the Alsietina was consumed entirely outside the city. At the time Frontinus wrote his treatise in A.D. 97, the Naumachia of Augustus had probably long ceased to function as a naval amphitheater, and the aqueduct's water was therefore distributed outside the city for irrigation and other industrial purposes. For a full discussion, see Taylor, "Torrent or Trickle," 471–74.

Celso Cittadini (1553–1627), whom Fabretti cites in his criticism of Ligorio, was a Sienese linguist who produced works on the Latin origins of the Tuscan language. See *DBI*, 26:71–75 (G. Formichetti). Cittadini's criticism of Ligorio here is quoted in Fioravante Martinelli's *Roma ex ethnica sacra* (429), from which Fabretti may have drawn the quotation in this passage.

6. THE WATER SOURCES LISTED IN THE REGIONARY CATALOGS

Fabretti now addresses the problem (introduced earlier) of making sense of the listing of water sources in the regionary catalogs in light of Procopius's notice of fourteen aqueducts in Rome. He bases his discussion on the summary of twenty aqueducts given in the listing of Publius Victor cited earlier in this dissertation (III.2b).

The first four sources listed in Victor (the **Appia, Marcia, Virgo,** and **Claudia**) present no problems. The fifth, the **Herculanea,** Fabretti argues, is to be identified with the Rivus Herculaneus tapped for supplementing the volume of the Anio Novus in the upper Anio Valley. His observation that the Herculanea is listed in the catalogs in place of the Anio Novus itself (nowhere listed in the regionary catalogs) is plausible; however, modern topographers identify the Herculanea with a branch of the Aqua Marcia called the Rivus Herculaneus, distributing water over the Caelian Hill within Rome: see Jordan, 2:224; Richardson, 17. Fabretti does cite the Rivus Herculaneus of the Marcia in this discussion, but he mentions it primarily to correct Alessandro Donati concerning its name and to introduce additional criticism of Ligorio.

Ligorio's *Collectanea* is a massive encyclopedic work on antiquity in eighteen manuscript volumes, now in the Italian National State Archives in Turin. There is also a copy made for Queen Christina of Sweden in the collection of the Vatican Library (Codices Ottoboniani latini 3364–77), but it is incomplete, lacking the first part of the A volume; in addition, the N–O volume is to be found in the Vatican Library's Barberini collection (Barb. lat. 5085), having apparently been stolen at some time from the library of Palazzo Altemps. For a full discussion, see T. Ashby, "The

Bodleian MS. of Pirro Ligorio," *JRS* 9 (1919): 172 n. 5. All citations of Ligorio's *Collectanea* in the notes are taken from the Vatican Library copy.

Fabretti then cites the **Tepula** and **Damnata** as the water sources listed sixth and seventh. The Tepula, cited by Frontinus, presents no problems, but Fabretti's identification of the Damnata with the Aqua Crabra discussed earlier (III.4c) is clearly in error.

The eighth water source listed, the **Traiana,** reintroduces points made in the first dissertation (I.4g) and criticism of earlier topographers, particularly Andrea Fulvio, who had associated it with a reworking of the Aqua Virgo, and Giovanni Pietro Bellori, who had identified it with a Trajanic extension of the Aqua Marcia to the Aventine. As in the first dissertation, Fabretti does not name Bellori here directly. Fabretti repeats his previous argument that the Aqua Traiana entered Rome over the Janiculum, but he now adds an observation omitted in the first dissertation, that Trajan's aqueduct also delivered to the eastern bank of the Tiber. He cites an inscription of a lead pipe purportedly found on the Aventine, attesting to supply by the Aqua Traiana to the Thermae Variane.

The inscription introduced here had an interesting history of its own; since Fabretti cites only Gruter as his source, he was apparently unaware that the inscription was a fabrication of Pirro Ligorio, included in the edition of the regionary catalogs published by Onofrio Panvinio in 1558; see Lanciani, 511 (no. 81). Like Gruter and other topographers, Fabretti appears to accept the inscription as genuine—indeed, as evidence that the Aqua Traiana was brought to the Aventine. He therefore argues that water of the aqueduct was carried through pipes over the Pons Aemilius, citing Frontinus (*Aq.* 11.2) and Statius (*Silv.* 1.3.67) for examples of similar bridge crossings.

The notice from Statius is not pertinent, since the passage from the *Silvae* cited here refers not to a bridge crossing but to underwater delivery by pipe of the Aqua Marcia to the villa of Manilius Vopiscus at Tibur; see H. B. Evans, "*In Tiburtium usum*," 452. However, Frontinus's notice that water was carried from the eastern bank over bridges to the Transtiber (*Aq.* 11.2) is clear evidence that the Romans used such arrangements within the ancient city.

Although based on a spurious inscription, Fabretti's reconstruction of a bridge crossing for the Traiana was quite farsighted. Until the discovery of inscribed pipes from the Aqua Traiana near the Thermae Traianae on the Oppian Hill and publication of the Fasti Ostienses citing distribution of the Traiana "omni parte urbis" [in every part of the city] in this century (see H. Bloch, "Aqua Traiana," *AJA* 48 [1944]: 337–41; A. DeGrassi, *Inscriptiones Italiae* 13.1 [Rome, 1947], 198–200), scholars were unaware that water from Trajan's aqueduct was delivered to the eastern bank of the Tiber. For the most recent discussion of the evidence, see R. Taylor, "A *Citiore Ripa Aquae:* Aqueduct River Crossings in the Ancient City of Rome," *BSR* 63 (1995): 91–102, especially 99–100. Although Taylor's treatment of the evidence cited by Fabretti and modern topographers is carefully presented, his argument that the Aqua Traiana crossed the river on a specially constructed aqueduct bridge south of the Aventine is not convincing.

Fabretti touches on two other points in this discussion of the Aqua Traiana. The first is that the extension of the Aqua Marcia to the Aventine cited by Frontinus (*Aq.* 87) was the work of Nerva, not Trajan, and is therefore not to be associated with the Aqua Traiana. Fabretti made this point earlier, in the first dissertation [I.4c, 4g], against the views of Bellori. But unlike in the earlier discussion, Fabretti argues here from the date of the publication of the *De aquaeductu* in A.D. 97 to demonstrate that this notice in Frontinus cannot refer to a Trajanic reworking of the Aqua Marcia. His second argument on this point, based on the spurious evidence of Ligorio's inscription, is less persuasive; even if the inscription were genuine, the practice and protocol of Roman pipe stamps are now known to be far more complicated than Fabretti represents here. For a detailed discussion of the problem, see Bruun, 20–95.

The second point is that the Traiana (as restored by Pope Paul V) was again in the seventeenth century, as in antiquity, being used for driving grain mills on the Janiculum. This observation prompts a complaint against the contemporary use of river mills on the Tiber River itself and a description of Belisarius's floating mills on the river. For the ancient mills on the Janiculum, see *LTUR*, 3:270–72 (M. Bell); Richardson, 258–59. For water mills in general, see O. Wikander, "The Water-Mill," in Wikander, *Handbook*, 371–400.

For his discussion of the ninth water source listed, the **Annia,** Fabretti follows Guido Panciroli in arguing that the name *Annia* is derived from *Anio* or *Ania*. This explanation is as persuasive as any put forth so far: see Jordan, 2:224; *LTUR*, 1:61 (D. Palombi); Richardson, 15. Fabretti's discussion also introduces another attack on Ligorio.

Fabretti identifies the tenth water source listed, the **Alsia,** with the Alsietina cited by Frontinus (*Aq.* 11.1), who also gives it the name *Augusta.* This reference prompts Fabretti to introduce a digression on three other lines also named Augusta: the branch supplementing the Aqua Marcia and Aqua Claudia in the upper Anio Valley, discussed in the second dissertation (II.4f); a branch supplementing the Aqua Appia; and the Aqua Virgo, referred to as the Aqua Augusta by Dio Cassius. On the notice in Dio, Fabretti is no doubt correct. Even if the name *Augusta* was given to the Virgo, as it was to other lines, it does not appear on extant *cippi* of the aqueduct or on the inscription on the Arco del Nazzareno commemorating Claudius's reconstruction of it (*CIL* VI, 1245). See Ashby, 168 (200).

The reference to the branch of the Aqua Appia also permits Fabretti to introduce a correction of Nardini's observations about the level of the Appia at Spes Vetus. Here, too, Fabretti is correct: the Appia runs underground when it enters the city at Porta Maggiore. See Ashby, 51 (67–68); Van Deman, 64. Fabretti's final observation concerns errors made by Francesco Angeloni in his discussion of Augustan aqueducts, to which Fabretti is sharply critical.

Fabretti identifies the eleventh water source listed, the **Caerulea,** with the Aqua Claudia, citing Frontinus (*Aq.* 14.8). The twelfth source listed, the **Julia,** is self-explanatory but introduces another attack on Ligorio.

The thirteenth water source listed, the **Algentiana,** described earlier in this dissertation (III.4e), now introduces a lengthy digression on the topography of the *ager Tusculanus*. Fabretti names the aqueduct, less tentatively than in his earlier discussion, from Mount Algidus, for which he cites literary references in great detail, in an attempt to fix the location of the mountain on the ridgeline east of Tusculum and to correct errors in Kircher's topographical study of Latium. **Algidus** is generally fixed today on the high ground southeast of the Alban Mount, directly north of Velletri: see *Barrington Atlas,* map 43 and Directory, 1:626; F. Melis and S. Quilici Gigli, "Votivi e luoghi di culto nella campagna di Velletri," *ArchClass* 35 (1983): 19–24.

Rocca Priora, which Fabretti describes as occupying the ridgeline of the Algidus, is well fixed topographically; for the **monastery of S. Silvestro,** near the modern town of Monte Compatri, see Tomassetti, 3:525–26. The **Castra Molaria** cited by Fabretti, in the valley between

Tusculum and the Alban Mount, is identified with the area of Molaria at the twenty-first kilometer of the modern Via Anagnina; see Tomassetti, 4:521–30.

Fabretti treats Kircher with relative restraint but stridently attacks **Jacob Gronovius,** for his emendation of Livy 26.9.11–12, cited earlier in this discussion. Gronovius (Gronov, 1645–1716), a student of his more illustrious father, Johann Friedrich Gronovius (1611–71), was professor of Greek at Pisa and at Leiden from 1679 until his death. In addition to producing new editions of his father's Tacitus, Gellius, and Senecan tragedy, he edited Herodotus and Polybius, Cicero, Livy, and Ammianus Marcellinus. For his career, see Sandys, 2:329.

Gronovius's emendation, which appeared in the third of his *Epistolae in quibus multa Titi Livii loca geographica emendantur* (1678) substituted "Pedo" for the manuscript reading "Algido" at Livy 26.9.11–12; in proposing it, Gronovius interpreted Livy's reference to Algidus as referring to a town, not to the ridgeline east of Tusculum. As Fabretti points out, the emendation makes little sense topographically, since the town of Pedum, the exact location of which is still uncertain, is probably to be placed further northeast, in the area of Gallicano, Zagarolo, and Corcolle. See L. Quilici, "Segnalazioni in margine alla costruenda autostrada Fiano-San Cesareo," *QArchEtr* 11 (1985): 165–66; Tomassetti, 3:596, 645; Ashby, *Campagna*, 137. The ferocity of Fabretti's criticism here, including his coarse pun on the terms *Pedum* and *oppedere*, led to the protracted scholarly battle described in chapter 2 of the present book.

Fabretti's topographical map entitled "Representation of the Double Ridge or Tusculan and Praenestine High Ground and the Course of the Via Latina, according to Strabo" (fig. 37), prepared for his work in progress on the Roman *suburbium*, was first published in his *Ad Grunnovium apologema* and was reprinted in the Barbiellini edition of the *De aquis* in 1788. In contrast to the other topographical maps in the *De aquis*, it is oriented northwest to southeast, to show the valley separating the Alban Mount from the high ground of the Algidus ridge, through which the Via Latina runs. The legend is much less detailed than those in the maps prepared for the *De aquis* itself, focusing on ancient and modern roads and a few of the more prominent ruins of the aqueducts, such as the Ponte Lupo, which Fabretti wrongly assigns to the Aqua Claudia and Aqua Anio Novus.

Marcus Meibom (Maybaum, 1630–1710), whom Fabretti cites in his attack on Gronovius, had published a *Liber de fabrica triremium* (= Graevius, 12:553–680) that provoked severe criticism from **Johann Scheffer** (1621–72), a professor at Uppsala, who issued a fierce response in his *Constantini Opelii De fabrica triremium meibomiana epistula perbrevis ad amicum* (= Graevius, 12:681–704).

The fourteenth water source listed, the **Ciminia,** presents a problem, since its name implies that its water came from the Lacus Ciminius north of Rome near Viterbo, a source from which no aqueduct is attested. Fabretti therefore equates the Ciminia with the Sabatina, which follows immediately in Victor's list, attributing this identification to **Fulvio Orsini** (Fulvius Ursinus, 1529–1600), canon of S. Giovanni in Laterano at Rome, whose large collection of manuscripts and printed books became the center of scholarly and antiquarian interests in Rome. See Sandys, 2:153–54. Fabretti also gently corrects Guido Panciroli for having erroneously associated the aqueduct with a nonexistent Ciminian forest near Tusculum.

The fifteenth water source listed, the **Sabatina,** is easily identified as one of the sources supplying the Aqua Alsietina, through Fabretti's citation of Frontinus (*Aq.* 71.1), with the emendation "Sabatino" for the manuscript reading "Abatino." Likewise, the sixteenth source listed, the **Aurelia,** is readily identified with the Traiana.

Fabretti introduces his discussion of the seventeenth water source listed, the **Septimiana,** with criticism of Nardini, citing the "enumerationem secundarum caducarumque aquarum" [listing of spillover waters] of the regionary catalogs, deliberately alluding to an adjective coined by Frontinus (*Aq.* 94). He then attempts to identify the line with the branch of the Anio Novus (cited in his earlier discussion of Procopius [III.3d]) that ran to the area of the Villa of the Quintilii on the Via Appia. His argument that the aqueduct was named from its delivery to the Thermae Severianae in Region I is not persuasive: Fabretti assumes that the bath complex was located outside the circuit of the Aurelian Wall, somewhere in the vicinity of the settlement discussed earlier, south of the Caffarella stream along the Via Appia, and that the aqueduct running to the Villa of the Quintilii at the fifth milestone of the Appia continued northwest along the line of the road, an assumption for which there is no supporting evidence.

Fabretti identifies the eighteenth water source listed, the **Severiana,** with a Severan conduit, probably that of the Alexandrina. He identifies the nineteenth, the **Antoniniana,** through Caracalla's inscription on the Porta S. Lorenzo, which he had also cited in the first dissertation (I.4f).

Fabretti also repeats his early erroneous arguments about the course of the ancient Via Tiburtina, correcting Kircher for a mistaken attribution of the Porta S. Lorenzo inscription to Marcus Aurelius and for his misidentification of the inscription of Titus on the Porta Maggiore. Fabretti concludes this discussion with a much sharper attack on Ligorio's explanation of the Aqua Antoniniana.

Fabretti presents the twentieth through the twenty-fourth water sources listed, the **Alexandrina, Anio Novus, Anio Vetus, Albudina,** and **Crabra,** with little discussion. Only the Alexandrina appears in the listing of Victor published in VZ (1:255–56), and the "new edition" that Fabretti cites as the source for this supplemental listing cannot be identified. In his discussion of these final conduits, Fabretti cites only Frontinus (*Aq.* 14.2) to support his identification of the Albudina with the Claudia. His identification of the Crabra with the Aqua Damnata (presented earlier in III.4c) is of course in error.

Finally, Fabretti turns to one discrepancy between Victor's list and that of the *Notitia*, the listing in the latter of an **Aqua Setina,** which he explains through an argument of scribal error proposed by Nardini. See Jordan, 2:224.

7. UNDERGROUND CONDUITS IN ROME

As his final topic, Fabretti discusses remains of several ancient drains within the city to demonstrate that they are not part of the aqueduct system. His topographical observations are limited, since no attempt at a systematic study of Rome's sewers and drains was undertaken until that of P. Narducci in the late nineteenth century (*Sulla fognatura della città di Roma* [Rome, 1889]).

Of the channels cited here, the first is a collector for the Cloaca Maxima in the vicinity of S. Giorgio in Velabro, not far from the Janus Quadrifrons, under which the Cloaca Maxima ran. For its course, see *LTUR*, 1:288–90 (H. Bauer); Lanciani, *FUR*, 29; C. Mocchegiani Carpano, "Le cloache dell'antica Roma," in *Roma sotterranea*, 171–72. The second is a drain under the Palazzo Grimani on the Via Rasella, just north of the Quirinal Palace, perhaps to be associated with the Fogna della Giuditta. For its course, see Lanciani, *FUR*, 16; Narducci, *Sulla fognatura*, 24–34; Mocchegiani Carpano, "Le cloache," 170. The third is a drain to be identified as part of the Chiavica dell'Olmo running south from the Piazza della Minerva to Piazza Mattei. Lanciani shows its course in *FUR*, 21, with the caption "Fabretti p. 183" (a reference to the Barbiellini edition of 1788). See also Nolli, no. 887 (Piazza dell'Olmo), indicating "cloache il di cui imbocco è coperto"; Narducci, *Sulla fognatura*, 34–39; Mocchegiani Carpano, "Le cloache," 170.

Fabretti correctly observes that none of these channels is an aqueduct conduit, although his comments about their appearance ("humili atque abiecta structura" [humble and poor construction]) are too vague to be useful. He correctly cites Festus's notice on the stream of the **Petronia Amnis** to demonstrate the abundance of natural water sources in the city. On the natural springs of ancient Rome, see Lanciani, 220–49; L. Lombardi and M. Polcari, "Acquedotti antichi e moderni," in *Roma sotterranea*, 31–35. However, given the limited knowledge in the seventeenth century about the sewers of ancient Rome, Fabretti was not aware that the Petronia Amnis was probably part of the system of the Chiavica dell'Olmo cited in this discussion. See Richardson, 289–90.

This final section of the dissertation introduces two more attacks on Ligorio, the first for an erroneous explanation of the Petronia Amnis, the second for his misidentification of the water supplied to the Fontana delle Tartarughe in Piazza Mattei, a point on which Fabretti also corrects Famiano Nardini.

CLOSING

In contrast to his introduction, full of literary citations, Fabretti closes the dissertation to Giulio with a single quote from Martial, without any specific personal reference to his addressee.

CHAPTER SIX

Conclusion

An assessment of Fabretti's aqueduct hunting as a whole is now in order. The contributions made in the *De aquis* are the direct result of the author's approach to his subject, one significantly different from those of earlier topographers and scholars. Building on his expertise in reading inscriptions and on his extensive firsthand knowledge of the Roman Campagna, as well as on his interest in construction techniques and ancient engineering, subjects that had become increasingly the focus of antiquarians during the seventeenth century, Fabretti moves beyond the literary evidence of Frontinus's *De aquaeductu* and other literary sources to make sense of physical remains on the ground, which had been either misread or neglected up to that time. Though his conclusions are sometimes faulty and his understanding of the ancient city was much less informed than our own, his approach and methodology are in many respects quite modern in their attempts to combine both archaeological and literary evidence.

In the *De aquis*, Fabretti tells us much about Rome's aqueducts but also reveals something about himself. Although the individual dissertations of his treatise were certainly the fruit of many years of investigation and study and were written separately over a two-year period, his decision to publish them as a single treatise seems to have been motivated in large part by a desire to disseminate the results of his research to a wider audience and to correct errors of earlier scholars. This is especially the case with Pirro Ligorio, the well-known antiquarian of the sixteenth century, who is severely criticized throughout Fabretti's dissertations (I.4f–g; II.4e; III.6–7). In some ways, of course, Ligorio, who had been dead for almost a century when the *De aquis* was published and who was already notorious in his own day for forgeries and falsifications, was an easy target; Fabretti's censure of him is therefore no surprise, despite its sharpness. Another sixteenth-century topographer, Georg Fabricius, is also subjected to fierce criticism in the dissertations (I.4a, 4c; 4g; III.3, 5b), for inattention to detail and ignorance of the Roman Campagna.

Fabretti also directs criticism at his contemporaries. The Jesuit polymath Athanasius Kircher is praised for his erudition and devotion to learning (I.2; III.3) but is also sharply censured for errors in his topographical work on Latium (I.2; II.4b; III.4b–d, 6). Lucas Holste, whose publications on topography Fabretti acknowledges as significant and indeed authoritative (I.7; II.2, 4c), is also corrected and criticized (II.3, 4c; III.5b), at times in highly adverse personal terms (I.7). The Dutch scholar Jacob Gronovius is dismissed as totally ignorant about the topography of the *ager Tusculanus* (III.6), in criticism so sharp that it provoked the prolonged scholarly battle discussed in chapter 2 of the present book.[1]

Giovanni Pietro Bellori's recently published work on Trajan's Column also seems to have been a source of special dissatisfaction for Fabretti. Indeed, the discussion of the "Trophies of Marius" *castellum* presented in Fabretti's first dissertation (I.4c), although pertinent to the problem of levels treated there, appears to have been prompted in large part to correct Bellori's interpretation of the "Trophies." In this dissertation, Fabretti severely censures not only Bellori himself but also Bellori's mentor Francesco Angeloni (also criticized elsewhere in the treatise), as well as the numismatist Jacques Oisel, who had recently published support for Bellori's argument that the "Trophies" were constructed as part of a Trajanic reworking of the Aqua Marcia.

Recent topographical guides to the city, those of Alessandro Donati and Famiano Nardini, are cited for both approval and correction. Donati is praised for his interpretation of the coin of L. Marcius Philippus (II.6)

1. Gronovius himself was equally fierce in his published reply to Fabretti, making reference to his Roman critic as a "faber rusticus" [rustic artisan]. This prompted Fabretti's uncomplimentary play on Gronovius's name in the title of his counterreply, *Ad Grunnovium apologema*, in which he derived the word *Grunnovium* from the Latin onomàtopoeic verb *grunnire*, used to describe the sounds of grunting animals. See Marotti, "Fabrettus," 195–99; *DBI*, 43:740 (M. Ceresa).

and the Navalia (III.5a[1]) but is corrected on the Rivus Herculaneus (III.6); Nardini is criticized on a number of points.[2] In addition, Robert Keuchen, who had published a recent edition of Frontinus, is severely censured, both for his "monstrum interpretationis" in misreading Frontinus's notice about elevation (*Aq.* 18) as a reference to the size and height of aqueduct conduits and for his theft of this idea from Thomas Dempster, without attribution (III.4c).

By modern standards, Fabretti's criticism of the work of his fellow scholars seems unusually harsh. Such polemics, however, appear to have been in keeping with the scholarly climate of his time. Topographical studies on the Roman *suburbium,* as well as the ancient city, were obviously a scholarly battleground in the seventeenth century, as they still are today. Fabretti's censure may strike us as fierce, even from an individual passionately devoted to his subject, but it is not unusual in a man devoted to learning and to correcting the errors of fellow learners to advance the understanding of antiquity. We may note Fabretti's own reference to the controversy over triremes between Marcus Meibom and Johann Scheffer, to which he alludes in the third dissertation (III.6), a scholarly battle that he certainly expected to have been familiar to his readers.

Certainly, the picture of Fabretti the man given by his contemporaries is that of a convivial and highly personable individual well established in Rome's scholarly community and on warm personal terms with the many colleagues whom he cites as friends and whose work he praises frequently in the *De aquis*.[3] Glimpses of his wit and humor appear throughout the work. It cannot be denied that Fabretti the scholar is often contentious, but his contemporaries within his learned circle and the larger *respublica litterarum* or scholarly community probably did not find his polemics objectionable, even if later scholars have reacted negatively to them.[4]

It is only fair to apply to Fabretti's work as a whole his own words introducing his discussion of the "Arco di Druso" in his first dissertation: "In

2. These corrections include Nardini's attribution of the brick-faced construct of the Arcus Caelimontani to late antiquity (I.4b), his assignment of the "Trophies of Marius" to the Aqua Marcia (I.4c), his reconstruction of a double *agger* (III.3), and his assignment of the Fontana delle Tartarughe Fountain to the Petronia Amnis (III.7)

3. See Marotti, "Fabrettus," 199–200. Among the colleagues Fabretti praises in the *De aquis* are Giovanni Ciampini, Francesco Gottifredi, Francesco Cameli, and Francesco Mezzabarba Birago.

4. See Poleni, 26 (on Fabretti's criticism of Aldus Manutius II), 29 (on his criticism of Keuchen); Cassio, 1:122 (on his criticism of Bellori, Angeloni, and Oisel).

rebus id genus, ipsa vestutate incertis et obscuris, aliena oppugnare, quam sua probare facilius semper visum fuit" [In matters of this sort, uncertain and obscure because of their very antiquity, it has always seemed easier to attack other people's positions than to prove one's own] (I.4d). For all the criticism of predecessors and contemporaries that Fabretti's three dissertations on the aqueducts present, they also comprise a scholarly publication that must be assessed on its own merits.

The dissertations are not without shortcomings and errors. Fabretti's descriptions of ancient remains, while frequently quite detailed, are also often disappointingly casual (although they are hardly to be dismissed for simply that reason). Indeed, the *De aquis* frequently falls short by modern standards of topographical and archaeological research and says nothing about some important questions. Such shortcomings are in many ways to be expected, given the time of the work's composition and the state of scholarly understanding of the ancient city in the late seventeenth century. However, a review of the individual dissertations will be instructive here.

DISSERTATION I

The first dissertation establishes conclusively the course of the Aqua Alexandrina and reviews the other aqueducts brought into ancient Rome (I.2–3), both the nine lines cited by Frontinus and the Aqua Traiana, which Fabretti proves was brought to the Janiculum as an independent conduit and was not to be identified with a Trajanic reworking of the Aqua Marcia (I.4g). This dissertation also demonstrates the different levels of the aqueducts entering the city at Spes Vetus (I.4a–b) and argues correctly that the "Trophies of Marius" *castellum* was supplied by one of the Claudian lines (I.4c). Fabretti also establishes the accurate site of the Porta Capena in the valley between the Caelian and Aventine (I.4d) and describes his exploration of the nearby Aqua Appia (I.4e).

But many related problems and issues are not discussed in this dissertation. We would like to learn far more, for example, about what significance water distribution in the *suburbium* had in determining the route of the Aqua Alexandrina into the city or whether the aqueduct served purposes in Rome in addition to its delivery to Alexander Severus's *thermae* in the Campus Martius. Of course, the *suburbium* was not Fabretti's main topic in this dissertation, since he was engaged in a separate study of it that was not completed at the time of his death.

Fabretti's treatment of monuments within the city itself is frequently disappointing. His discussion of the "Trophies of Marius" (I.4c) does not attempt to make architectural or hydraulic sense of the *castellum* itself but seems to have been included rather for the purpose of attacking the recent interpretation of the monument by Angeloni and Bellori. Even more unfortunate is Fabretti's long digression on the "Arco di Druso" at the Porta S. Sebastiano (I.4f). In this discussion, Fabretti pays far too much attention to perceived similarities between the arch itself and monuments represented on Augustan coinage and far too little to conduit levels; his comparisons and arguments in this section rightly prompted the criticism of Jacob Gronovius in the years immediately following. In addition, Fabretti's attempts to identify the Specus Octavianus in this dissertation (I.4d) are not persuasive: in no way can the *specus* be reasonably connected with the conduit carried by the "Arco di Druso." However, to be fair, no topographer to date has persuasively made sense of Frontinus's notice about the Specus Octavianus, because of the limited evidence available.

DISSERTATION II

The second dissertation, focused primarily on the topography of the upper Anio Valley, attempts to fix the sources of the Aquae Marcia, Claudia, and Augusta—all identified by Frontinus—with natural springs in the area. This was a highly ambitious undertaking that was, perhaps for precisely that reason, never systematically attempted by earlier scholars. As Panimolle has correctly observed, the task would be relatively easy if the upper Anio Valley did not abound with natural springs, but because the opposite is the case, the problem is one of making sense of the textual evidence in light of the many springs in the region.[5] Fabretti's task was all the more challenging because the area was poorly documented, aside from the recent topographical work of Lucas Holste, which Fabretti frequently criticizes throughout the dissertation.

Fabretti's major focus of investigation here, his attempt to identify

5. Panimolle, 71. Panimolle (63–69) also provides a summary of the many springs in the upper Anio Valley.

securely the sources of the Marcia and Claudia, is flawed because of fundamental errors in his reconstruction of the course of the ancient Via Valeria. Although Fabretti correctly posits the existence of an earlier original route for this Roman road, a course running northeast from the Osteria della Ferrata directly past Riofreddo to the Piana dei Cavalieri (II.4b), he was ignorant of the subsequent, much more important course of the ancient Via Valeria, which followed a route along the Anio River, then turned north to S. Giorgio. This route is now confirmed by the discovery in 1889 of the *bivium* or intersection of the Via Valeria and Via Sublacensis, along with three *cippi* of the Via Valeria itself. As is frequently the case in the history of archaeology, this important discovery of the late nineteenth century has reduced much of Fabretti's argument in this dissertation to a curiosity.

The second is the longest and most expansive of the three dissertations. Fabretti begins it with a lengthy digression on the measurement of the Roman foot and mile, partly because his readers would have found the topic interesting in itself, but primarily to throw doubt on the distances recorded by Holste in his earlier topographical research on the region (II.3). Fabretti also includes within it a lengthy digression on a possible iconographical representation of the Aqua Claudia, with comparisons to reliefs of other female divinities, a discussion that has little to do with his main subject (II.4d). In his discussion of the source of the Marcia—now securely fixed through archaeological evidence at the location he posits for the Claudia—he also presents a lengthy refutation of Gerolamo Mercuriale for a misreading of Galen (II.4f). Throughout the dissertation, there are many references to inscriptions in which Fabretti displays his wide knowledge of epigraphical conventions and practice.

Two additional digressions complete the dissertation. The first is a detailed discussion of a distribution *castellum* on the ancient Via Latina, which Fabretti seeks to identify as the settling tank of the Aqua Marcia, along with an attempt to reconstruct its hydraulic operation (II.5a). Although he was able to observe and document this structure in far better condition than it is in today, Fabretti's reading of its evidence is clearly in error. The second digression is a lengthy criticism of Benedetto Castelli's seventeenth-century treatise on hydraulics, in which Fabretti is not entirely fair to Castelli (II.5c). A short discussion of the famous coin of L. Marcius Philippus depicting the arcade of the Aqua Marcia closes the treatise (II.6).

This dissertation must be judged as uneven at best. Fabretti is mistaken in his primary conclusions here, because (through no fault of his own) he did not know the most compelling archaeological evidence in fixing the courses of the Via Valeria and Via Sublacensis so that he might make sense of Frontinus's references to them. His numerous digressions, while giving valuable insights into his methodology and approaches to literary and archaeological evidence and into his wide knowledge of inscriptions, do not contribute substantially to the overall discussion. His account of the settling tanks is confused and in error. In this discussion, as well as in his remarks about Frontinus's measurement of aqueduct capacity and his earlier account of the "Trophies of Marius" in the first dissertation, he shows that his interest in the problems of hydraulic engineering must be considered superficial by modern standards.

We must conclude that Fabretti was unusually expansive in this dissertation because it was dedicated not to a scholarly colleague who might have expected closer focus on the topic but to his patron Carpegna, a prominent churchman and a famous antiquarian and collector, whom the author wished to impress through a broad display of knowledge. Certainly, he makes reference several times in the dissertation to antiquities in Carpegna's own collection. However, the dissertation cannot be dismissed as a total failure: in it, Fabretti not only presents far more detailed cartography of the area of the upper Anio than had been previously available but also fixes an original course for the ancient Via Valeria, which he documents in detail (II.4b). He also provides an extremely valuable discussion of aqueduct *cippi*, presenting for the first time the persuasive argument that their regular intervals of 240 feet used for measuring the conduits from their distribution points in the city represent the lengths of Roman *iugera* (II.4e). In addition, his illustrations of two other structures he identifies as settling tanks—the first a large cistern that he assigns to the Aqua Claudia/Anio Novus, the second a settling tank of the Aqua Virgo in the Campus Martius (II.5b)—have been repeated by later scholars.

DISSERTATION III

In contrast to the first two dissertations, the third treats primarily literary evidence, specifically the problem of making sense, in light of Frontinus and the archaeological remains, of Procopius's citation of fourteen aqueducts and the listing of twenty-four water sources in the regionary catalogs. As Fabretti approaches it, the problem becomes archaeological as well as literary, since his discussion raises a question that had engaged the attention of many antiquarians of his day, that of the size of the ancient city—specifically, how far the city of ancient Rome must be thought to have extended into its *suburbium*.

In his interpretation of the evidence, Fabretti adopts a position similar to that of most modern topographers, that the Romans considered their city as an entity extending as far as the continuous buildings (III.5). He introduces this argument by a lengthy discussion of the eastern gates of the republican "Servian" Wall and the roads issuing from them (III.3). He next turns to Procopius's notice citing fourteen aqueducts in the ancient city, arguing that Frontinus's listing of nine conduits can be supplemented by five more, including three that he identifies as the Aqua Crabra, Aqua Septimiana, and Aqua Algentiana.

His arguments are less than persuasive. Fabretti attempts to identify the Aqua Crabra with the papal Marrana Mariana (III.4c), the Septimiana with a conduit now known to be a branch of the Aqua Claudia supplying the Villa of the Quintilii on the Via Appia (III.4d), and the Algentiana with remains of an unidentified conduit northeast of the modern Via Tuscolana (III.4e). From his manner of presentation of the individual points in order, it seems that the author himself recognizes each argument as more tentative than the last.

Fabretti then discusses the limits of ancient Rome, reviewing settlements located on the major roads emanating from the city (III.5a), with a lengthy digression on the Emporium and the structure now commonly referred to as the "Porticus Aemilia," southwest of the Aventine; full discussion of the Vicus Alexandri on the Via Ostiensis (III.5a); and much more summary treatment of archaeological evidence he identifies as remains of settlements on the other roads (III.5b). His next topic is the famous topographical puzzle presented by Pliny's description of the size of Rome (*HN* 3.66–67), based on distances from the Golden Milestone to the edge of the built-up area. Fabretti's attempt to explicate this passage by a review of the settlements already discussed shows ingenuity more than cogency.

Finally, Fabretti turns to the water sources listed in the regionary catalogs (III.6), reviewing at length arguments he had made about the Aqua Traiana earlier in the first dissertation and expanding these arguments with two new points, on the use of the Traiana for powering mills on the Janiculum and its possible transport by bridge to the eastern bank of the Tiber. His argument on delivery of the Traiana across the Tiber is based primarily on the evidence of a spurious inscription but was confirmed in the mid–twentieth century by the discovery of lead pipes of the Aqua Traiana on the Oppian Hill. A short discussion of ancient drains and sewers in seventeenth-century Rome closes the dissertation (III.7).

Fabretti's own words in his discussion of the Porta Capena and Specus Octavianus in the first dissertation should be applied to his major thesis in this dissertation as well: "Ea veritatis praerogativa est, ut illi omnia concinant" [This is the special claim of truth, that all things agree with it] (I.4d). Fabretti's attempt to make sense of the evidence of Procopius and the problematic listings of the regionary catalogs in light of the archaeological evidence makes too many assumptions, begs too many questions, and must ultimately be judged ingenious but unconvincing. Of course, the regionary catalogs remain an enormously difficult source for Roman topography even today, and no one to date has made persuasive sense of their aqueduct listings.[6]

Yet Fabretti's final dissertation, for all its shortcomings, is not without its contributions. In addition to his observations about the distribution of the Aqua Traiana, Fabretti provides more detailed cartography of much of the same area of the Roman Campagna treated in the first dissertation; a full presentation of the Emporium and structure commonly known as the "Porticus Aemilia," which seems to be our earliest plan of the complex; additional insights into inscriptions, especially one documenting staggered distribution of aqueduct water; and valuable observations about ancient sewers and drains in seventeenth-century Rome.

FABRETTI'S CONTRIBUTIONS

Fabretti's first dissertation must be judged the most successful of the three, the other two being much less so. His documentation of the Aqua Alexandrina and his systematic treatment of the other major aqueducts of ancient Rome are far more valuable to scholarship than his presentations in the

6. For the problems in assessing the catalogs' evidence on lacus, baths, and aqueducts, see Bruun, 73–74.

second and third dissertations. But even with all of their omissions and errors, all three of Fabretti's dissertations comprise a gold mine of information on the aqueduct system of the ancient city.

Fabretti's contributions to the study of Roman aqueducts include a detailed presentation of the remains of the Aqua Alexandrina (repeated almost without correction by all later topographers), along with a carefully drafted map of its course; extremely detailed plans of the conduits and arches of the aqueducts, as well as the ancient monuments described; vivid accounts of his personal explorations; and insights on the topography of the upper Anio Valley and the Roman *suburbium* based on autoptic observation and keen familiarity with the region, accompanied by three stunningly presented, well-designed, and highly detailed topographical maps, far surpassing those of contemporary cartographers in their accuracy and coverage. Fabretti's frequent references to his maps throughout his text show that he clearly considered them integral to his overall work and to the arguments presented in it. Certainly they became the most accurate and detailed cartographic rendering of the territory represented and courses of the aqueducts outside the city until the publication of Ashby's maps in the early twentieth century. For the orientation of Fabretti's maps with the modern topography of the Roman Campagna and upper Anio Valley, see fig. 38.

Fabretti includes in his discussion of the settling tanks in the second dissertation an apt quotation from Horace, "Est aliquid prodire tenus, si non datur ultra" [It is no small thing to advance so far, even if it is not allowed to go further] (*Epist.* 1.1.32, quoted by Fabretti at II.5b). Certainly, when he published the *De aquis*, Fabretti did not intend the three dissertations to be his final word on Roman aqueducts. At the end of the second dissertation (II.6), he promises to treat elsewhere a list of problems he calls "doubtful" matters, including the capacities of individual aqueducts given in Frontinus's *De aquaeductu* (along with the obvious mathematical errors in the statistics presented in that text) and the levels of the conduits. In the final dissertation, he declares that he will devote a separate treatise to the routes or courses of the aqueducts (III.4d). These investigations were never published during Fabretti's lifetime, probably because he was determined to finish his massive epigraphical corpus before his death.

Whether Fabretti would have made major contributions to our understanding of Frontinus's distribution statistics in the *De aquaeductu* is prob-

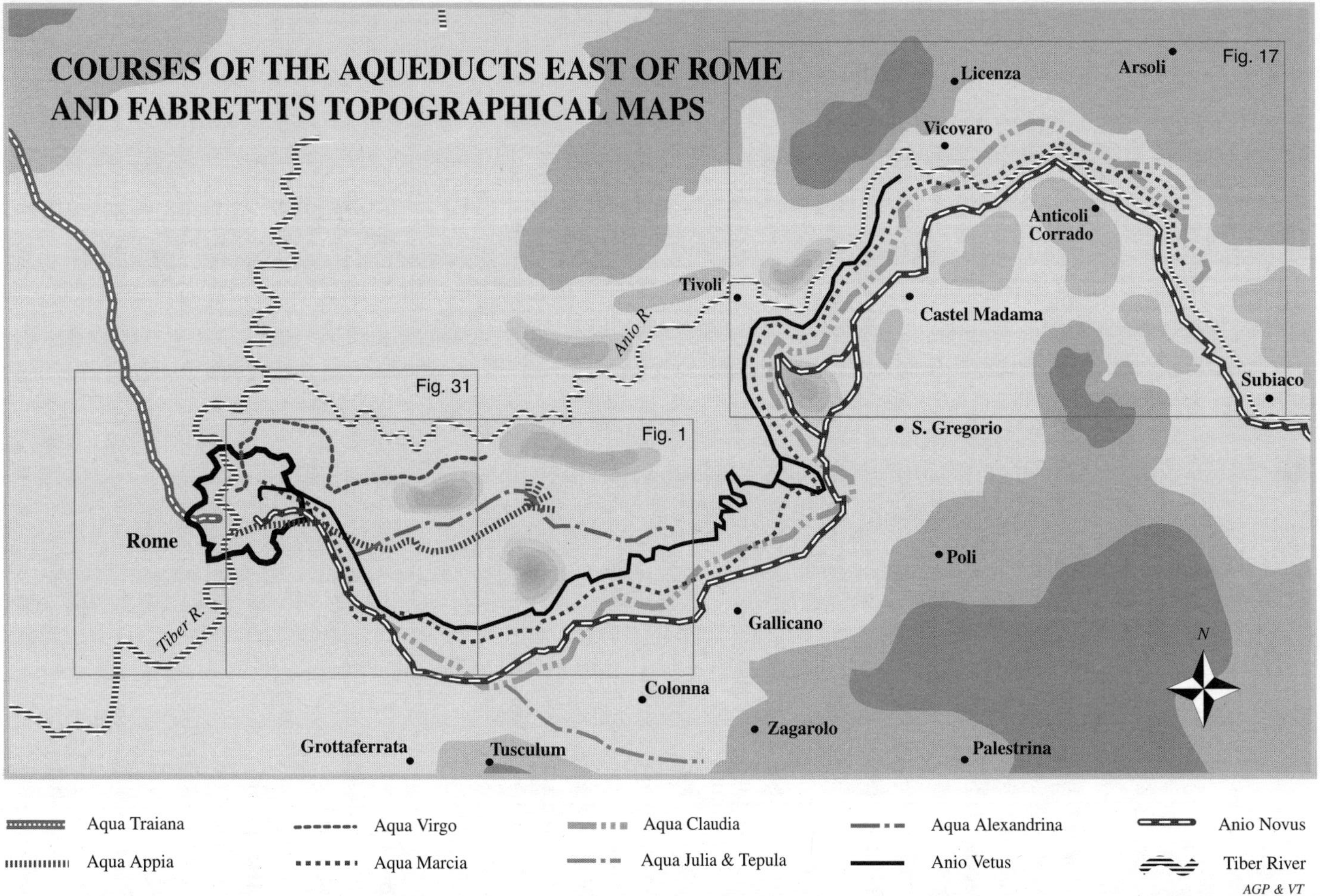
COURSES OF THE AQUEDUCTS EAST OF ROME
AND FABRETTI'S TOPOGRAPHICAL MAPS
Fig. 17
Fig. 31
Fig. 1
Arsoli
Licenza
Vicovaro
Anticoli Corrado
Tivoli
Castel Madama
Anio R.
Subiaco
S. Gregorio
Rome
Poli
Tiber R.
Gallicano
N
Colonna
Zagarolo
Grottaferrata
Tusculum
Palestrina
Aqua Traiana
Aqua Appia
Aqua Virgo
Aqua Marcia
Aqua Claudia
Aqua Julia & Tepula
Aqua Alexandrina
Anio Vetus
Anio Novus
Tiber River
AGP & VT

lematic. Although Fabretti displays an impressive familiarity with Frontinus's text, as well as with the other literary sources he cites frequently, his real strengths as a scholar and topographer lie primarily in his firsthand knowledge of the *ager Romanus* and his passionate interest in inscriptions. It cannot be denied, however, that Fabretti's familiarity with the Campagna more than qualified him to address the problems of the levels of the aqueducts and the routes of their conduits outside Rome. It was precisely this unfinished business to which Thomas Ashby and Esther Van Deman directed their focus in their magisterial studies of the aqueducts in the early twentieth century.

Within the overall framework of his publications and career, Fabretti can in some respects be considered a scholar typical of his day—producing commentaries on antiquities already collected and documented and drawing on the contributions of the last two centuries—rather than a pioneer introducing new critical methods or approaches to the study of ancient Rome. One scholar has recently given the following general assessment of his scholarly work: "Infatti non si osserva nei contributi fabrettiani un effettivo progresso rispetto alle concezioni concurrenti nel campo dell'antiquaria durante il Seicento e, in definitiva, anche se condotti con un padronanza della materia allora esemplare, essi sembrano chiudere un'epoca piuttosto che aprirne una nuova" [In fact, in Fabretti's contributions, one does not observe real progress with respect to ideas prevalent in antiquarian study during the seventeenth century, and in conclusion, even if carried out with a mastery of the material exemplary for that time, his contributions seem to close an era rather than to begin a new one].[7]

Such a verdict is unreasonably dismissive, however. It might well be valid in the case of Fabretti's longer monumental works on Trajan's Column and his corpus of inscriptions updating that of Gruter, but it cannot be applied with fairness to the *De aquis*. The three dissertations cannot claim to introduce a new methodology, but Fabretti's research does break important new ground, marking a significant advance beyond all earlier study of the aqueduct system in the sixteenth and seventeenth centuries

7. Mennella, *Museo Lapidario*, 17. Pfeiffer (143) offers a general assessment of classical scholarship in seventeenth-century Europe: "the characteristic scene in most Continental countries was that of self-satisfied polymaths filling enormous volumes with collected antiquities and reproducing in their editions the accumulated notes of the last two centuries." Cf. Sandys, 2:370: "On the whole, it was a century of multifarious erudition rather than minute and accurate scholarship . . ." Fabretti's work on the aqueducts is an exception to this overall view.

and contributing far more than simply a rehash of ancient sources and earlier topographical findings. Having gone over the ground of the Roman Campagna and upper Anio Valley, Fabretti presents new material and observations, many of which have never been superseded, in an effort to contribute a coherent and convincing overview of the aqueduct system.

Perhaps even more interesting is the vivid picture that the *De aquis* provides of a serious scholar at work. We encounter our author meticulously weighing literary and epigraphical evidence against that of the physical remains (many of which he was documenting for the first time), in lively dialogue with his contemporaries and predecessors, vigorously correcting their errors when appropriate and making significant contributions of his own to the understanding of the ancient city in his time.

Most important, of course, are the long-term results of this pioneering work. Fabretti's aqueduct hunting effectively demonstrated as early as the seventeenth century that Frontinus's text, while an indispensable source for understanding the water system of the ancient city—indeed, still the starting point for any serious study of it—neither was for Fabretti nor can be for us the last word on the subject. Fabretti's approach to the problem of making sense of the aqueduct system of the ancient city therefore not only became authoritative in his own day and in the years immediately following but also had a profound influence on all later investigations of the subject. Aqueduct hunting has continued throughout the last three centuries, distinguished by the contributions of such giants as Rodolfo Lanciani, Thomas Ashby, and Esther Van Deman, but Raffaello Fabretti started it all.

Works Cited

Abbreviations in the works cited follow closely those listed in *AJA* 95 (1991): 1–16. The abbreviation *Trionfo* is used in entries for essays from *Il trionfo dell'acqua: Acque e acquedotti a Roma, IV. sec. a.C.–XX. sec.*, ed. G. Pisani Sartorio and A. M. Liberati Silverio (Rome, 1986).

Aicher, P. J. *Guide to the Aqueducts of Ancient Rome*. Wauconda, Ill., 1995.

Almagià, R. *L'opera geografica di Luca Holstenio*. Studi e Testi 102. Vatican City, 1942.

———. *Monumenta cartographica vaticana*. Vol. 2. Vatican City, 1948.

Anastasius bibliothecarius. *De vitis pontificum romanorum*. Rome, 1718.

Angeloni, F. *L'istoria augusta da Giulio Cesare a Costantino il Magno illustrata con la verità dell'antiche medaglie*. Rome, 1641.

Aringhi, P. *Roma subterranea*. Rome, 1651.

Ashby, T. *The Aqueducts of Ancient Rome*. Oxford, 1935. Published in Italian as *Gli acquedotti dell'antica Roma* (Rome, 1991).

———. "The Bodleian MS. of Pirro Ligorio." *JRS* 9 (1919): 170–201.

———. "The Classical Topography of the Roman Campagna-I." *BSR* 1 (1902): 127–285.

———. "The Classical Topography of the Roman Campagna-III." *BSR* 4 (1907): 1–159.

———. *The Roman Campagna in Classical Times*. London, 1927. Reprint, 1970.

———. "Sul vero sito del Lago Regillo." *RendLinc* 7 (1898): 103–26.

Ashby, T., and G. Lugli. "La villa dei Flavi cristiani 'ad duas Lauros.'" *MemPontAcc* 2 (1928): 158–92.

Barbetta, S. *Via Labicana*. Rome, 1995.

Barbiellini, N., pub. *Raph. Fabretti Gasparis F. Urbinatis de aquis et aquaeductibus veteris Romae dissertationes tres, editio secunda romana adnotationibus illustrata*. Rome, 1788.

Bardon, H. *La letterature latine inconnue*. Paris, 1956.

Bartoli, P. S. *Colonna Traiana . . . accresciuta di medaglie, iscrittioni e trofei da Gian Pietro Bellori*. Rome, 1672.

Bellori, G. P. *Fragmenta vestigii veteris Romae xx. tabulis comprehensa*. Rome, 1673.

Bembo, P. *Opere*. 12 vols. Venice, 1728.

Berriman, A. E. *Historical Metrology*. London, 1953.

Biondo, F. *Roma instaurata: de origine et gestis venetorum*. Verona, 1481–82.

Blackman, D. R., and A. T. Hodge. *Frontinus' Legacy: Essays on Frontinus' "de aquis urbis Romae."* Ann Arbor, 2001.

Bloch, H. "Aqua Traiana." *AJA* 48 (1944): 337–41.

———. *I bolli laterizi di Roma e la storia edilizia romana. Studi e materiali del Museo dell'Impero Romano* 4 (1947).

Bosio, L. *La Tabula Peutingeriana*. Rimini, 1983.

Bracciolini, P. *Historiae de varietate fortunae libri quatuor*. Rome, 1448. Reprint, Forni, 1969.

British Museum, Department of Coins and Medals. *Coins of the Roman Empire in the British Museum*. 6 vols. to date. London, 1965–.

British Museum, Department of Coins and Medals. *Coins of the Roman Republic in the British Museum*. 3 vols. London, 1970.

Bruun, C. *The Water Supply of Ancient Rome: A Study in Roman Imperial Administration*. Commentationes Humanarum Litterarum 93. Helsinki, 1991.

Buechner, C., ed. *Fragmenta Poetarum Latinorum*. Leipzig, 1982.

Burmann, P., ed. *Publii Ovidii Nasonis opera omnia*. 4 vols. Amsterdam, 1727.

Calci, C. "L'acquedotto alessandrino alla tenuta della Mistica." *BC* 89 (1994): 105–6.

Cancellieri, C. "L'acquedotto Paolo (sec. XVII–XX)." In *Trionfo*, 225–31.

Cardano, N. "La mostra dell'acqua Felice." In *Trionfo*, 243–50.

Carnabuci, E. *Via Aurelia*. Rome, 1992.

Caruso, G. "L'Aqua Alexandriana." In *Trionfo*, 120–23.

Caruso, G., and P. Giusberti. "Acquedotto Alessandrino: Restauro del tratto tra via del Fosso di Centocelle e via dei Pioppi." *BC* 95, no. 2 (1993): 117–21.

Cassio, A. *Corso dell'acque antiche*. 2 vols. Rome, 1756–57.

Castelli, B., O.S.B. *Della misura dell'acque correnti*. 3d ed. Bologna, 1660.

Chacon, A. *Historia utriusque belli Dacici a Traiano Caesare gesti*. Rome, 1576.

Champlin, E. "The *Suburbium* of Rome." *AJAH* 7 (1982): 97–117.

Cherubini, L., ed. *Magnum bullarium romanum: ab Leone Magno usque ad S.D.N. Clementem X opus absolutissimum*. 5 vols. in 4. Lyons, 1673–92.

Ciofano, E. *In omnia P. Ovidii Nasonis opera observationes*. Antwerp, 1581–83.

Clüver, P. *Italia antica, auctoris methodo, verbis et tabulis geographicis retentis*. Wolfenbüttel, 1659.

Coarelli, F. *Dintorni di Roma*. Guide archeologiche Laterza 7. Rome, 1981.

———. "La situazione edilizia di Roma sotto Severo Alessandro." In *L'urbs: Espace urbain et histoire (Ier siècle avant J-C.–IIIe siècle après J-C.)*, 429–56. CÉFR 98. Rome and Paris, 1987.

———. "L'*urbs* e il *suburbium*." In *Società romana e impero tardoantico*, ed. A. Giardina, 2:1–58, 395–412. Rome, 1986.

Coste, J. "I casali della Campagna di Roma all'inizio del '600." *Archivio della Società romana di Storia patria* 92 (1969): 41–115.

Courtney, E., ed. *The Fragmentary Latin Poets*. Oxford, 1993.

Crainz, F., and C. F. Giuliani. "I due tracciati della Via Valeria fra *ad Lamnas* e *Carseoli*." AMST 58 (1985): 71–88.

Crescimbene, G. M. *Le vite degli arcadi illustri*. 5 vols. Rome, 1708.

Dausque, C., S.J. *Ortographia Latini sermonis vetus et nova*. Paris, 1677.

DeGrassi, A. *Inscriptiones Italiae* 13.1. Rome, 1947.

de Grummond, N. T., ed. *An Encyclopedia of the History of Classical Archaeology*. 2 vols. Westport, Conn., 1996.

del Re, N. *Antonio Massa da Gallese, giurista e litterato*. Naples, 1992.

———. "Luca Peto Giuresconsulto e magistrato Capitolino (1512–1581)." In *Scritti in onore di Filippo Caraffa*, ed. G. Giammaria, 307–39. Rome, 1986.

Dempster, T. *Antiquitatum romanarum corpus absolutissimum in quo praeter ea, quae Johannes Rosinus delineaverat, infinita supplentur, mutantur, adduntur*. Cologne, 1620.

de Rossi, G. M. *Bovillae*. Forma Italiae 1.15. Florence, 1979.

de Strada à Rosberg, O. *Vitae imperatorum caesarumque Rom. orient. et occid. uxorum et liberorum eorundem*. Frankfurt, 1629.

Dilke, O. A. W. *Greek and Roman Maps*. Ithaca, N.Y., 1985.

———. *Mathematics and Measurement*. Berkeley and London, 1987.

Donati, A., S.J. *Roma vetus ac recens utriusque aedificiis ad eruditam cognitionem expositis*. Rome, 1638.

Dorcey, P. F. *The Cult of Silvanus: A Study in Roman Folk Religion*. Leiden, 1992.

Drakenborch, A., ed. *T. Livi Patavinii Historiarum libri, qui supersunt, omnes*. Stuttgart, 1828.

Dumolinet, C. *Historia summorum pontificum a Martino V ad Innocentium XI per eorum numismata*. Paris, 1679.

Evans, H. B. "*In Tiburtium usum:* Special Arrangements in the Roman Water System (Frontinus, *Aq.* 6.5)." AJA 97 (1993): 447–55.

———. *Water Distribution in Ancient Rome: The Evidence of Frontinus*. Ann Arbor, 1994.

Ewald, P., and L. M. Hartmann, eds. *Gregorii I papae Registrum epistolarum* (Monumenta Germaniae Historica Epistolae 1–2). 2 vols. Munich, 1978.

Fabretti, R. *De aquis et aquaeductibus veteris Romae dissertationes tres*. Rome, 1680. Reprint, 1788. Original edition reprinted in the series The Printed Sources of Western Art, ed. T. Besterman, no. 3 (Portland, Or., 1972).

———. *De columna Traiani syntagma*. Rome, 1683.

———. *Inscriptionum antiquarum quae in aedibus paternis asservantur explicatio et additamentum*. Rome, 1699.

———. *Isathei ad Grunnovium apologema, in eiusque Titilivitia sive somnia de Tito Livio animadversiones*. Naples, 1686.

Fabricius, G. [Georgius Fabricius Chemnicensis]. *Descriptio urbis Romae*. Rome, 1550.

Fabroni, A., ed. *Vitae Italorum doctrina excellentium qui saeculis XVII et XVIII floruerunt*. Vol. 6. Pisa, 1780.

Fiore Cavaliere, M. G., and Z. Mari. "Acquisizioni lungo il tracciato degli acquedotti aniensi." *QArchEtr* 24 (1995): 463–73.

Folkes, M. "An Account of the Standard Measures preserved in the Capitol at Rome." *Philosophical Transactions of the Royal Society in London* 442 (1736): 262–66.

Frazer, A. "The Iconography of the Emperor Maxentius' Buildings in Via Appia." *Art Bulletin* 48 (1966): 385–92.

Frézouls, E. "Rome ville ouverte: Réflexions sur les problémes de l'expansion urbaine d'Auguste à Aurélian." In *L'urbs: Espace urbain et histoire (Ier siècle avant J-C.–IIIe siècle après J-C.)*, 373–92. CÉFR 98. Rome and Paris, 1987.

Frutaz, A. M. *Le carte del Lazio*. 3 vols. Rome, 1972.

———. *Le piante di Roma*. 3 vols. Rome, 1962.

Fulvio, A. *Dell'antichità della città di Roma*. Venice, 1543.

Garbrecht, G., and H. Manderscheid. "*Etiam fonte novo Antoniniano*." *ArchClass* 44 (1992): 193–234.

Gaston, R. W., ed. *Pirro Ligorio: Artist and Antiquarian*. Florence, 1998.

Gatti, G. "Il viadotto della via Aurelia nel Trastevere." *BC* 58 (1940): 129–41.

Gatti, S. "Un nouvo cippo dell'Aqua Marcia." In *Il trionfo dell'acqua: Atti del convegno "Gli antichi acquedotti di Roma: Problemi di conoscenza, conservazione e tutela,"* ed. G. Pisani Sartorio and A. M. Liberati Silverio, 93–104. Rome, 1992.

Gelen, S., trans. *Dionysii Halicarnassei antiquitatum sive originum Romanarum*. Basel, 1549.

Ghisalberti, Alberto M., ed. *Dizionario biografico degli Italiani*. Rome, 1960–.

Gillispie, C. C., ed. *Dictionary of Scientific Biography*. New York, 1970–.

Giuliani, C. F. *Tibur, pars altera*. Forma Italiae 10.7. Rome, 1966.

Graevius, J. G. *Thesaurus antiquitatum romanarum congestus*. 2d ed. 12 vols. Venice, 1732.

Grafton, A. *The Footnote: A Curious History*. Cambridge, Mass., 1997.

———. *Joseph Scaliger: A Study in the History of Classical Scholarship*. Vol. 1, *Textual Criticism and Exegesis*. Oxford, 1983.

———, ed. *Rome Reborn: The Vatican Library and Renaissance Culture*. Washington, D.C., 1993.

Greaves, J. *A Discourse on the Romane Foot and Denarius*. London, 1647.

———. *The Miscellaneous Works of Mr. John Greaves*. London, 1737.

Gronovius, J. *Epistolae in quibus multa Titi Livii loca geographica emendantur et illustrantur*. Amsterdam, 1678.

———. *Responsio ad cavillationes Raphaelis Fabretti*. Leiden, 1684.

Gross, H. *Rome in the Age of Enlightenment*. Cambridge, 1990.

Gruter, J. *Inscriptiones antiquae totius orbis Romani in corpus absolutum simul redactae*. Heidelberg, 1603.

Guasti, C., ed. *Scritti vari di Lorenzo Panciatichi*. Florence, 1856.

Helbig, W. *Führer durch die öffentlichen Sammlungen klassicher Altertümer in Rom*. Ed. H. Speier. 4th ed. 4 vols. Tübingen, 1963–72.

Hodge, A. T. "Purity of Water." In *Handbook of Ancient Water Technology*, ed. O. Wikander, 95–99. Leiden, 2000.

———. *Roman Aqueducts and Water Supply*. London, 1992.

Hoefer, J. C. F., ed. *Nouvelle biographie générale*. 46 vols. in 23. Paris, 1852–66.

Holste, L. [Lucas Holstenius]. *Annotationes in Geographiam sacram Caroli a S. Paolo, Italiam antiquam Cluverii, et Thesaurum geographicam Ortelii*. Rome, 1666.

Jordan, H. *Topographie der Stadt Rom im Alterthum*. 2 vols. in 4. Berlin, 1871–85. Reprint, Rome, 1970.

Keuchen, R. ed. *Sexti Julii Frontini viri consularis quae extant*. Amsterdam, 1661.

Kircher, A., S.J. *Latium: Id est, nova et parallela Latii tum veteris tum novi descriptio*. Amsterdam, 1671.

Kleiner, F. S. *The Arch of Nero in Rome*. Rome, 1985.

Krautheimer, R. *The Rome of Alexander VII*. Princeton, 1985.

La Bigne, Marguerin de, ed. *Maxima bibliotheca veterum patrum et antiquorum scriptorum ecclesiasticorum*. Lyons, 1677.

Lanciani, R. *Forma Urbis Romae*. Milan, 1893–1901. Reprint, Rome, 1990.

———. *I commentarii di Frontino intorno le acque e gli acquedotti*. MemAccLinc 3d ser., 4 (1881): 213–616. Reprint, Rome, 1975.

———. "Miscellanea topografica." *BC* 19 (1891): 217–22.

———. *Storia degli scavi di Roma e notizie intorno le collezioni romane di antichità*. 4 vols. Rome, 1902–12.

Leighton, R., and C. Castelino. "Thomas Dempster and Ancient Etruria: A Review of the Autobiography and *De Etruria Regali*." *BSR* 58 (1990): 337–52.

Lexicon Iconographicum Mythologiae Classicae. 9 vols. in 18. Zurich, 1981–97.

Lemerle, F. *Les "Annotations" de Guillaume Philandrier sur le "De architectura" de Vitruve: livres I a IV*. Paris, 2000

Lewis, M. "Theoretical Hydraulics, Automata, and Water Clocks." In *Handbook of Ancient Water Technology*, ed. O. Wikander, 343–69. Leiden, 2000.

Ligorio, P. *Collectanea*. 18 vols. Codices Ottoboniani latini, 3364–77. Vatican City, n.d.

———. *Libro delle antichità di Roma, nelquale si tratta de circi, theatri e anfitheatri, con le paradosse del medesimo autore, quai confutano la commune opinione sopra varii luoghi della città di Roma*. Venice, 1553.

Lipsius, J. *Admiranda sive de magnitudine Romana libri quattuor*. Antwerp, 1617.

———. *Opera omnia postremum ab ipso aucta et recensita*. 3 vols. Antwerp, 1637.

Lombardi, L., and M. Polcari. "Acquedotti antichi e moderni." In *Roma sotterranea*, ed. R. Luciani, 31–35. Rome, 1984.

Lo Sardo, E., ed. *Athanasius Kircher, S.J.: Il Museo del Mondo*. Exhibition catalog. Rome, 2001.

Luciani, R., ed. *Roma sotterranea*. Exhibition catalog. Rome, 1984.

Luni, M., and G. Gori. *1756–1986: Il Museo Archeologico di Urbino*. Urbino, 1986.

Lupu, N. "La villa dei Setti Bassi sulla via Latina." *Ephemeris Daco-romana* 7 (1937): 117–88.

L'urbs: Espace urbain et histoire (I[e] siècle avant J-C.–IIIe siècle après J-C). CÉFR 98. Rome and Paris, 1987.

Manutius, Aldus II. *De aquis, in urbem Romam olim influentibus*. In *De quaesitis per epistulam libri III*, 26–46. Venice, 1576.

Margarini, C., O.S.B., ed. *Bullarium casinense, seu constitutiones summorum pontificum*. Vol. 1: Venice, 1650. Vol. 2: Todi, 1670.

Mari, Z. "La valle dell'Aniene nell'antichità." *AMST* 68 (1995): 25–52.

———. *Tibur, pars quarta*. Forma Italiae 1.35. Florence, 1991.

Marliani, G. *Antiquae Romae topographia libri septem*. Rome, 1534.

Marotti, G. "Raphael Fabrettus." In *Vitae Italorum doctrina excellentium qui saeculis XVII et XVIII floruerunt*, ed. A. Fabroni, 6:174–229. Pisa, 1780.

Martinelli, F. *Roma ex ethnica sacra exposita*. Rome, 1653.

Martyrologium Romanum. 4th ed. Vatican City, 1956.

Massa, A., S.J. *De origine et rebus Faliscorum liber*. Rome, 1546.

Mattingly, H., E. A. Sydenham, and R. A. G. Carson. *The Roman Imperial Coinage*. 6 vols. London, 1923–61. Rev. ed. by C. H. V. Sutherland and R. A. G. Carson, London, 1981.

Mayhoff, K., ed. *C. Plini Secundi Naturalis historiae libri XXXVII*. 6 vols. 1906. Reprint, Leipzig, 1933.

Melis, F., and S. Quilici Gigli. "Votivi e luoghi di culto nella campagna di Velletri." *ArchClass* 35 (1983): 19–24.

Mennella, G. *Il Museo Lapidario del Palazzo Ducale di Urbino: Saggio storico su documenti inediti*. Genoa, 1973.

Meogrossi, P. "Villa dei Quintilii: Il restauro dell'acquedotto." In *Il trionfo dell'acqua: Atti del convegno "Gli antichi acquedotti di Roma: Problemi di conoscenza, conservazione e tutela,"* ed. G. Pisani Sartorio and A. M. Liberati Silverio, 211–15. Rome, 1992.

Mercuriale, G. *Variarum lectionum in medicinae scriptoribus et aliis, libri quinque*. Paris, 1585.

Mocchegiani Carpano, C. "Le cloache dell'antica Roma." In *Roma sotterranea*, ed. R. Luciani, 164–78. Rome, 1984.

Montfaucon, B. de, O.S.B. *Diarium italicum sive Monumentorum veterum, bibliothecarum, musaeorum, etc. notitiae singulares in itinerario italico collectae*. Paris, 1702.
Motta, R. "L'acquedotto Felice." In *Trionfo*, 220–25.
———. "L'acquedotto Vergine (sec. XVI–XVIII)." In *Trionfo*, 205–8.
———. "La decadenza degli acquedotti antichi e la conduzione dell'acqua Mariana." In *Trionfo*, 203–5.
Nardini, F. *Roma antica*. Rome, 1666.
Nardini, N. *La catedra vescovile di S. Timoleo in Nepi: La pentapoli Nepesina e il vero sito degli antichi Veienti, Falisci, e Capenati*. Rome, 1677.
Narducci, P. *Sulla fognatura della città di Roma*. Rome, 1889.
Nash, E. *A Pictorial Dictionary of Ancient Rome*. 2d ed. 2 vols. New York, 1967.
Negri, G., S.J. *Istoria de Fiorentini scrittori*. Ferrara, 1722.
Nestori, A. "La Catacomba di Calepodio all III miglio dell'Aurelia vetus e i sepulchri dei Papi Callisto I e Giulio I (I parte)." *RACrist* 57 (1971): 169–278.
———. "Osservazioni sulla topografia cimiteriale dell'Aurelia Antica." *RACrist* 40 (1964): 112–22.
Nolli, G. *Rome, 1748: The Pianta grande di Roma of Giambattista Nolli in facsimile*. Highmount, N.Y., 1984.
Occo, A. *Imperatorum romanorum numismata ab Adolfo Occone olim congesta, illustrata a Francisco Mediobarba Birago et expurgata curante Philippo Argelato*. Milan, 1730.
Oisel, J. *Thesaurus selectorum numismatum*. Amsterdam, 1677.
Pace, P. *Gli acquedotti di Roma e il "De aquaeductu" di Frontino: Con testo critico, versione, e commento*. Rome, 1983.
———. "Tecniche di conduzione e distribuzione dell'acqua in epoca romana." In *Trionfo*, 138–51.
Panciroli, G. *De quattuordecim regionibus urbis Romae earumdem que aedificiis tam publicis quam privatis libellus*. Rome, 1602.
Panimolle, G. *Gli acquedotti di Roma antica*. Rome, 1963.
Panvinio, O. *Fastorum libri V a Romolo rege usque ad imp. Caesarem Carolum V*. Venice, 1557.
Pareti, L. "Sulla battaglia del Lago Regillo." *StRom* 7 (1959): 18–30.
Parker, J. H. *The Aqueducts of Ancient Rome*. Oxford, 1876.
Peto, L. *De mensuris et ponderibus Romanis et Graecis cum his quae hodie sunt collatis libri V*. Venice, 1573.
———. *De restitutione ductus Aquae Virginis*. Rome, 1570.
Petrocchi, M. *Roma nel Seicento*. Storia di Roma 14. Bologna, 1970.
Pfeiffer, R. *History of Classical Scholarship from 1300 to 1850*. Oxford, 1976.
Pighius, S. V. *Annales Romanorum*. 3 vols. Antwerp, 1599–1615.
Piranesi, G. B. *Del castello dell'Acqua Giulia*. Rome, 1761.
Pisani Sartorio, G. *La villa Massenzio sulla via Appia*. Rome, 1976.
———. "Punto di derivazione dell'acqua Marcia alla cisterna della c.d. villa delle

Vignacce sulla via Latina; strada di manutenzione e cippo terminale delle acque Marcia, Tepula, e Iulia." In *Trionfo*, 55–56.

———. "Tomba detta di Alessandro Severo a Monte del Grano." In *Piranesi nei luoghi di Piranesi*, 5:65–71. Rome, 1979.

Pisani Sartorio, G., and A. M. Liberati Silverio, eds. *Il trionfo dell'acqua: Acque e acquedotti a Roma, IV. sec. a.C.–XX. sec.* Rome, 1986.

———. *Il trionfo dell'acqua: Atti del convegno "Gli antichi acquedotti di Roma: Problemi di conoscenza, conservazione, e tutela."* Rome, 1992.

Pökel, W. *Philologisches Schriftsteller-Lexicon*. Leipzig, 1882.

Poleni, G. *Sex. Iulii Frontini de Aquaeductibus Urbis Romae Commentarius*. Padua, 1722.

Prado, J., S.J., and J. B. de Villalpando, S.J. *In Exechielem explanationes*. 3 vols. Rome, 1604.

Preller, L. *Die Regionen der Stadt Rom*. Jena, 1846.

Puliti, C., A. Bergioli, and C. Terzano. "Studio chimico-fisico su formazioni calcaree prelevate da antichi acquedotti romani." In *Trionfo*, 195–98.

Quilici, L. *Collatia*. Forma Italiae 1.10. Rome, 1974.

———. "Gli acquedotti di Roma." *Archeo. attualità del passato* 53 (July 1989): 51–97.

———. "Il sistema di captazione delle sorgenti." In *Il trionfo dell'acqua: Atti del convegno "Gli antichi acquedotti di Roma: Problemi di conoscenza, conservazione e tutela,"* ed. G. Pisani Sartorio and A. M. Liberati Silverio, 47–58. Rome, 1992.

———. "La Campagna romana come suburbio di Roma." *PP* 29 (1974): 410–38.

———. "La posterula di Vigna Casali nella pianificazione urbanistica dell'Aventino e sul possibile prospetto del Tempio di Diana." In *L'urbs: Espace urbain et histoire (Ier siècle avant J-C.–IIIe siècle après J-C.)*, 713–45. CÉFR 98. Rome and Paris, 1987.

———. *La Via Latina da Roma a Castel Savelli*. Passeggiate nel Lazio 4. Rome, 1978.

———. *La Via Prenestina: I suoi monumenti e i suoi paesaggi*. Passeggiate nel Lazio 2. Rome, 1969.

———. "La villa dei Quintilii sulla via Appia." *Bolletino Unione Storia ed Arte* 1–2 (1974): 15–26.

———. "Segnalazioni in margine alla costruenda autostrada Fiano-San Cesareo." *QArchEtr* 11 (1985): 165–66.

———. "Sull'acquedotto Vergine dal Monte Pincio alle sorgenti." *QITA* 5 (1968): 125–60.

Quilici Gigli, S. *La Via Salaria da Roma a Passo Corese*. Passeggiate nel Lazio 3. Rome, 1977.

Reeve, M. D. "Frontinus." In *Texts and Transmission: A Survey of the Latin Classics*, ed. L. D. Reynolds, 166–70. Oxford, 1983.

Riccioli, G. B., S.J. *Almagestum novum astronomiam veterem novamque complectens*. Bologna, 1651.

———. *Geographiae et hydrographiae reformatae nuper recognitae et auctae libri duodecim*. Venice, 1672.

Richardson, L., jr. *A New Topographical Dictionary of Ancient Rome*. Baltimore and London, 1992.

Richmond, I. A. *The City Wall of Imperial Rome*. Oxford, 1930.

Rodgers, R. H., ed. and comm. *Sextus Julius Frontinus, "De aquaeductu urbis Romae."* Cambridge, forthcoming.

———. "The Mystery of 62 Miles: *CIL* VI 1256." *ZPE* 63 (1986): 157–60.

Roncaioli Lamberti, C. "Il percorso dell'*Anio Vetus* e la Porta Trebana di Tivoli." *Giornale italiano di filologia* 40 (1988): 87–100.

Säflund, G. *Le mura di Roma repubblicana*. Uppsala, 1932.

Sandys, J. E. *A History of Classical Scholarship*. 3d ed. 3 vols. Cambridge, 1921.

———. *Latin Epigraphy*. Cambridge, 1919.

Schoonhoven, H., ed. *The Pseudo-Ovidian "Ad Liviam de morte Drusi."* Groningen, 1992.

Simili, A. *Gerolamo Mercuriale lettore e medico a Bologna*. Bologna, 1966.

Smet, M. *Inscriptionum antiquarum quae passim in Europam liber*. Leiden, 1588.

Spanheim, E. *Dissertationes de praestantia et usu numismatum antiquorum*. Amsterdam, 1671.

Spini, G. "Italy after the Thirty Years War." In *The Ascendancy of France, 1648–88: The New Cambridge Modern History*, ed. F. L. Carsten, 5:458–73. Cambridge, 1964.

Steinby, E. M., ed. *Lexicon Topographicum Urbis Romae*. 6 vols. Rome, 1993–2000.

———. "L'industria laterizia di Roma nel tardo impero." In *Società romana e impero tardoantico*, ed. A. Giardina, 2:99–163. Rome, 1986.

Sylburg, F., ed. *Dionysii Halicarnassei opera omnia*. Leipzig, 1774–77.

Talbert, R. J. A., ed. *"Barrington Atlas of the Greek and Roman World"* and *"Map-By-Map Directory."* 3 vols. Princeton and Oxford, 2000.

Taylor, Rabun. "A *Citiore Ripa Aquae:* Aqueduct River Crossings in the Ancient City of Rome." *BSR* 63 (1995): 75–103.

———. "Torrent or Trickle? The Aqua Alsietina, the Naumachia Augusti, and the Transtiberim." *AJA* 101 (1997): 465–93.

Taylor, René. "Hermeticism and Mystical Architecture in the Society of Jesus." In *Baroque Art: The Jesuit Contribution*, ed I. B. Jaffe, 63–97. New York, 1972.

Tedeschi Grisanti, G. *I "trofei" di Mario: Il ninfeo dell'Acqua Giulia sull Esquilino*. Rome, 1977.

———. "Primo contributo ad una livellazione urbana sistematica degli antichi acquedotti di Roma." In *Il trionfo dell'acqua: Atti del convegno "Gli antichi acquedotti di Roma: Problemi di conoscenza, conservazione e tutela,"* ed. G. Pisani Sartorio and A. M. Liberati Silverio, 59–69. Rome, 1992.

Tolomeo, M. G. "L'Acquedotto Vergine (sec. XVI–XVII)." In *Trionfo*, 205–8.

Tomassetti, G. *La campagna romana antica, medioevale e moderna*. 4 vols. Rome, 1910–26. New edition, ed. L. Chiumenti and F. Bilancia. 7 vols. Florence, 1979.

Tuck, S. L. "A New Identification for the 'Porticus Aemilia.'" *JRA* 13 (2000): 175–82.

Vacca, F. *Memorie di varie antichità trovate di diversi luoghi dell'alma città di Roma nell'anno 1594*. Rome, 1704.

Valentini, R., and G. Zucchetti. *Codice topografico della città di Roma*. 4 vols. Rome, 1940–53.

Van Deman, E. B. *The Building of the Roman Aqueducts*. Washington, D.C., 1934.

Vico, E. *Augustarum imagines aereis formis expressae*. Venice, 1588.

———. *Discorsi sopra le medaglie de gli antichi*. Venice, 1558.

———. *Le imaggini con tutti i reversi trovati e le vite degli imperatori*. [Venice?], 1548.

Volpe, R. *Aqua Marcia: Lo scavo di un tratto urbano*. Florence, 1996.

Vossius, G. J. *De origine et progressu idololatriae*. Amsterdam, 1668.

Ward-Perkins, J. B. "The Via Gabina." *BSR* 40 (1972): 91–126.

Wasserstein, A. "Aqua Damnata." *Hermes* 103 (1975): 382–83.

Wikander, O. "The Water-Mill." In *Handbook of Ancient Water Technology*, ed. O. Wikander, 371–400. Leiden, 2000.

———, ed. *Handbook of Ancient Water Technology*. Technology and Change in History 2. Leiden, 2000.

Winslow, E. M. A *Libation to the Gods: The Story of the Roman Aqueducts*. London, 1963.

General Index

Index of Ancient Authors